# Regional Congress on Membrane Technology and Regional Conference on Environmental Engineering (RCOM & RCEnvE 2020)



## Enhancing Translational Regional into Sustainable Environmental Technology

IOP Conference Series: Materials Science and Engineering
Volume 1142

Johor, Malaysia
16 - 17 January 2020

**Editors:**

**Farhana Aziz**
**Hasrinah Hasbullah**
**Norhaniza Yusof**
**Wan Norharyati Wan Salleh**

ISBN: 978-1-7138-3415-1
ISSN: 1757-8981

**Printed from e-media with permission by:**

Curran Associates, Inc.
57 Morehouse Lane
Red Hook, NY 12571

**Some format issues inherent in the e-media version may also appear in this print version.**



No changes have been made to the content of these proceedings. There may be changes to pagination and minor adjustments for aesthetics.

Printed with permission by Curran Associates, Inc. (2021)



**Additional copies of this publication are available from:**

Curran Associates, Inc.
57 Morehouse Lane
Red Hook, NY 12571 USA
Phone: 845-758-0400
Fax: 845-758-2633
Email: curran@proceedings.com
Web: www.proceedings.com

# TABLE OF CONTENTS

# Regional Congress on Membrane Technology 2020 (RCOM 2020) and Regional Conference Environmental Engineering (RCEnvE 2020)

Edited by:

1) Dr Farhana Aziz
AMTEC, Universiti Teknologi Malaysia

2) Dr Hasrinah Hasbullah
AMTEC, Universiti Teknologi Malaysia

3) Dr Norhaniza Yusof
AMTEC, Universiti Teknologi Malaysia

4) Dr Wan Norharyati Wan Salleh
AMTEC, Universiti Teknologi Malaysia

## PREFACE

AMTEC and Universiti Teknologi Malaysia (UTM) successfully held a two-day joint online event on 16-17 January 2021 with the involvement of the Regional Conference on Environmental Engineering (RCEnvE2020) and the Regional Congress on Membrane Technology 2020 (RCOM2020) under the theme 'Enhancing Translational Research into Sustainable Environmental Technology.' RCOM 2020 & RCEnvE 2020 is a premier forum for the presentation of new advances and research results in the field of membrane and environment. The conference has brought together networks of professionals and researchers worldwide in various disciplines to discuss current technology development in different field and valuable information of the latest research activities in maintaining the sustainability of the environment for the future.

All papers were presented on-line on Webex platform in about 15 minutes plus 5 minutes for Q&A, managed by the Universiti Teknologi Malaysia, in Johor Bahru, Malaysia, divided in 5 parallel sessions. In addition, four plenary speeches on interesting issues in membrane manufacturing and environmental engineering inspired the participants and provided a platform for valuable discussions. More specifically, the plenary sessions were the following: Professor Dr Ishihara Keiichi from the Kyoto University, Japan speaking on greenhouse gases mitigation by agrivoltaic technology, Prof Dr Bunsho Ohtani from Hokkaido University, with a lecture on novel approach on efficient photocatalyst design, Professor Emeritus Dr Takeshi

IOP Conf. Series: Materials Science and Engineering 1142 (2021) 011001 doi:10.1088/1757-899X/1142/1/011001

Matsuura from University of Ottawa, Canada talking about recent progress on membrane transport theory and design and last but not least, Dato' Norhalim Yunus from Malaysian Technology Development Corporation with a speech on the enhancing translational research towards sustainable environmental technologies.

**List of Topics:**

- Environmental Engineering
- Membrane Technology
- Advanced Separation Technology
- Absorption & Adsorption Technology
- Catalyst and Photocatalytics
- Other Separation Technology
- Biotechnology
- Nanotechnology
- Materials and advanced Materials
- Water and Wastewater Treatment
- Air pollution control
- Renewable Energy and Energy Efficiency Materials
- Other Energy applications

**List of Organizing Committee:**

| | Committee | Head of Unit |
|---|---|---|
| 1 | Patron | Prof. Datuk Ts. Dr. Ahmad Fauzi Ismail |
| 2 | Advisor I<br><br>Advisor II | Prof. Dr. Madzlan Aziz<br><br>Assoc. Prof. Ts. Dr. Mohd Hafiz Dzarfan Othman |
| 3 | Chair<br><br>Co-Chair | Assoc. Prof. Dr. Juhana Jaafar<br><br>Assoc. Prof. Dr. Lau WoeiJye |
| 4 | Focal Person AUN/SEED | Dr. Mohd Zamri Bin Mohd Yusop |
| 5 | Secretary<br><br>Co-Secretary | Dr. Norhaniza Binti Yusuf<br><br>Ts. Dr. Mohd Nazri Bin Mohd Sokri |
| 6 | Treasurer<br><br>Co-Treasurer | Dr. Hasrinah Hasbullah<br><br>Assoc. Prof. Dr. Mukhlis |
| 7 | Protocol Unit | Assoc. Prof. Ts. Dr. Goh Pei Sean |
| 8 | Publicity & Promotion Unit | Dr. Zulhairun Abdul Karim |

IOP Conf. Series: Materials Science and Engineering 1142 (2021) 011001 doi:10.1088/1757-899X/1142/1/011001

| | | |
|---|---|---|
| 9 | Registration Unit | Dr. Wan Norharyati Wan Salleh |
| 10 | Publication Unit | Dr. Farhana Aziz |
| 11 | Logistic Unit | Mr. Mohammad Abdul Razis Saidin |
| 12 | Technical Unit | Ts. Dr. Ng. Be Cheer |
| 13 | Sponsorship Unit | Ts. Mr. Sohaimi Abdullah |
| 14 | Award Unit | Assoc. Prof. Dr. Lau Woei Jye |

# Peer review declaration

All conference organisers/editors are required to declare details about their peer review. Therefore, please provide the following information:

- **Type of peer review: Single-blind ~~/ Double-blind / Triple-blind / Open / Other~~**
  **In a single blind peer review, reviewers' identities are kept hidden from authors.**

- **Conference submission management system: EasyChair**

- **Number of submissions received: 21**

- **Number of submissions sent for review: 20**

- **Number of submissions accepted: 19**

- **Acceptance Rate (Number of Submissions Accepted / Number of Submissions Received X 100): 90.5%**

- **Average number of reviews per paper: 2**

- **Total number of reviewers involved: 30**

- **Any additional info on review process:-**
  At least two reviewers were assigned to each submission. The reviewers must give the following scores and include a detailed review, providing reasoning for the scores:
  3: strong accept
  2: accept
  1: weak accept
  0: borderline paper
  -1: weak reject
  -2: reject
  -3: strong reject
  The editor sent the reviewers' comments to the author, and the author was given the opportunity to revise and resubmit their manuscript.

- **Contact person for queries: Farhana Aziz, farhana@petroleum.utm.my**

# The influence of calcination temperature on the optical, morphological properties and photocatalytic activity of lanthanum orthoferrite

**Nur Aisyah Jamaludin[a], Farhana Aziz[a,b*], Norhaniza Yusof[a,b], Juhana Jaafar[a,b], Wan Norharyati Wan Salleh[a,b]**
[a]Advanced Membrane Technology Research Centre (AMTEC), Universiti Teknologi Malaysia, 81310, Johor Bahru, Johor
[b]*School of Chemical and Energy Engineering, Faculty of Engineering, Universiti Teknologi Malaysia, 81310, Johor Bahru, Johor*

*Corresponding author: farhana@petroleum.utm.my

**Abstract.** In this study, the effect of calcination temperature towards the optical, morphological properties and photocatalytic degradation performance of the synthesized $LaFeO_3$ were investigated. The precursor and synthesized $LaFeO_3$ calcined at 500-900 ºC were characterized using X-ray diffractometer (XRD), Thermogravimetric analysis (TGA), UV-Vis spectrophotometer, Fourier Transform Infrared Spectroscopy (FTIR) and BET. The photocatalytic activity of the synthesized $LaFeO_3$ was performed on oily wastewater. The photocatalytic degradation performance of $LaFeO_3$-600 is 67%, 72% and 63% for 1000ppm, 10000ppm and 20000ppm respectively. This study verified that calcination temperatures have a major impact on the optical, structural and morphological properties of nanoparticles.

## 1. Introduction

Countless harmful hydrocarbon mixture, chemical components and heavy metal present in oily wastewater [1]. There is a typical limit for oil and grease discharges depend on the types of the oils. For instance, the discharge limits for mineral and synthetic oils are 10-15 mg/L while for those of vegetable and animal based from 100-150 mg/L. Oily wastewater need to be treated because it can affecting drinking water and groundwater resources, endangering aquatic resources and human health, affecting crop production and destructing the natural landscape [2].

Various semiconductors have been applied as photocatalysts for photo-degradation of organic and inorganic contaminants such as $TiO_2$, $WO_3$, ZnO, CuS, BiOBr, and $BiFeO_3$ [3-5]. Among them, titanium dioxide ($TiO_2$) is the most widely studied since it is non-toxic, chemically stable, commercially available and inexpensive [6]. However, $TiO_2$ can only be activated using ultraviolet light (<390nm) as it cannot be activated by visible light because of the high band gap (3.2eV) [7]. The other barrier using $TiO_2$ as the photocatalyst is the rapid recombination of electron-hole pair. Due to this limitations, this method is still not practically being used in large industry [8]. In order to activate photocatalyst using visible light, various ways and solution came out such as altering the photocatalyst band gap using dye sensitization, metal, non-metal doping and synthesizing new narrow band gap photocatalyst [4, 7, 9, 10].

Perovskite with typical $ABO_3$ formula such as $NaTaO_3$, $KTaO_3$, $LaFeO_3$ have a higher photocatalytic activity due to its narrow band gap, unique crystal structure and electronic properties [11]. Citric acid and glucose based sol-gel method were used in previous studied and successfully synthesized $LaFeO_3$. Glucose is more favoravble as complexing agents as it is eco-friendly nature, low cost, facile and reproducibility. The use of glucose as a chelating agent often leads to convenience that requires less energy to synthesize high-purity nanoparticles. Based on previous studies, the temperature used in calcining while synthesizing was undoubtly affected the structure, particle size and properties of the end product but rarely explored and reported [12, 13].

In this study, lanthanum based photocatalyst, Lanthanum Orthoferrite ($LaFeO_3$) with narrow bandgap has been synthesized via sol-gel method in order to be activated under visible light irradiation. Study on the effect of calcination temperature towards the physicochemical properties and photocatalytic degradation performance of the synthesized $LaFeO_3$ were investigated.

## 2. Experimental

### *2.1 Materials*

In the typical synthesis of perovskite $LaFeO_3$, lanthanum nitrate, La $(NO_3)_3.6H_2O$, iron (III) nitrate nonahydrate, $Fe(NO_3)_3.9H_2O$, glucose $C_6H_{12}O_6$ were used as starting materials to prepare $LaFeO_3$ nanostructures by sol-gel route. Oil-in-water emulsion was made using *Bagus* cooking oil together with sodium dodecyl sulphate (SDS) for photocatalytic activity test.

### *2.2 Preparation of photocatalyst*

Perovskite type of lanthanum orthoferrite ($LaFeO_3$) nanoparticles materials was synthesized according to a method which has been described elsewhere (T. Liu & Xu, 2011). Glucose solution with 80% concentration was prepared using 500ml beaker. 150ml of distilled water then was added into the beaker with stirrer bar. Bear in mind that glucose powder must be added after adding water in the beaker to avoid rock candy. Put the beaker on the hot plate and heat the water to become warm water using the temperature of 50-60°C. While waiting for the water, weight the glucose powder, 80% glucose solution refers to the 80% of glucose concentration to the water. 80% from 500ml, 400g glucose powder was added to the warm water. Additional water will be added after the solution become homogeneous until 500ml. The glucose solution needs must be stirred each time before use.

La $(NO_3)_3.6H_2O$ and glucose solution was mixed together in the 250ml beaker on the hot plate with magnetic stirrer in it and at 55-65°C. Using the mol ratio of glucose to metal ions of 3:5, Fe $(NO_3)_3.9H_2O$ was added to the solution under constant stirring and temperature around 70°C until it dissolves completely. Decomposition of ions nitrate gave out brownish gas after dissolving Fe $(NO_3)_3.9H_2O$. The sample then was removed from hot plate and heated in the oven at 250°C for 2 hours. After 2 hours, the sample was let to cool down and out of the oven. The sample was then be ground into slightly ground powder and precursor was obtained. Precursor was then be calcined at various temperatures ranging from 500-900°C for 2 hours. Final product should be ground again to get finer powder.

### *2.3 Characterization*

The precursor and calcined samples were characterized using various characterization techniques. The TGA analysis is conducted to determine the weight loss and the phase change of the sample precursor and calcined samples. Crystallite phase and size were analyzed by X-ray diffraction (XRD) that were performed on a X'Pert PRO PANanalytica from from 10° to 90° using Cu $K_{\alpha\ radiation}$ of $\lambda$=0.1540nm. UV-Vis absorption spectrophotometer using Perkin Elmer Lambda 25 were used in this study to analyzed the optical properties of synthesized samples. The morphology of the synthesized samples was observed using Low Voltage FE-SEM (Hitachi FE-SEM SU8000 EDX (Bruker) with an acceleration voltage of 3.0 kV. Specific surface area and pore volume were measured and calculated by $N_2$ adsorption using Brunauer, Emmet and Teller (BET) (Micromeritics Instrument Corporation TriStarII 3020) analyzer based on nitrogen gas adsorption.

### *2.4 Photocatalytic activity*

The synthetic oily wastewater was prepared using cooking oil brand Bagus, by Sime Darby. The synthetic oil-in-water emulsion was prepared according to method by Gohari et al., (2014) in which the synthetic oil of different concentrations, i.e. 10, and 2000 ppm, was mixed with DI water under vigorous stirring at 250 rpm for about 30 min at room temperature while adding sodium dodecyl sulphate (SDS) to produce homogenous mixture between oil and water. Once the process was completed, a solution with uniform yellowish colour was obtained. Considering coalescence of oil droplets that may happen during an extended period of storage, synthetic oily wastewater was prepared a day before experiment to keep the feed characteristics consistent.

The photocatalytic degradation of oily wastewater using synthesized $LaFeO_3$ as photocatalyst was conducted in a closed stainless-steel box as a batch reactor. 300 mL synthetic oil emulsion were put in the beaker. 0.3g of $LaFeO_3$ calcined at 600°C was added into the beaker. The mixture then was being stirred for 30 minutes for adsorption process in a dark place with continuous stirring. After 30 minutes of adsorption process, the photocatalytic activity was then tested under visible-light irradiation for 150 minutes. 10mL of the mixture was aliquot using 10mL syringe and was filtered using syringe filter (brand, size) at 30 minutes interval for 2 hours and 30 minutes.

Stir the mixture for another 30 minutes under visible light irradiation (100 watt). After 30 minutes, an aliquot was taken again. This activity continues for 2 hours and 30 minutes, for every 30 minutes 10mL aliquot was taken as sample and stored in dark place to avoid further photocatalytic activity. The concentration of the synthetic oil was varied for 1000 ppm, 10 000 ppm, 20 000 ppm.

## 3. Results and Discussion

### *3.1 Thermal decomposition analysis*

The TGA analysis is conducted to determine the weight loss and the phase change of the sample precursor and calcined samples. Figure 1 shows the thermal decomposition curves of the $LaFeO_3$ precursor and nanoparticles that can be described as follows. The weight loss for precursor shows two weight loss segments. The first segment is attributed to the elimination of water molecules and the crystallization process of the materials. The second significant weight loss observed between 360-629.5°C mainly due to the decomposition of organic compound.

The weight loss for $LaFeO_3$ calcined at 500°C shows only a segment and there's no further weight loss was observed after that. The segment observed between 59.5 - <629.5 °C is due to the decomposition remaining organic compound and formation of the disordered $La_2O_3$ [14]. There is no weight loss observed for sample calcined at 600°C-900°C confirmed that all organic matter has been burnt out and there is no change in phase [15]. This finding shows that $LaFeO_3$ precursor and nanoparticles calcined at 500°C were completely in crystal form after 630°C. On the other hand, $LaFeO_3$ calcined at 600°C-900°C were thermally stable and completely crystallized.

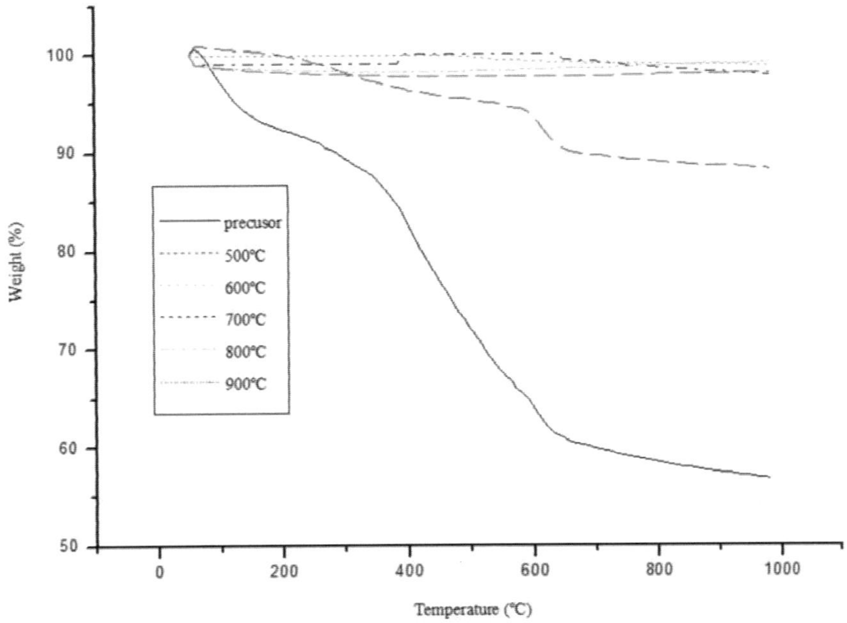


Figure 1 TGA Curves of precursor and $LaFeO_3$ calcined at 500 – 900 °C

*3.2 Crystallinity properties*

Figure 2 shows the pXRD pattern of $LaFeO_3$ calcined at various temperatures range between 500°C and 900°C for 2h in air. Lattice parameter for $LaFeO_3$ synthesized by sol-gel route is α=3.926. The observed patterns all indexed with JCPDS card no.37- 1493. The precursor and sample synthesized at 500°C is in amorphous structure as it can be seen by wide and broad continuum. $LaFeO_3$ was completely crystallized into perovskite phase at calcination temperature of 600°C. Further calcined the precursor at 700°C to 900°C for 2h yield more crystalline pure $LaFeO_3$ with orthorhombic structure.

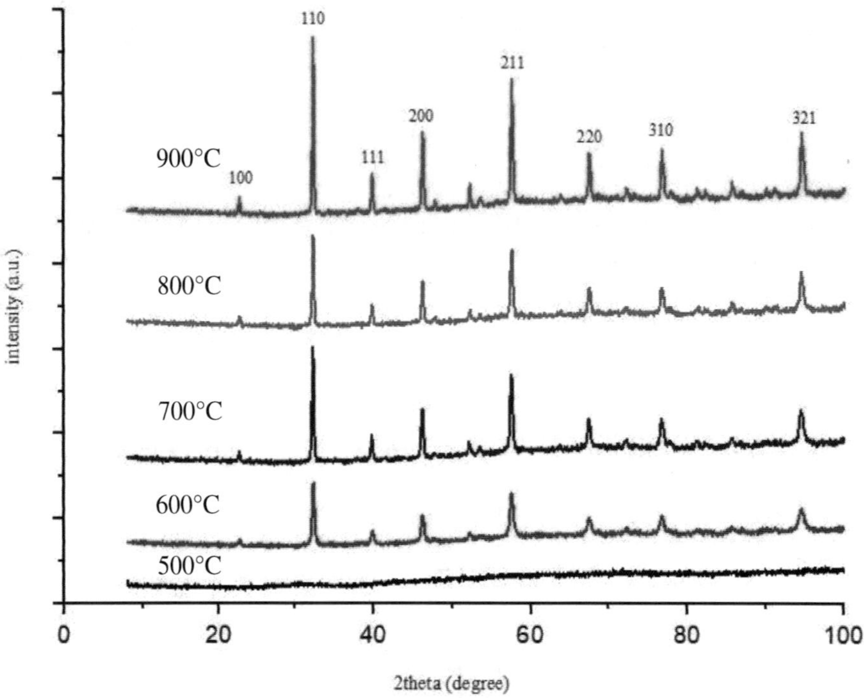


Figure 2 pXRD pattern of $LaFeO_3$ precursor, 500°C, 600°C, 700°C, 800°C, 900°C

Increasing the calcination temperature will result in sharper and stronger peaks indicating greater crystallization. The average crystallite size of all samples was calculated from X-ray line broadening of the (1 1 0) diffraction peak using Debye-Scherrer's equation. The crystallite size of particles has been determined according to the x-ray line-broadening at (1 1 2). Crystallite size of 23.82 nm, 31.17 nm, 32.0 nm, and 38.3 nm were obtained for the samples calcined at 600°C, 700°C, 800°C, 900°C respectively. It can be seen clearly that crystallite size of each sample gradually increases with the increasing of the calcination temperature (T. Liu & Xu, 2011). Narrow band gap is needed to ensure the photocatalyst can be activated by visible light to carry the photocatalytic activity. Crystallite size of the particle gives strong influence to the value of the band gap as smaller crystallite size give rise to the band gap value [16]. This result is consistent with the band gap value as shown in Table 2.

*3.3 Morphological properties*

The surface morphology of the precursor and pure $LaFeO_3$ were analysed by field emission scanning electron microscopy (FESEM). FESEM micrographs of $LaFeO_3$ precursor and powders calcined at various temperature ranges from 500°C-900°C are presented in Figure 3. Powder calcined at 500°C consists of agglomerated particles but porous. At 600°C, the nanoparticle seems to be in regular porous morphology and connect with each other to form a large network system. Further calcined the precursor to 700°C-900°C, changes the particle morphology. From FESEM images, it can clearly be seen that the particle size increase with the calcination temperature. The formation of sponge-like structure and "chess-like" structure can be observed in Figure 3(b) and 3(c) for samples calcined at 500°C and 600°C. Figure 3(d) and 3(e) are the images of $LaFeO_3$ calcined at 700°C and 800°C shows the pseudohexagonal-like structure of nanoparticles. While uniform nano hexagonal shape for the powder calcined at 900°C can be seen through the FESEM image. All sample calcined at 500-900°C exhibit porous morphology, due to the charring reaction causing the evaluation of a large amount of gasses (Feng et al., 2011). It is well known that smaller particles size with higher surface area will provide more active sites thus promising for photocatalytic activity. It is also understood that the particle size and surface area are inversely related to each other. The surface area for each particle will be discussed in next section. As the particle size increase with calcination temperature, $LaFeO_3$ nanoparticles calcined at 600°C exhibit more favorable condition with smaller particle size and high surface area.

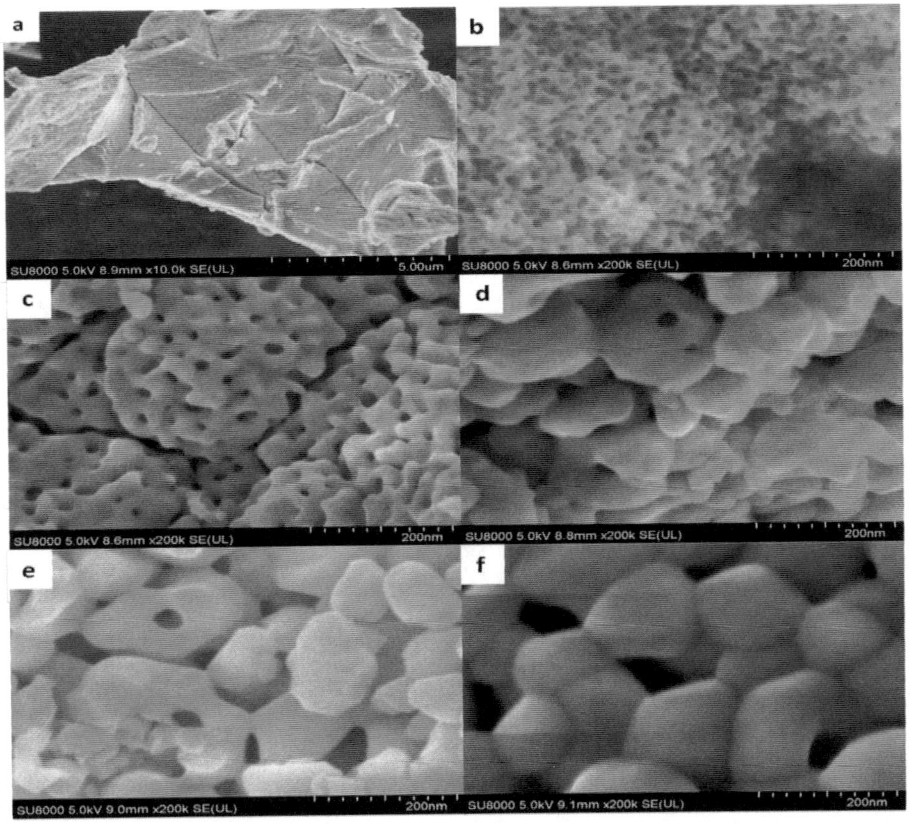


Figure 3 FESEM micrographs of $LaFeO_3$ (a)precursor and sample calcined at (b)500°C, (c)600°C, (d)700°C, (e)800°C, (f) 900°C for 2h.

*3.4 Structural Analysis*

FTIR spectra of $LaFeO_3$ precursor and the powder calcined at different temperatures are illustrated in Figure 4. it can be observed that all spectra exhibit two dominant absorption regions at 500-600cm$^{-1}$ and at band ~3500cm$^{-1}$. All the calcined samples show a peak around 550-570 cm$^{-1}$ are attributed to Fe-O stretching vibrational [17-20]. This new band appeared at nearly 600 cm$^{-1}$ is the band for the formation of $LaFeO_3$ attributed to the Fe-O stretching band and being characteristic of the octahedral $FeO_6$ group in $LaFeO_3$ [20-22]. The O-Fe-O deformation vibration ($V_2$ mode) obviously can be seen by the strong absorption band less and near 500 $cm^{-1}$ for all the calcined samples except for sample calcined at 500°C and precursor [12, 17, 23].

The smaller peak exists at 850 and 1080 $cm^{-1}$ were resemble to the principal vibration of $CO_3^{-2}$ group due to the contact to the ambient atmosphere, demonstrating that La-Carbonate exist at the surface of $LaFeO_3$ nanoparticles. Two vibrational bands can also be observed at 1570 $cm^{-1}$ and 1385 $cm^{-1}$ that are assigned to the splitting of the asymmetric stretching of metal carbonate (T. Liu & Xu, 2011). Almost complete disappearance of this band for the sample calcined 600°C and complete disappearance for the samples calcined at 700-900°C showing that the $CO_3^{-2}$ group were completely removed leaving behind only La. The wide band observed at ~3400 $cm^{-1}$ was assigned for O-H stretch for precursor and 500-700°C samples (Lebid & Omari, 2013)(T. Liu & Xu, 2011). The reduction or disappearance of all the stated peaks (1080 $cm^{-1}$, 1570 $cm^{-1}$ and 1385 $cm^{-1}$, 3400 $cm^{-1}$) indicating relatively pure phase formation for the samples calcined at 600-900°C.

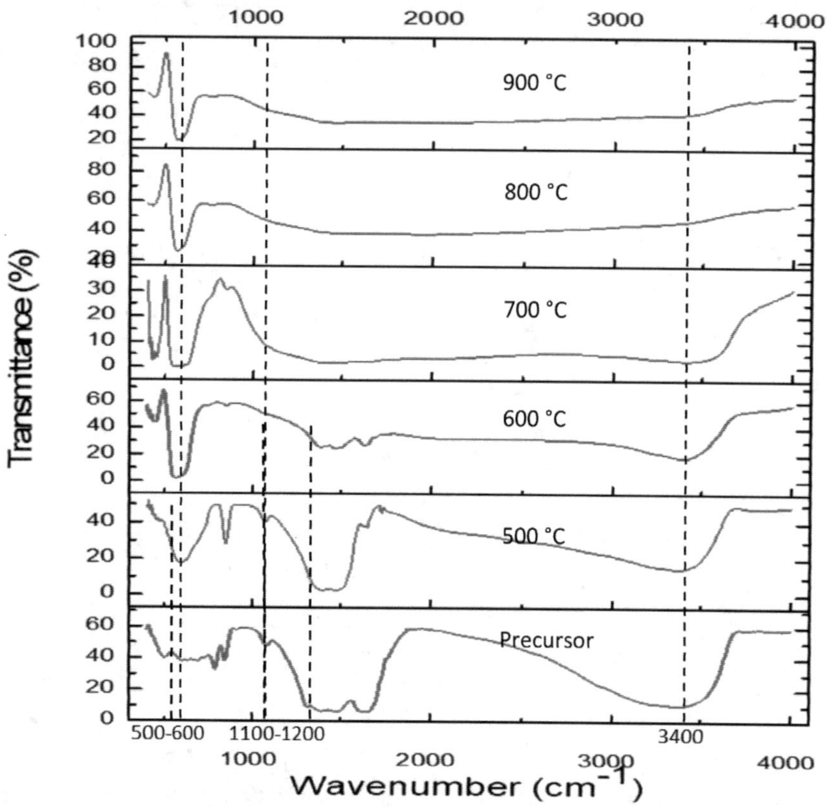


Figure 4 FTIR spectra of $LaFeO_3$ precursor and the powder calcined at different temperature

Based on the above FTIR results, Figure 5 shows the expected molecular structure of the metal ions and glucose complex. The formation of the chelation process can be described during the hydrolysis and condensation phase, in which the xerogel was fully formed. The chelation mechanism could be further explained from the structure of glucose, which consists of five hydroxyl groups (OH-) that can chelate to form metal complexes with metal ions.

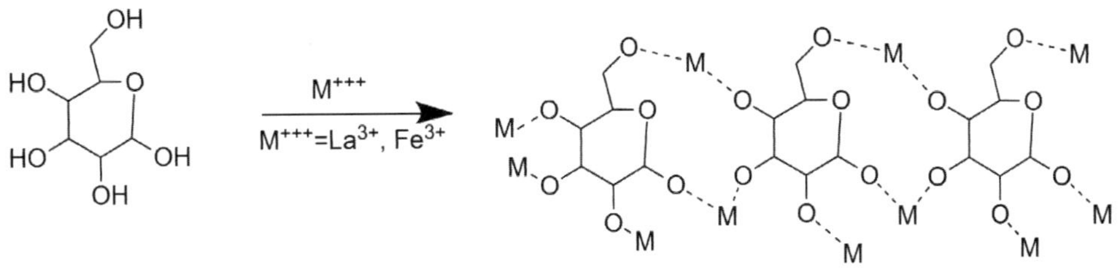


Figure 5 Possible complex of metal ions and glucose in gel precursor of $LaFeO_3$ nanoparticles

*3.5 Surface area analysis*

In this study, BET surface area and pores size distribution were analyzed to determine the specific surface area of the synthesized $LaFeO_3$ at different calcination temperature. For the photocatalysis process where the reaction happens on the surface of the catalyst, surface areas of the nanoparticles are the main concern. In general, small particle size is more favorable as its lead to higher surface area. The surface area values for the calcined samples are tabulated in Table 1. It can be seen that the surface area values were decreasing with the calcination temperature except for the sample calcined at 600°C. Besides, this result is also in agreement with the crystallite size and particles size of $LaFeO_3$ that can be observed in XRD pattern and SEM images, respectively. For $LaFeO_3$-500, lower surface area could be related to the agglomeration of the particles.

Comparison our results with those reported in the literature is shown in Table 1. Based on this table, it can be concluded that our photocatalyst surface area is comparable with others at 600°C calcination temperature. Besides, it is clear that calcination time and complexing agents does have significant effects on the surface area of the photocatalyst. The surface area of the $LaFeO_3$ nanoparticles was found to be $15.68 m^2/g$ for sample calcined at 600°C which is higher than the reported value $8.5 m^2/g$ [19] and $14.1 m^2/g$ [24] . This result shows that $LaFeO_3$ synthesized in this study have enough surface area to promote the degradation of oily wastewater. Even tough samples prepared using PVP K-30 have high surface area, their calcination time is longer which is not favorable as it is time and energy consuming. For citric acid, a lot study has demonstrated that the used of citric acid as complexing agents leads to less purity photocatalyst

Table 1 Surface areas of $LaFeO_3$ synthesized using different complexing agents and calcination temperature

| **Sample** | **Calcination temperature °C** | **Calcination time (h)** | **Complexing agents** | **Surface area ($m^2/g$)** | **References** |
|---|---|---|---|---|---|
| 1 | 500 | 2 | Glucose | 3.89 | This study |
| 2 | 600 | 2 | Glucose | 15.68 | |
| 3 | 700 | 2 | Glucose | 6.43 | |
| 4 | 800 | 2 | Glucose | 4.63 | |
| 5 | 900 | 2 | Glucose | 2.40 | |
| 6 | 600 | n/a | Glucose | 8.5 | [19] |
| 7 | 600 | n/a | Glucose | 14.1 | [24] |
| 8 | 600 | 5 | Citric acid | 5.37 | [24] |
| 9 | 600 | 2 | Citric acid | 22.55 | [25] |
| 10 | 600 | 3 | PVP K-30 | 26.62 | [26] |

*3.6 UV–vis diffuse reflectance spectra*

Figure 6 represents the UV-Vis absorption spectrum of $LaFeO_3$ calcined at 500-900°C in the wavelength range from 200-800 nm. Lambda max from UV-Vis analysis should have peak more than

400nm and above to show the active photocatalytic activity in visible wavelength region. The absorption of $LaFeO_3$ precursor and $LafeO_3$ calcined at 600-900°C was characterized with a sharp absorption edge at about 600-700 nm with band gap range 1.63-2.15 eV. The absorption edge in this region indicates that $LaFeO_3$ nanoparticles prepared by sol-gel method could serve as a potential visible-light-driven photocatalytic material. Figure 7 shows the results of band gap obtained by applying the Kubelka-Munk method to $LaFeO_3$ powder samples calcined at the temperature of 500-900°C whose values are tabulated in the Table 2.

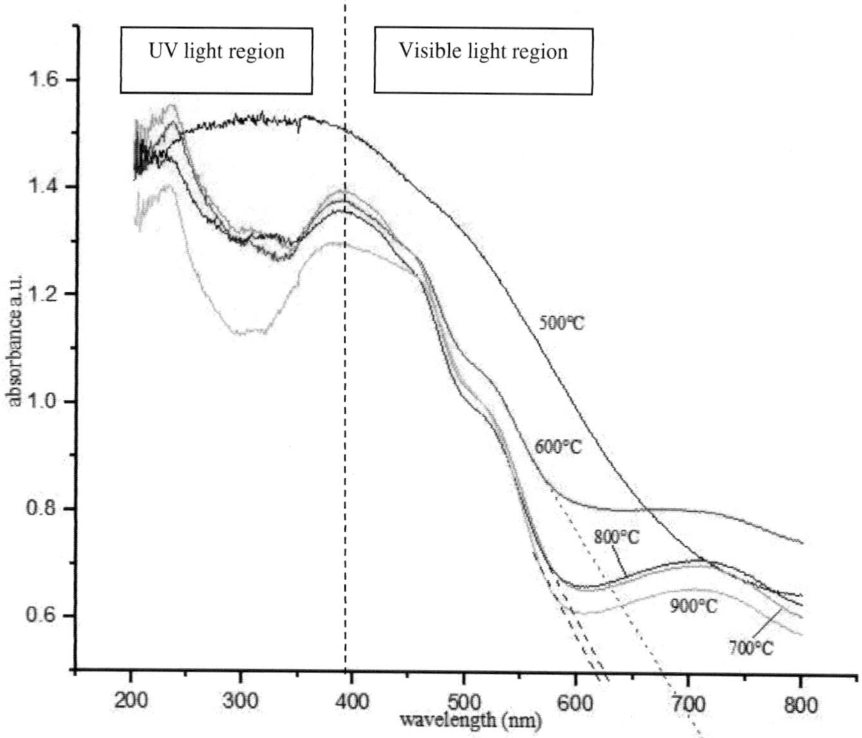


Figure 6 UV-Vis spectrum of the perovskite $LaFeO_3$ calcined at different temperature, 600-900°C

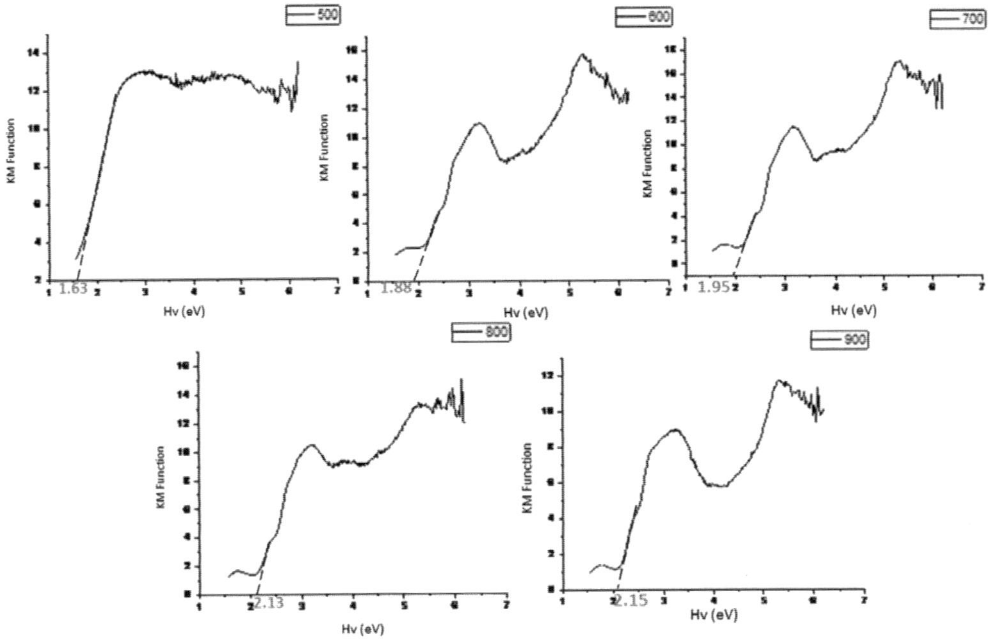


Figure 7 Kubelka-Munk Function of LaFeO3 Calcined at 500-900°C

According to Figure 7, the estimated value of band gap using Kubelka-Munk Function for $LaFeO_3$ calcined at 500°C is ~1.63 eV, which is relatively lower than P25 at 3.24 eV. The low band gap value of $LaFeO_3$ supported the visible light-driven characteristics of the synthesized photocatalyst. In Table 2 shows all the criteria that outstandingly has been encountered to find the finest sample calcined. The band gap value is increasing with the crystal size of nanoparticles. This is related to the quantification effect obtained due to morphological variations in the nanoparticles. As noted in the FESEM images (see Figure 3), the calcination treatment of $LaFeO_3$ material results in the formation of sponge-like structure, "chess-like" structure, pseudohexagonal-like structure, and hexagonal-like structure. The development of the structure of variance is responsible for quantum confinement.

Table 2 The value of the band gap, surface area and crystal size for $LaFeO_3$ calcined at 500-900°C.

| **Samples** | **Band gap (eV)** | **Surface area ($m^2/g$)** | **Crystal Size (nm)** |
|---|---|---|---|
| $LaFeO_3$-500 | 1.63 | 3.89 | Amorphous |
| $LaFeO_3$-600 | 1.88 | 15.68 | 7.29 |
| $LaFeO_3$-700 | 1.95 | 6.43 | 11.55 |
| $LaFeO_3$-800 | 2.13 | 4.63 | 12.60 |
| $LaFeO_3$-900 | 2.15 | 2.40 | 15.43 |

The finest photocatalyst among sample in this study is $LaFeO_3$-600 because of its properties. This $LaFeO_3$-600 has the outstanding surface area (15.68$m^2/g$) with low band gap value (1.88 eV) and small crystal size (7.29 nm). This $LaFeO_3$-600 were served as a potential visible-light-driven photocatalytic material and has been chosen to undergo photocatalytic activity to treat synthetic oily wastewater.

IOP Conf. Series: Materials Science and Engineering 1142 (2021) 012001 doi:10.1088/1757-899X/1142/1/012001

*3.7 Photocatalytic activity*

UV-Vis spectrometer analysis on the degradation of the synthetic oily wastewater in dark (adsorption) and under visible light irradiation are shown in Figure 8. The photocatalytic activity of $LaFeO_3$ samples was evaluated by observing the degradation of synthetic oily water driven by visible light as the model pollutant. Only $LaFeO_3$-600C was selected to study the photocatalytic activity due to its large surface area. It is clearly shown that $LaFeO_3$ powder calcined at 600°C exhibit strong visible light absorption and more stable than the others. After absorption and desorption time have passed, at t=0s, the photocatalytic activity was proceeded under visible light irradiation and constant aeration. After half (0.5) hour adsorption equilibrium, we can find initial characteristic peak of synthetic oily wastewater decreased different from each other, which indicate the adsorption effect of each sample with an adsorption amount of 20% (1000ppm), 27% (10000ppm) and 55% (20000ppm). This results is comparable with previous study (Moslehyani et al., 2015). Under the irradiation of visible light for the next 150 minutes, the removal efficiency increases up to 67%, 72% and 63% for 1000ppm, 10000ppm and 20000ppm respectively. It was also showed that without any photocatalyst (photolysis) the degradation of synthetic oily wastewater was low, where removal efficiency of less than 5% was observed proving the stability of the pollutant under visible light irradiance.

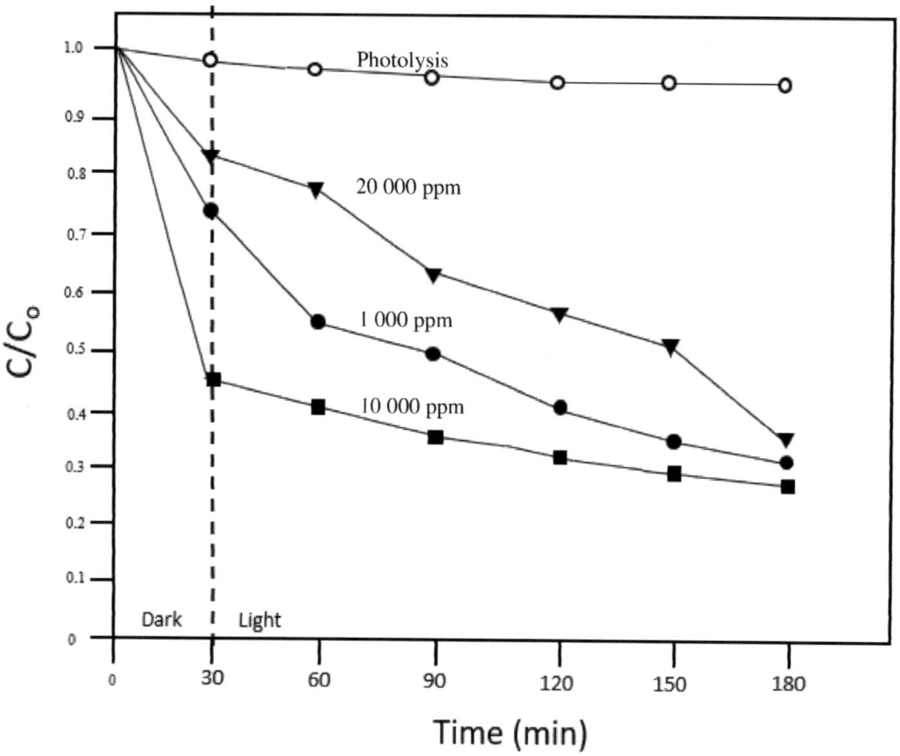


Figure 8 Degradation of synthetic oily wastewater using $LaFeO_3$-600 for different initial concentration

Figure 9 illustrates the possible photocatalytic process at different initial concentrations. Limited interaction between oil droplet and photocatalyst result in less degradation for 1000 ppm initial concentration. 10 000 ppm is considered as optimum initial concentration for 1g/L of $LaFeO_3$ photocatalyst. Possibility of oil interact and absorb on the surface of photocatalyst was increased thus, increasing the degradation rate. Turbidity is high at 20 000 ppm, which leads to less light penetration and therefore less hydroxyl radical production, which reduces degradation. Besides, the excessive contaminant adsorbed on the surface of the photocatalyst is believed can block the active sites from receiving light for photoactivation and lead to low photocatalytic performance.

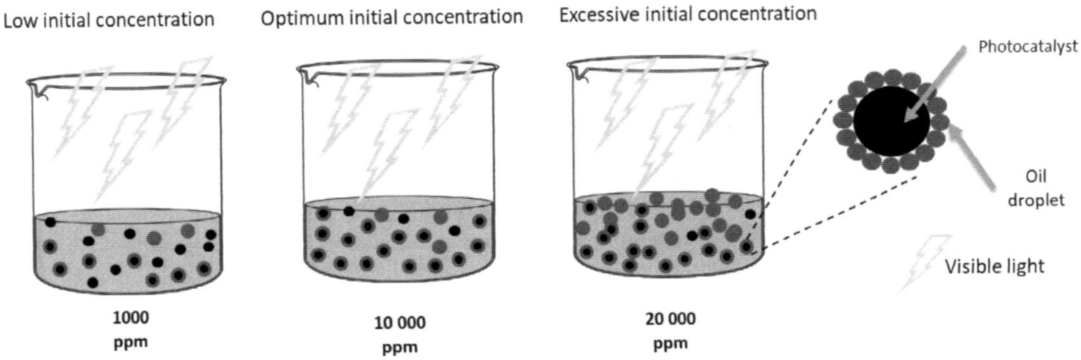


Figure 9 A schematic diagram of photocatalytic process with different initial concentration.

## 4. Conclusions

In summary, $LaFeO_3$ perovskite has been successfully synthesized using sol-gel method with varied calcination temperature between 500-900°C. XRD analysis confirmed that $LaFeO_3$ fully crystallized for calcination temperature of 600°C and above. The crystallite size of 23.82 nm, 31.17 nm, 32.0 nm, and 38.3 nm were obtained for the samples calcined at 600°C, 700°C, 800°C, 900°C respectively. The $LaFeO_3$ photocatalyst with the highest surface area is $LaFeO_3$-600 with a surface area of 15.68 $m^2/g$. The synthesized $LaFeO_3$-600 able to degrade oily wastewater up to 72% in 150 min. This study confirmed that the calcination temperatures significantly affects the optical, structural and morphological properties of nanoparticles.

**Acknowledgement**

This work was supported by the Malaysia Ministry of Education (MOE) under FRGS (Grant No. R.J130000.7851.5F007) and MRUN (Grant no: R.J130000.7813.4L874). The authors would also like to acknowledge Universiti Teknologi Malaysia for financial support and instruments used for characterizations of samples within this study.

## References

1. Ong, C.S., et al., *Investigation of submerged membrane photocatalytic reactor (sMPR) operating parameters during oily wastewater treatment process.* Desalination, 2014. **353**: p. 48-56.
2. Yu, L., M. Han, and F. He, *A review of treating oily wastewater.* Arabian Journal of Chemistry, 2013.
3. Mohamed, M.A., et al., *Incorporation of N-doped TiO2 nanorods in regenerated cellulose thin films fabricated from recycled newspaper as a green portable photocatalyst.* Carbohydrate Polymers, 2015. **133**: p. 429-437.
4. Mohamed, M.A., et al., *Photodegradation of phenol by N-Doped TiO2 anatase/rutile nanorods assembled microsphere under UV and visible light irradiation.* Materials Chemistry and Physics, 2015. **162**: p. 113-123.
5. Saleh, R. and N.F. Djaja, *UV light photocatalytic degradation of organic dyes with Fe-doped ZnO nanoparticles.* Superlattices and Microstructures, 2014. **74**: p. 217-233.
6. Leong, S., et al., *TiO2 based photocatalytic membranes: A review.* 2014. p. 167-184.
7. Banerjee, S., et al., *New Insights into the Mechanism of Visible Light Photocatalysis.* The journal of physical chemistry letters, 2014. **5**(15): p. 2543-54.

8. Shen, S., et al., *Visible-light-driven photocatalytic water splitting on nanostructured semiconducting materials Shaohua Shen *, Jinwen Shi.* Int. J. Nanotechnol., 2011. **8**: p. 523-591.
9. Yao, L., et al., *Fabrication of catalytic membrane contactors based on polyoxometalates and polyvinylidene fluoride intended for degrading phenol in wastewater under mild conditions.* Separation and Purification Technology, 2013. **118**: p. 162-169.
10. Chen, X., et al., *Semiconductor-based Photocatalytic Hydrogen Generation.* Chemical Reviews (Washington, DC, United States), 2010. **110**(11): p. 6503-6570.
11. Grabowska, E., *Selected perovskite oxides: Characterization, preparation and photocatalytic properties-A review.* 2016. p. 97-126.
12. Lebid, M. and M. Omari, *Synthesis and Electrochemical Properties of LaFeO3 Oxides Prepared Via Sol–Gel Method.* Arabian Journal for Science and Engineering, 2013. **39**(1): p. 147-152.
13. Sadrzadeh, M., E. Gorouhi, and T. Mohammadi, *Oily Wastewater Treatment Using Polytetrafluoroethylene (Ptfe) Hydrophobic Membranes.* Water-Observatory.Net, 2008: p. 1-11.
14. Shabbir, G., A.H. Qureshi, and K. Saeed, *Nano-crystalline LaFeO3 powders synthesized by the citrate–gel method.* 2006. p. 3706-3709.
15. Farhadi, S. and F. Siadatnasab, *Perovskite-type LaFeO3 nanoparticles prepared by thermal decomposition of the La[Fe(CN)6]??5H2O complex: A new reusable catalyst for rapid and efficient reduction of aromatic nitro compounds to arylamines with propan-2-ol under microwave irradiation.* Journal of Molecular Catalysis A: Chemical, 2011. **339**(1-2): p. 108-116.
16. Kim, D.S., S.J. Han, and S.-Y. Kwak, *Synthesis and photocatalytic activity of mesoporous TiO2 with the surface area, crystallite size, and pore size.* Journal of Colloid and Interface Science, 2007. **316**(1): p. 85-91.
17. Gosavi, P.V. and R.B. Biniwale, *Pure phase LaFeO3 perovskite with improved surface area synthesized using different routes and its characterization.* Materials Chemistry and Physics, 2010. **119**(1-2): p. 324-329.
18. Tang, P., et al., *Microwave-assisted synthesis of nanoparticulate perovskite LaFeO3 as a high active visible-light photocatalyst.* Current Applied Physics, 2013. **13**(2): p. 340-343.
19. Thirumalairajan, S., et al., *Controlled synthesis of perovskite LaFeO3 microsphere composed of nanoparticles via self-assembly process and their associated photocatalytic activity.* Chemical Engineering Journal, 2012. **209**: p. 420-428.
20. Liu, T. and Y. Xu, *Synthesis of nanocrystalline LaFeO3 powders via glucose sol-gel route.* Materials Chemistry and Physics, 2011. **129**(3): p. 1047-1050.
21. Fujii, T., I. Matsusue, and J. Takada, *Superparamagnetic Behaviour and Induced Ferrimagnetism of LaFeO3 Nanoparticles Prepared by a Hot-Soap Technique.* Advanced Aspects of spectroscopy, 2012.
22. Thirumalairajan, S., et al., *Novel synthesis of LaFeO3 nanostructure dendrites: A systematic investigation of growth mechanism, properties, and biosensing for highly selective determination of neurotransmitter compounds.* Crystal Growth and Design, 2013. **13**(1): p. 291-302.
23. Zheng, W., et al., *Hydrothermal synthesis of LaFeO 3 under carbonate-containing medium.* 2000(March): p. 20-23.
24. Li, Y., et al., *Sol–gel combustion synthesis and visible-light-driven photocatalytic property of perovskite LaNiO3.* Journal of Alloys and Compounds, 2010. **491**(1): p. 560-564.
25. Parida, K.M., et al., *Fabrication of nanocrystalline LaFeO3: An efficient sol-gel auto-combustion assisted visible light responsive photocatalyst for water decomposition.* International Journal of Hydrogen Energy, 2010. **35**(22): p. 12161-12168.
26. Shang, M., et al., *Bi 2 WO 6 Nanocrystals with High Photocatalytic Activities under Visible Light.* The Journal of Physical Chemistry C, 2008. **112**: p. 10407-10411.

IOP Conf. Series: Materials Science and Engineering 1142 (2021) 012002 doi:10.1088/1757-899X/1142/1/012002

# Photocatalytic disinfection of bacteria under visible light irradiation by $BiFeO_3$ photocatalyst

**N A Daub[1,2], F Aziz[1,2*], N A Mohd Zain[3], W J Lau[1,2], N Yusof[1,2], W N W Salleh[1,2], J Jaafar[1,2]**

[1]School of Chemical and Energy Engineering, Faculty of Engineering, Universiti Teknologi Malaysia, 81310 Johor Bahru, Johor, Malaysia

[2]Advanced Membrane Technology Research Centre (AMTEC), Universiti Teknologi Malaysia,81310 Johor Bahru, Johor, Malaysia

[3]Department of Biosciences, Faculty of Science, Universiti Teknologi Malaysia, 81310, Johor Bahru, Johor, Malaysia.

*Corresponding author: farhanaaziz@utm.my

**Abstract.** $BiFeO_3$ nanoparticles was synthesized by sol gel auto combustion. The as-synthesized $BiFeO_3$ were characterized by X-ray diffraction (XRD), UV-Vis diffuse reflectance spectra (DRS) and Brunauer-Emmett-Teller (BET) analysis. The disinfection activities towards Gram-positive *Staphylococcus aureus* (*S.aureus*) were examined under visible light irradiation. The results showed a complete inactivation of 3 x $10^6$ CFU/mL *S.aureus* was achieved within 20 min. The disruption of bacterial cell structure was observed by Transmission electron microscope (TEM). The cells were severely damaged after being exposed to $BiFeO_3$ under visible light irradiation. Hence, the results demonstrated the potential application of perovskite-type photocatalyst, $BiFeO_3$ in photocatalytic disinfection of various microorganisms.

## 1. Introduction

Clean and safe water access is necessary for the substance of all living organisms. The presence of pathogenic microorganisms play a major role in threating water safety. Removal of pathogenic microorganisms from contaminated water is crucially important to prevent potential negative effects on both human health and environment. Currently, various traditional disinfection techniques such as chlorination, ozonation, ultraviolet irradiation and advance filtration processes have been used for contaminated water treatment. These conventional techniques were found to suffer from some drawbacks such as formation of carcinogenic and hazardous byproducts (DBPs), regrowth of harmful bacteria [1], [2]. Escherichia coli (*E.coli*) are known as an indicator of water decontamination, have potential to many health problems such as serious diarrhea and can lead to fatality in humans [2]. It is of critical important searching for an alternative technology that are safe, low-cost and environmental-friendly methods in order to tackle the abovementioned problem. Photocatalysis have been extensively studied to circumvent the environmental problems including the inactivation of microorganisms by utilizing solar energy or visible light source [2]. Many studies have been conducted in photocatalytic disinfection of various microorganisms by light-induced nanomaterials such as Cu-$TiO_2$ [3], Fe-doped $BiVO_4$ [2], Cd, Mo and W sulphide [4], Vanadium tetrasulfide ($VS_4$) [5], g-$C_3N_4$-AgBr [1], Pb-$BiFeO_3$/rGO [6].

Among various visible-light-driven photocatalyst, perovskite-type bismuth iron oxide ($BiFeO_3$) have been recognized as promising and efficient photocatalyst due to their attractive properties. $BiFeO_3$ has a small band gap (~2.0 eV) which appears to be sensitive to visible light and exhibit high efficiency in photocatalytic activity. $BiFeO_3$ also demonstrated a good thermal stability [7], high chemical stability [8], has magnetic properties [9], multiferroic at room temperature which helps in separation of charge carriers [10], showing a great potential in photocatalytic activity. Thus, this study will focus on photocatalytic disinfection of *Staphylococcus aureus (S.aureus)* under visible light irradiation.

## 2. Experimental

### 2.1. Materials

Bismuth nitrate pentahydrate, $Bi(NO_3)_3.5H_2O$, iron (III) nitrate nonahydrate, $Fe(NO_3)_3.9H_2O$ and 25% ammonia solution, $NH_3$ were purchased from Sigma Aldrich, while citric acid monohydrate, $C_6H_8O_7.H_2O$ was purchased from QRec Chemicals. Nutrient agar and Nutrient broth were provided by Merck. *Staphylococcus aureus* was obtained from DSMZ- German Collection of Microorganisms and Cell Cultures. The experiment will take place in a customized photocatalytic reactor under visible light irradiance using 100 W LED light source.

### 2.2. Preparation of $BiFeO_3$ photocatalyst

The synthesis of $BiFeO_3$ was carried out by gel-combustion method based on the previous study carried out by Hassan & Islam (2015). According to stoichiometric preparation of bismuth ferrite, BiFeO3, the calculated amount of bismuth nitrate pentahydrate, $Bi(NO_3)_3 \cdot 5H_2O$ and iron (III) nitrate nonahydrate $Fe(NO_3)_3 \cdot 9H_2O$ were dissolved in citric acid solution at 60 °C of constant stirring for 30 minutes. The amount of citric acid needed was equal to the molar amount of metal nitrates in the solution. Subsequently, ammonia solution was added dropwise to adjust the pH to 7 and subsequently stabilize the solution. The nitrate-citrate solution was then heated to 130 °C for 2 hours to obtain viscous gel. The resultant gel evaporates all water molecules from the solution and vigorous fuming was occurred which subjected to combustion reaction. The powders were annealed at temperature of 400°C overnight to obtain homogenous $BiFeO_3$ nanoparticles. This activation process will transform the precursor into loose powder. The obtained powder was then collected as the $BiFeO_3$ nanoparticles.

### 2.3. Characterization of $BiFeO_3$ photocatalyst

The crystallinity and the phase purity of the samples were determined by powder X-ray diffraction (XRD) using an X-ray diffractometer (PANanalytical X'Pert PRO, Japan) with Cu Kα radiation ($\lambda$ = 0.154056 nm). The prepared materials were characterized by UV–vis Diffuse Reflectance Spectroscopy (DRS) measurements using a UV–vis-NIR spectrophotometer (Lambda 1050 PerkinElmer). Brunauer-Emmett-Teller (BET) surface area were determined from nitrogen adsorption-desorption method.

### 2.4. Photocatalytic disinfection experiment

The strain *S.aureus* used in this study was purchased from DSMZ- German Collection of Microorganisms and Cell Cultures. All apparatuses utilized were sterilized by an autoclave at 121°C. *S.aureus* was grown in nutrient broth and incubated overnight at 37°C under constant shaking. The bacterial cells were centrifuged and washed with saline solution (0.9% NaCl). After that, the bacterial stock solutions were re-suspended in sterilized saline solution and initial bacterial concentration was approximately 3 X $10^6$ CFU/mL. The experiment was conducted in a photocatalytic reactor under irradiation of 100 W visible light. Upon disinfection experiment, bacterial suspension was stirred in the dark for 30 min to attain the equilibrium. The photocatalyst (1 mg) was dispersed in bacterial suspension under continuous stirring. The samples (1 mL) were collected at predetermined time intervals and cultivated onto the Nutrient agar plate to assess the viability of bacteria. Prior to colony count, the plates were incubated overnight at 37°C. Three replicate plates were used in each sampling time. Three control experiments procedure was conducted as stated: (i) light control (without photocatalyst), (ii) dark control (without visible light) and negative control (without light and photocatalyst).

### 2.5. *Characterization of the morphology of S.aureus*
The morphology of treated *S.aureus* by $BiFeO_3$ was examined by TEM analysis (Hitachi).

## 3. Results and discussion

### 3.1. *Characterization of $BiFeO_3$ photocatalyst*
The crystal structure of the prepared samples were investigated by XRD analysis. Figure 1 shows the XRD patterns of the $BiFeO_3$ nanoparticles. The synthesized $BiFeO_3$ demonstrated the formation of rhombohedral perovskite structure as all the diffraction correspond to JCPDS card no 82-2327. The most prominent peak of $BiFeO_3$ can be seen at 32.08° indexed correspond to (110) diffraction plane. The characteristic of this peak indicates that $BiFeO_3$ has the perovskite-type rhombohedral structure [12]. From the XRD pattern, sharp and high intensity with narrow width of the observed peaks signifies that the $BiFeO_3$ nanoparticles are well crystallized [13][14]. The average of crystallite size of nanoparticles was calculated using Scherrer's formula. The calculated crystallite size of $BiFeO_3$ was about 12 nm.

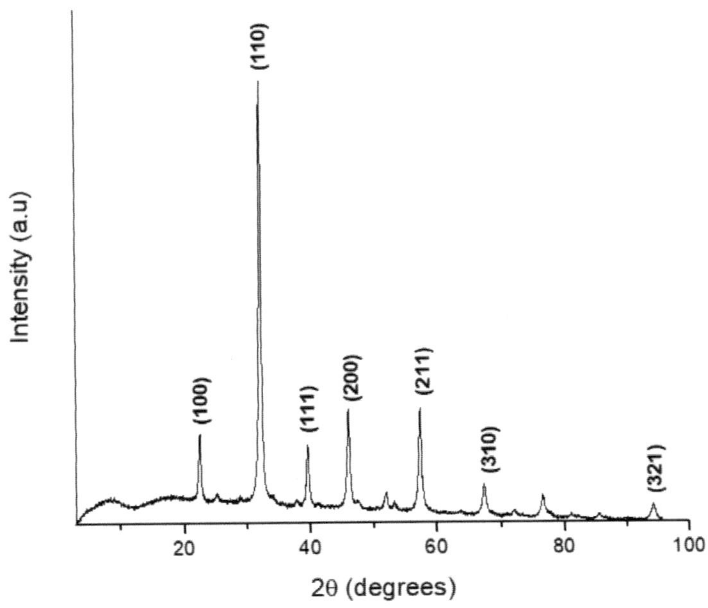


Figure 1. XRD pattern of $BiFeO_3$ nanoparticles.

Understanding optical properties of photocatalyst is one of the important key factor in photocatalytic studies. The UV-Vis diffuse reflectance spectra was performed to observe the optical absorption of prepared $BiFeO_3$ nanoparticles. As shown in Figure 2, the absorption edge of $BiFeO_3$ was found at approximately 580 nm implying that $BiFeO_3$ can be activated under visible-light source. Visible light wavelength range is between 400 nm to 700 nm [15]. Strong absorption edge in perovskite-type oxide can mainly be ascribed to the electronic transition from the valence band to conduction band [16][17]. The optical band gap energy ($E_g$) is a crucial parameter to assess the activity of prepared photocatalyst, which the low band gap values of materials often exhibit the best performances in photocatalytic activity [18]. Generally, the common acceptable band gap value of the most photocatalyst are below 3 eV, making it desirable to be activated under visible-light irradiation [19]. The band gap value of the materials was observed by plotting $(\alpha hv)^2$ versus photon energy ($hv$), where $\alpha$, $h$ and $v$ were absorption coefficient, Planck constant and light frequency, respectively. The calculated band gap value of the $BiFeO_3$ was shown in Figure 2 (b) below. The band gap value for the synthesized samples was 2 eV, thus making it desirable materials to be activated under visible light. Hence, $BiFeO_3$ photocatalyst can be considered for further study as a great visible light photocatalyst.

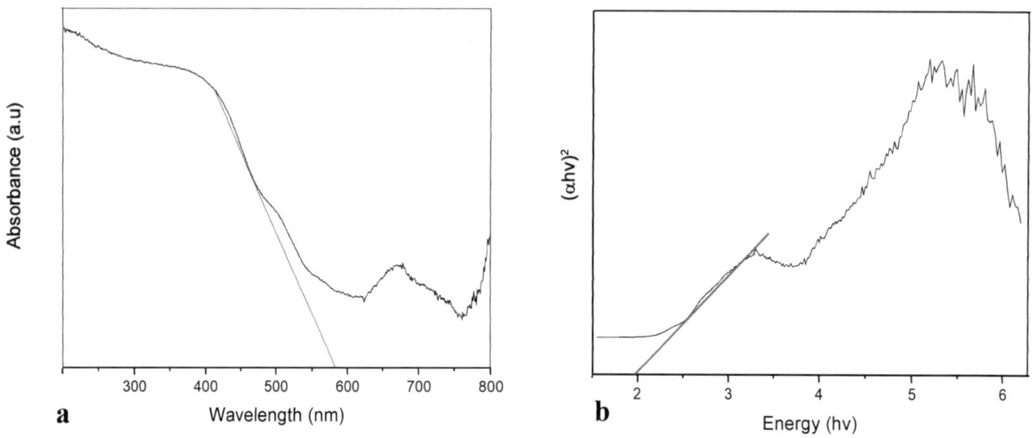


Figure 2. (a) UV-Vis absorption spectra of $BiFeO_3$ (b) Kubelka-Munk function $BiFeO_3$ to estimate band gap value.

The specific surface area and pore size distribution are of important role in determining its capabilities to absorb and degrading pollutants or microbial inactivation. $BiFeO_3$ exhibited BET surface area, pore size and total pore volume of 27.61 $m^2/g$, 16.40 nm and 0.11 $cm^3/g$, respectively. These findings are consistent with previous reported studies [8]. High surface area provides more active adsorption sites and promotes better adsorption capacities in photocatalytic activity [20]. Figure 3 showed nitrogen adsorption-desorption isotherms and the pore size distribution curves of $BiFeO_3$. According to IUPAC classification, $BiFeO_3$ shows a typical type III isotherm with H3 hysteresis loops, implying the presence of certain amount of macroposity in the materials.

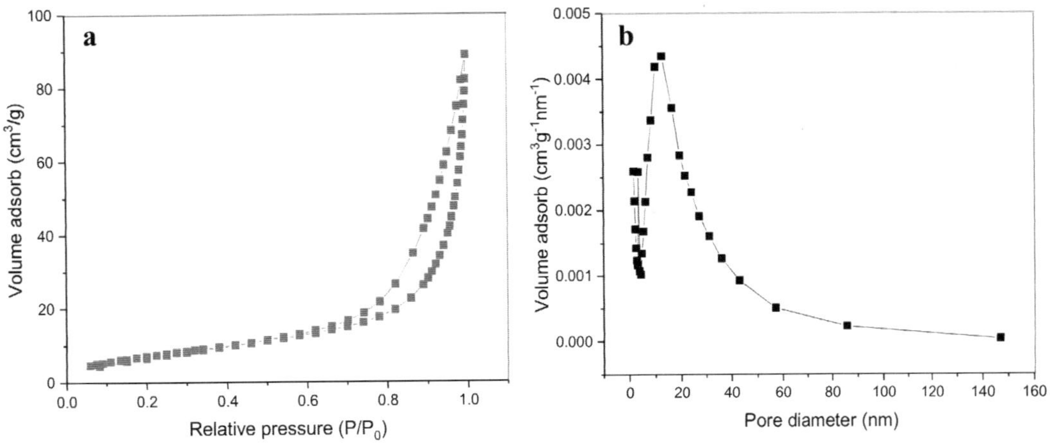


Figure 3. (a) Nitrogen adsorption-desorption isotherm (b) pore size distribution curves of $BiFeO_3$.

### 3.2. *Morphological change of S.aureus exposed to $BiFeO_3$*

The change of morphological bacterial cells during the disinfection process were observed under TEM images. Figure 4 (a) shows a smooth morphology of cells which indicated a normal display of bacterial cells. After 10 min reaction with $BiFeO_3$ under visible light, the cells become lysed as shown in Figure 4 (b). After 20 min of light irradiation, cells were irregularly shaped and become severely damaged as shown in Figure 4 (c). The outer membrane of the cells was ruptured which then results in release of intracellular components of the cells. Cells that have been exposed to photocatalyst demonstrated the production of reactive oxygen species during photocatalytic disinfection and interrupted all the unsaturated lipids as well as organic covalent bonds of the cells [1]. This eventually lead to the cellular death.

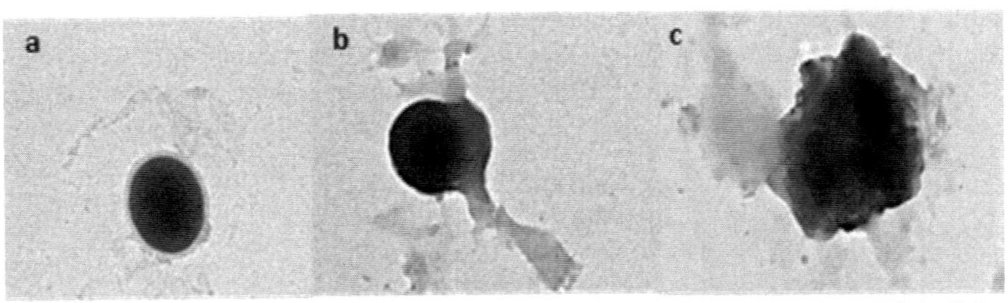


Figure 4. TEM images of S.aureus exposed to $BiFeO_3$ photocatalyst under visible light irradiation at time (a) 0 min (b) 10 min (c) 20 min.

### 3.3. *Disinfection efficiency of photocatalyst*

Photocatalytic disinfection activity of $BiFeO_3$ towards *S.aureus* was investigated at 1 mg of photocatalyst under visible light irradiation as shown in Figure 5. The results clearly show that no noticeable decrease in the viable cell density for all three control experiments, indicating that the bacteria cells were not affected by the absence of photocatalyst and light irradiation (negative control). Also, the presence of photocatalysts without light irradiation (dark control) and the presence of visible light without photocatalyst (light control) were not toxic to the cells. The inactivation of *S.aureus* can be clearly seen when the cells were exposed to $BiFeO_3$ nanoparticles under visible light irradiation. The cell density of *S.aureus* significantly decreased, suggesting that the death of bacterial cells.

Complete inactivation of *S.aureus* at initial concentration 3 x $10^6$ CFU/mL was successfully observed in 20 min by $BiFeO_3$ nanoparticles under visible light irradiation.

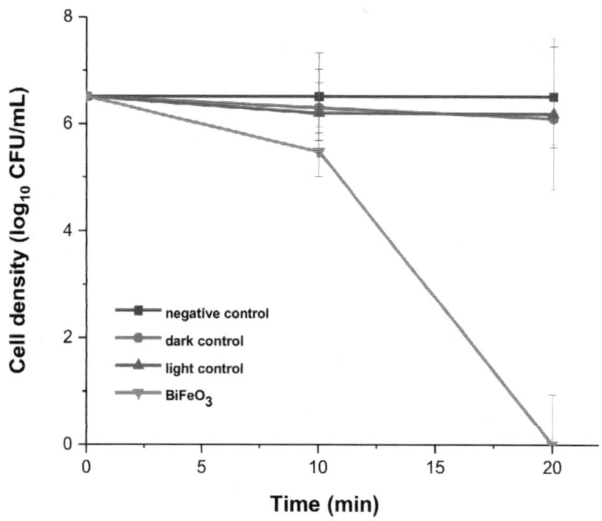


Figure 5. Disinfection efficiencies of $BiFeO_3$ under visible light irradiation with bacterial density 3 × $10^6$. Error bars represent standard deviations from triplicate experiments (n=3).

## 4. Conclusion

In this study, $BiFeO_3$ nanoparticles was synthesized by sol-gel auto combustion method. Visible-light driven photocatalyst, c The absorption intensity of $BiFeO_3$ was found to be in the visible light range. Moreover, high surface of $BiFeO_3$ helps to increase the efficiency of photocatalytic disinfection activity. The as-synthesized $BiFeO_3$ exhibited strong antibacterial activity towards *S.aureus.* Direct contact between the nanoparticles and bacterial cells play a crucial role for the cell inactivation under visible light irradiation. Complete inactivation of *S.aureus* was achieved within 20 min and the damage of cell membrane of the cells led to bacterial inactivation. Taking into consideration, $BiFeO_3$ nanoparticles would be a great potential photocatalyst for an excellent photocatalytic disinfection of microbes.

Acknowledgement
This work was supported by the Universiti Teknologi Malaysia under Collaborative Research Grant [Grant number: Q.J130000.2451.08G85], Fundamental Research Grant Scheme (FRGS) [Grant number: R.J130000.7851.5F007] and Japan International Cooperation Agency (JICA) under Special Program for Research Against Covid-19 (SPRAC) Aun/Seed-Net [Grant number: RJ130000.7351.4B610].

## References

[1] J. Deng, J. Liang, M. Li, and M. Tong, "Enhanced visible-light-driven photocatalytic bacteria disinfection by g-C3N4-AgBr," *Colloids Surfaces B Biointerfaces*, vol. 152, pp. 49–57, 2017.

[2] C. Regmi, Y. K. Kshetri, T. Kim, R. Prasad, and S. Wohn, "Visible-light-induced Fe-doped BiVO 4 photocatalyst for contaminated water treatment," *Mol. Catal.*, vol. 432, pp. 220–231,

2017.
[3] X. Zheng, Z. peng Shen, C. Cheng, L. Shi, R. Cheng, and D. hai Yuan, "Photocatalytic disinfection performance in virus and virus/bacteria system by Cu-TiO2 nanofibers under visible light," *Environ. Pollut.*, vol. 237, pp. 452–459, 2018.
[4] E. Shang, J. Niu, Y. Li, Y. Zhou, and J. C. Crittenden, "Comparative toxicity of Cd, Mo, and W sulphide nanomaterials toward E. coli under UV irradiation," *Environ. Pollut.*, vol. 224, pp. 606–614, 2017.
[5] B. Zhang, S. Zou, R. Cai, M. Li, and Z. He, "Highly-efficient photocatalytic disinfection of Escherichia coli under visible light using carbon supported Vanadium Tetrasulfide nanocomposites," *Appl. Catal. B Environ.*, vol. 224, no. October 2017, pp. 383–393, 2018.
[6] Y. Li, J. Zhao, G. Zhang, L. Zhang, S. Ding, and E. Shang, "Visible-light-driven photocatalytic disinfection mechanism of Pb-BiFeO3/rGO photocatalyst," *Water Res.*, vol. 161, pp. 251–261, 2019.
[7] S. Chakraborty and M. Pal, "Highly efficient novel carbon monoxide gas sensor based on bismuth ferrite nanoparticles for environmental monitoring," *New J. Chem.*, vol. 42, no. March, pp. 7188–7196, 2018.
[8] S. Irfan, Z. Zhuanghao, F. Li, Y. X, Chen, G. X. Liang, J. T. Luo and F. Ping, "Critical review: Bismuth ferrite as an emerging visible light active nanostructured photocatalyst," *J. Mater. Res. Technol.*, vol. 8, no. 6, pp. 6375–6389, 2019.
[9] N. Wang, X. Luo, L. Han, Z. Zhang, R. Zhang, and H. Olin, "Structure , Performance , and Application of - Nanomaterials," no. 0123456789, 2020.
[10] X. Shi, S. Quan, L. Yang, G. Shi, and F. Shi, "Chemosphere Facile synthesis of magnetic Co 3 O 4 / BFO nanocomposite for effective reduction of nitrophenol isomers," *Chemosphere*, vol. 219, pp. 914–922, 2019.
[11] M. Hasan, M. F. Islam, R. Mahbub, M. S. Hossain, and M. A. Hakim, "A soft chemical route to the synthesis of BiFeO3 nanoparticles with enhanced magnetization," *Mater. Res. Bull.*, vol. 73, pp. 179–186, 2016.
[12] L. Jiquan, Y. Wang, H. Ling, Y. Qiu, J. Lou, X. Hou, S.P. Bag, J. Wang, H. Wu and G. Chai. "Significant Enhancement of the Visible Light Photocatalytic Properties in 3D BiFeO 3 / Graphene Composites," *Nanomaterials*, vol. 9, no. 1, p. 65, 2019.
[13] H. Wu, W. Xia, and X. Zhu, "Synthesis of single-crystalline BiFeO3 nanoparticles and their magnetic properties," *Ferroelectrics*, vol. 519, no. 1, pp. 1–8, 2017.
[14] S. Godara, N. Sinha, G. Ray, and B. Kumar, "Combined structural, electrical, magnetic and optical characterization of bismuth ferrite nanoparticles synthesized by auto-combustion route," *J. Asian Ceram. Soc.*, vol. 2, no. 4, pp. 416–421, 2014.
[15] B. H. Mahmoud *et al.*, "Impact of Long-Wavelength UVA and Visible Light on Melanocompetent Skin," *J. Invest. Dermatol.*, vol. 130, no. 8, pp. 2092–2097, 2010.
[16] R. D. K. R. Jayavel, "Facile hydrothermal synthesis and characterization of LaFeO 3 nanospheres for visible light photocatalytic applications," pp. 3953–3961, 2014.
[17] S. Thirumalairajan, K. Girija, N. Y. Hebalkar, D. Mangalaraj, C. Viswanathan, and N. Ponpandian, "Shape evolution of perovskite LaFeO3nanostructures: A systematic investigation of growth mechanism, properties and morphology dependent photocatalytic activities," *RSC Adv.*, vol. 3, no. 20, pp. 7549–7561, 2013.
[18] A. C. Martins, A. L. Cazetta, O. Pezoti, J. R. B. Souza, T. Zhang, E. J. Pilau, T. Asefa and V.C. Almeida, "Sol-gel synthesis of new TiO2/activated carbon photocatalyst and its application for degradation of tetracycline," *Ceram. Int.*, vol. 43, no. 5, pp. 4411–4418, 2017.
[19] M. M. Mahlambi, C. J. Ngila, and B. B. Mamba, "Recent Developments in Environmental Photocatalytic Degradation of Organic Pollutants : The Case of Titanium Dioxide Nanoparticles — A Review," *J. Nanomater.*, vol. 2015, pp. 1–29, 2015.
[20] J. Wei, J. Yang, Z. Wen, J. Dai, Y. Li, and B. Yao, "Efficient photocatalytic oxidation of methane over β-Ga2O3/activated carbon composites," *RSC Adv.*, vol. 7, no. 60, pp. 37508–37521, 2017.

# Estimation of energy consumption during cross-flow membrane filtration of oil/water emulsion

**N U Barambu[1], M R Bilad[1,2], N A H Nordin[1], and M A Bustam[1]**

[1] Department of Chemical Engineering, Universiti Teknologi PETRONAS, 32610, Bandar Seri Iskandar, Perak, Malaysia; barambunafiu@gmail.com (N.U.B); mroil.bilad@utp.edu.my (M.R.B); nahadi.sapiaa@utp.edu.my (N. A. H); azmibustam@utp.edu.my (M.A.B)
[2] HICoE-Centre for Biofuel and Biochemical Research, Institute of Self-Sustainable Building, Universiti Teknologi PETRONAS, 32610, Seri Iskandar, Perak, Malaysia

**Abstracts:** Pressure driven membrane filtration has been widely adopted for oil and water recovery from oil/water emulsion. The technology offers numerous economic and environmental advantages. However, specific energy consumption remains a critical factor for its economic assessment. The specific energy consumption depends on several factors, among which is the fouling filtration period/time. Therefore, the fouling filtration time indirectly affects the overall economic viability of the system. Herein, the effect of fouling filtration time on the specific energy consumption during cross-flow filtration of 1000 ppm oil/water emulsion was evaluated. During the first 20 min of oil/water emulsion filtration, the energy consumption was found to be 0.0225 KWh/$m^3$. While the filtration was continued to 90 min, the system requires 0.0301 KWh/$m^3$ energy input. The tradeoff between fouling filtration time and energy consumption was found to be just before 30 mins. Suggesting that at 30 mins fouling filtration time, clean water flushing/backwashing would restore significant hydraulic performance and thus, less increase in energy input would be required. At the tradeoff point 0.0254 KWh/$m^3$ energy was required to drive the filtration. This represents up to ~19 % energy saving compared to the straight nonstop 90 min fouling filtration time. The overall results demonstrated that fouling filtration time affects the overall economics of cross-flow membrane filtration.

**Keywords:** Oil/water emulsion; Pressure driven membrane filtration; fouling filtration time; energy consumption

## 1. Introduction

A global challenge has been the proper handling of oil/water emulsion produced by oil and gas, pharmaceuticals, and the food industry, etc. [1–3]. Conventional techniques such as dissolved air floatation and many others are inefficient to handle oil/water emulsion properly [4,5]. Pressure-driven membrane processes provide the best economic and eco-friendly option for the treatment of oil/water emulsion [6]. However, pressure-driven membrane systems require energy input to drive their overall separation performances [7–9]. Therefore, an assessment of the energy consumption of the system remains one of the critical factors for its continuous widespread adoption [10]. The energy consumption depends on several factors, among which are the feed property, membrane properties, and operating parameters [10–12]. Those factors directly or indirectly affect the overall membrane performance and thus, affect the energy consumption of the system. Since energy consumption sorely depends on membrane performance [7,13].

Feed property involves the concentration of the model foulant in the feed, which directly affects the membrane performances and therefore affects the energy consumption as well [9]. Membrane properties such as surface wetting property, membrane thickness, and pore size and porosity have a significant effect on the hydraulic and fouling performances of the membrane [14–16]. The operating parameters constitute the operating pressure, filtration set-up, and fouling filtration time [14]. Those factors contributed to the

overall energy consumption of the membrane filtration system. And therefore, should be considered during the economic assessment of the technology [7,11]. However, many of those are rarely reported in energy consumption assessment such as fouling filtration time [14].

Recently, Li et al. [17] investigated the effect of rotating disk with vanes on energy consumption during linseed oil/water emulsion filtration. The system recorded up to ~140% energy saving against the smooth disk. The smooth disk and disk with vanes exhibited 1070 kWh/m$^3$ and 445 kWh/m$^3$ specific energy consumption, respectively. Substantial reduction in energy input was achieved thanks to the improved hydrodynamic to suppress membrane fouling. Krsti´c et al. [18] reported energy saving of up to ~40 % using the static mixer as turbulence promoter during filtration of cutting oil/water emulsion. Similarly, Vatai et al. [19] also reported tremendous energy saving and attributed it to the hydrodynamic effect induced by the static mixer and the sparged air bubbles.

However, the recently reported energy assessment during oil/water emulsion treatment does not take into consideration the effect of fouling filtration time during the energy assessment [17–19]. Therefore, this manuscript found it necessary to evaluate it effect on the overall energy consumption. This manuscript reported the effect of fouling filtration time on energy consumption during cross-flow filtration of oil/water emulsion. The oil concentration and the trans-membrane pressure (ΔP) of 1000 ppm and 0.2 bar respectively were used throughout the filtration period. The fouling filtration time range 10 to 90 mins, with 10 mins intervals were used, and their respective energy consumptions were evaluated.

## 2. Material and Methods

### *2.1 Membrane properties and filtration set-up*

The polysulfone ultra-filtration membrane fabricated in the previous study [20] was adopted in this research. The membrane properties are summarized in **Table 1**. Whereas the cross-flow filtration set-up shown in **Figure 1** was used in this study.

**Table 1**. Physical properties of the PSF membrane.

| Membrane | Pore size (μm) | Porosity (%) | Thickness (μm) | Contact angle (°) | Rejection (%) |
|---|---|---|---|---|---|
| PSF | 0.057 | 66.1 ± 0.2 | 234.3±1.3 | 67.1 ± 0.5 | 97.1 ± 2.2 |

The filtration set-up is a simplified one that constituted only the feed pump as the primary source of energy input. The pump outlet pressure was enough to recirculate the retentate to the feed tank, and therefore a recirculation pump was not included in the set-up.

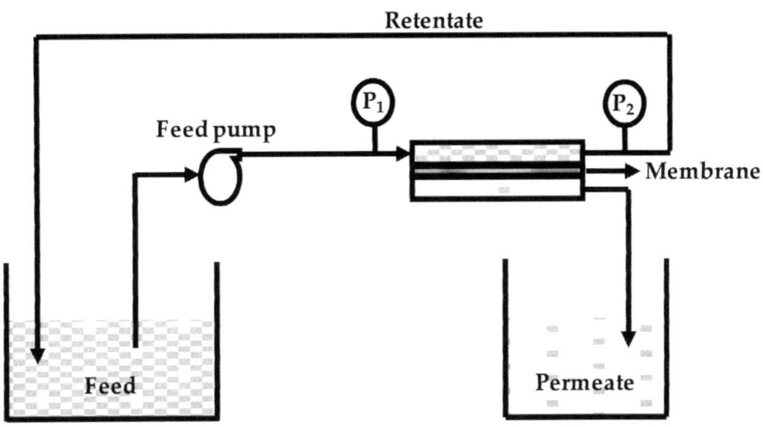

**Figure 1**. Cross-flow filtration set-up for flat and wavy membrane.

*2.2 Oil/water emulsion filtration*

The fouling filtration test was conducted according to the previous study [20]. Briefly, the membrane sample was cut and installed into the filtration cell with a 0.0037 $m^2$ effective area. The membrane was compacted using clean water at $\Delta P$ of 0.2 bar until nearly three same consecutive values were recorded and reported as clean water permeability. The filtration was continued for 90 mins with oil/water emulsion as the feed, and the respective permeabilities were reported every 10 mins. The clean water and oil/water emulsion permeabilities were evaluated using **Equation 1**. The oil/water emulsion was prepared by mixing crude oil obtained from one of the Petronas oilfields in Malaysia with water that contains 1% surfactant (Sodium Dodecyl Sulfate, Sigma Aldrich, USA). The mixing speed of 3500 rpm was used for a period of 24 h.

$$L = \frac{V}{A.t.\Delta P} \quad (1)$$

Where V is the volume of the permeate collected (L), t is the filtration time (mins.), $\Delta P$ is the trans-membrane pressure (bar), and A is the effective membrane area ($m^2$).

*2.3 Estimation of energy consumption*

In cross-flow filtration set-up, the feed pump is the only energy-consuming component and therefore, is the main component for energy consumption assessment. For the cross-flow filtration set-up, the energy consumption per unit volume of the permeate produced ($P_p$, $KWh/m^3$) was evaluated using **Equation 2.**

$$P_p = \frac{\dot{m}W_p}{V_P} \quad (2)$$

Where $\dot{m}$ is the mass rate (kg/s) defined by **Equation 3**, $W_p$ is the work done by the feed pump (J/kg) defined by **Equation 4**, and $V_P$ is the volume of the permeate collected ($m^3$).

$$\dot{m} = \rho \mathcal{V} w L \quad (3)$$

$$W_p = \frac{\Delta P}{\rho} + \frac{\mathcal{V}^2}{2} + F \quad (4)$$

Where $\rho$ is the density of water ($kg/m^3$), $\mathcal{V}$ is the cross-flow velocity (m/s), $w$ is the width of the membrane (m), L is the depth of the flow area (m), $\Delta P$ is the trans-membrane pressure (Pa), and F is the laminar flow friction loss ($m^2/s^2$).

## 3. Results and Discussion

*3.1 Oil/water emulsion filtration and energy consumption*

As shown in **Figure 2**, the membrane performances declined rapidly. In the first 30 min of the fouling filtration, the performance declined by 27 % while after 90 min, the performance declined by ~39 %. The continuous decline in the performance was due to the continuous increase in hydraulic resistance caused by membrane fouling [21]. Therefore, the declined membrane performance contributed to the increase in the overall power consumption of the system.

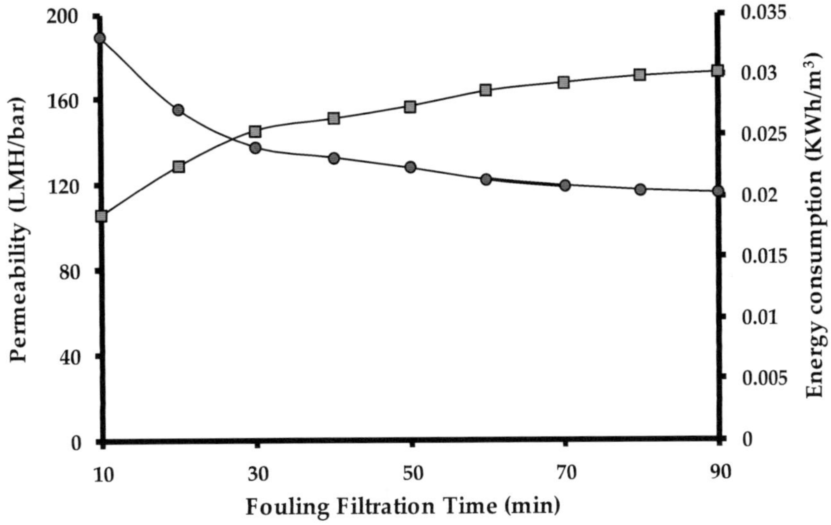


**Figure 2**. Membrane performances and energy consumptions during oil/water filtration.

As the membrane performances decline with time, the energy consumption increases due to the increased hydraulic resistance [7]. To maintain the membrane performance and compensate for the increase in hydraulic resistance, more energy input would be required [12]. In the first 10 min of the fouling filtration, the system requires only 0.0185 KWh/m$^3$ to drive the filtration. However, as the filtration was continued, the system requires up to 0.0301 KWh/m$^3$ to maintains its performance, which is ~63 % more energy input. This is very huge most especially when considering a full-scale operation.

However, the performances and the energy consumption intercepted just before 30 min of operation as shown in **Figure 2**. The point of interception locates the appropriate time where clean water flushing/backwash would restore significant membrane performance. And therefore, the membrane performances would be prolonged, and less increase in the energy input would be required. 0.0254 KWh/m$^3$ energy was required to drive the filtration at the point of interception. This accounted for ~19 % energy saving compared to the straight nonstop 90 min fouling filtration. The interception point is the tradeoff between filtration time and energy consumption, it gives a guide most especially in auto mode operation to set the backwash setting.

## 4. Conclusion

This study explores the influence of fouling filtration time on energy consumption during cross-filtration of 1000 ppm oil/water emulsion. Continuous performance decline coursed by membrane fouling was observed with an increase in the filtration time. Moreover, the energy consumption increases with an increase in the fouling filtration time. The trade-off point between the fouling filtration time and energy consumption was found to be at 30 min of fouling filtration time. At the trade-off point, the system demonstrated up to ~19 % energy saving compared to the straight nonstop 90 min fouling filtration. Thanks to the clean water flushing that restored significant membrane performances and restricted the hydraulic resistance buildup. And therefore, more energy saving.

## References

[1] Sun Y, Yang Z, Li L, Wang Z and Sun Q 2019 Facile preparation of isotactic polypropylene microporous membranes with bioinspired hierarchical morphology for nano-scale water-in-oil

emulsion separation *J. Memb. Sci.* **581** 224–35

[2] Ding L, Gao J and Chung T S 2019 Schiff base reaction assisted one-step self-assembly method for efficient gravity-driven oil-water emulsion separation *Sep. Purif. Technol.* **213** 437–46

[3] Paixão M V G and Balaban R de C 2018 Application of guar gum in brine clarification and oily water treatment *Int. J. Biol. Macromol.* **108** 119–26

[4] Ismail N H, Salleh W N W, Ismail A F, Hasbullah H, Yusof N, Aziz F and Jaafar J 2020 Hydrophilic polymer-based membrane for oily wastewater treatment: A review *Sep. Purif. Technol.* **233** 1–18

[5] Zoubeik M, Ismail M, Salama A and Henni A 2018 New Developments in Membrane Technologies Used in the Treatment of Produced Water: A Review *Arab. J. Sci. Eng.* **43** 2093–118

[6] Wan Ikhsan S N, Yusof N, Aziz F, Misdan N, Ismail A F, Lau W J, Jaafar J, Wan Salleh W N and Hayati Hairom N H 2018 Efficient separation of oily wastewater using polyethersulfone mixed matrix membrane incorporated with halloysite nanotube-hydrous ferric oxide nanoparticle *Sep. Purif. Technol.* **199** 161–9

[7] Massé A, Thi H N, Legentilhomme P and Jaouen P 2011 Dead-end and tangential ultrafiltration of natural salted water: Influence of operating parameters on specific energy consumption *J. Memb. Sci.* **380** 192–8

[8] Krzeminski P, Van Der Graaf J H J M and Van Lier J B 2012 Specific energy consumption of membrane bioreactor (MBR) for sewage treatment *Water Sci. Technol.* **65** 380–92

[9] Liu C, Rainwater K and Song L 2012 Calculation of energy consumption for crossflow RO desalination processes *Desalin. Water Treat.* **42** 295–303

[10] Eliseus A, Bilad M R, Nordin N A H M, Putra Z A and Wirzal M D H 2017 Tilted membrane panel: A new module concept to maximize the impact of air bubbles for membrane fouling control in microalgae harvesting *Bioresour. Technol.* **241** 661–8

[11] Fenu A, Roels J, Wambecq T, de Gussem K, Thoeye C, de Gueldre G and van de Steene B 2010 Energy audit of a full scale MBR system *Desalination* **262** 121–8

[12] Stephens B, Novoselac A and Siegel J A 2010 The effects of filtration on pressure drop and energy consumption in residential hvac systems (rp-1299) *HVAC R Res.* **16** 273–94

[13] Stillwell A S and Webber M E 2016 Predicting the specific energy consumption of reverse osmosis desalination *Water (Switzerland)* **8** 1–18

[14] Scott K, Mahmood A J, Jachuck R J and Hu B 2000 Intensified membrane filtration with corrugated membranes *J. Memb. Sci.* **173** 1–16

[15] Prince J A, Bhuvana S, Anbharasi V, Ayyanar N, Boodhoo K V K and Singh G 2016 Ultra-wetting graphene-based PES ultrafiltration membrane – A novel approach for successful oil-water separation *Water Res.* **103** 311–8

[16] Long Y, Shen Y, Tian H, Yang Y, Feng H and Li J 2018 Superwettable Coprinus comatus coated membranes used toward the controllable separation of emulsified oil/water mixtures *J. Memb. Sci.* **565** 85–94

[17] Li L, Ding L, Tu Z, Wan Y, Clausse D and Lanoisellé J L 2009 Recovery of linseed oil dispersed within an oil-in-water emulsion using hydrophilic membrane by rotating disk filtration system *J. Memb. Sci.* **342** 70–9

[18] Krsti D M and Wilhelm H 2007 Energy-saving potential of cross-flow ultrafiltration with inserted static mixer : Application to an oil-in-water emulsion ଓ **57** 134–9

[19] Vatai G N, Krstic D M, Höflinger W and Koris A K 2007 Combining air sparging and the use of a static mixer in cross-flow ultrafiltration of oil / water emulsion **204** 255–64

[20] Barambu N U, Bilad M R, Wibisono Y, Jaafar J, Mahlia T M I and Khan A L 2020 Development of Polysulfone Membrane via Vapor-Induced Phase Separation for Oil / Water *Polymers (Basel).* **12** 1–16

[21] Barambu N U, Bilad M R, Wibisono Y, Jaafar J, Mahlia T M I and Khan A L 2019 Membrane

surface patterning as a fouling mitigation strategy in liquid filtration: A review *Polymers (Basel).* **11** 1–14

# Evolution of Photobioreactors: A Review based on Microalgal Perspective

**Imran Ahmad[1*], Norhayati Abdullah[1*], Iwamoto Koji[1], Ali Yuzir[1], Shaza Eva Muhammad[1]**

[1]Algae and Biomass Research Laboratory, Malaysia-Japan International Institute of Technology (MJIIT), Universiti Teknologi Malaysia, Jalan Sultan Yahya Petra, 54100, Kuala Lumpur, Malaysia

*mustafwibinqamar@gmail.com; norhayati@utm.my

**Abstract**. Energy is indispensable for bringing competence, resilience, and comfort for the ever-advancing human lifestyle. Therefore, to satisfy the growing energy requirement researchers are trying to delve in exploring sustainable and renewable energy sources that will contribute to the reduction of carbon footprint and ultimately help to check the issue of global warming caused by the increased emission of CO2, due to the excessive usage of non-renewable and exhaustive fossil fuels. Microalgae are having the potential which can be exploited to produce biofuels (substitute for fossil fuels) and other value-added compounds. Microalgae is a form of unicellular photoautotrophic microorganism which can attain higher rate of proliferation. Microalgae is having merits over other terrestrial crops and fossil fuels as they have higher productivity of oil per hectare of land. Various other advantages of microalgae include their tolerance in different types of environment because of their ubiquitous nature (in terms of pH, salinity, and temperature). Furthermore, they can be cultivated in nonarable land and wastewater which will resolve the food-fuel duel and problem of freshwater usage, therefore leading towards environmental sustainability. The main factors governing the growth and cultivation of microalgae are light, CO2/air, nutrients, process parameters (pH, temperature, growth medium etc) and most importantly the type of system used for cultivation. The systems usually employed for the cultivation of microalgae are open system (open ponds, raceway ponds, scrubbers), closed system also commonly known as photobioreactors (flat plate and tubular photobioreactors) and hybrid system in which separation of biomass growth and lipid accumulation is achieved in two stages. All the systems have some merits and demerits but photobioreactors are widely accepted and used because they are having an upper hand over open system because of the optimised control of the growth conditions, contamination evasion and efficient productivity of microalgal biomass. This review will provide an insight to different parameters which govern the growth of microalgae and various types of photobioreactors with their advantages and disadvantages. This study will help in the optimized selection of the photobioreactors for a particular species of microalgae because despite the continuous and intensive research going on the cultivation systems it is a challenge for the researchers to achieve a suitable and economically viable system.

**Keywords**: Microalgae; Photobioreactors; biomass; open ponds; energy and environment

## 1. Introduction

Algae are considered as one of the common and impactful living organisms which are having various applications. They can be used as feedstock to produce biofuels, are a rich source of pigments and a tool for environmental bioremediation. Various useful products obtained from algal biomass are shown in Figure 1. That is why scientists are emphasizing on algal research to find efficient and viable technologies to be used in the bulk production of algal biomass to match the ever-increasing demand. Figure 2 shows the increasing interest of researchers for the microalgal aspect of photobioreactors in last three decades. Since the natural environmental bodies are lagging in the production of but of algal biomass thus it led to the development of manmade cultivation systems to increase the yield of algal biomass. The initialization off mono algal cultivation was made in 1890 using microalgae *chlorella vulgaris,* the study focused on the physiology of the algae. It is reported that approximately 5000 tons of dry algae are produced every year on the global scale [1]. It is considered as green gold of future. There are different technologies which are employed for the mass production of algal biomass i.e., the open systems and the closed systems. Open systems are simple and economic while the closed systems have higher biomass yield as well as they have better control of the operational parameters. Open ponds were used for the cultivation of *chlorella pyrenoidosa* and *scenedesmus* and the diatom *nitschea palea* during World War II in Germany for the purpose of studying their lipid content [2]. Initially the photobioreactors (PBRs) were developed in 1950s. The application of PBRs for the removal of pollutants was first studied at Carnegie Institute, USA in 1953 and the main objective was $CO_2$ sequestration [3].

PBRs are robust cultivation systems which employ the effective cultivation techniques this providing all the essential elements i.e., light, nutrients, temperature and mixing for the healthy growth of microalgae. the operating conditions are monitored, regulated, and controlled for the high yield of algal biomass [4]. PBR can be defined as a closed vessel for the photoautotrophic cultivation of microalgae where energy is supplied by artificial light or sunlight. PBRs can be installed outdoor or indoor depending upon the light collection and distribution together with their commercial feasibility. There are different types of PBRs based on their configuration and operation modes but the most common are flat plate, tubular, and column PBRs. Every microalga based PBRs are having their advantages and drawbacks in terms of operation mode, biomass yield, remediation potential and upscaling level [5]. The research is going on and the PBRs are continuously being modified and improved for the optimization of biomass yield, economic aspects, efficiency in bioremediation, space factors and resilience under varying environmental conditions [6]. This review will focus on the different growth factors and growth modes of microalgae and detailed classification of open and closed system of microalgal cultivation with greater emphasis on the closed PBRs.

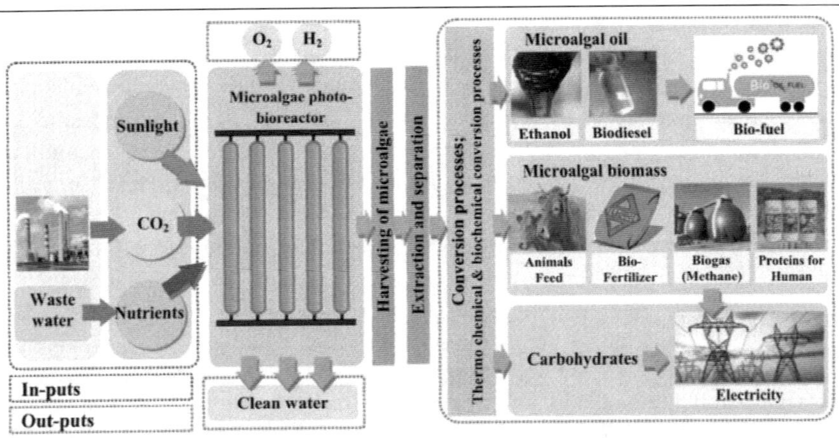


**Figure 1.** Microalgae the *green gold* of future [7]

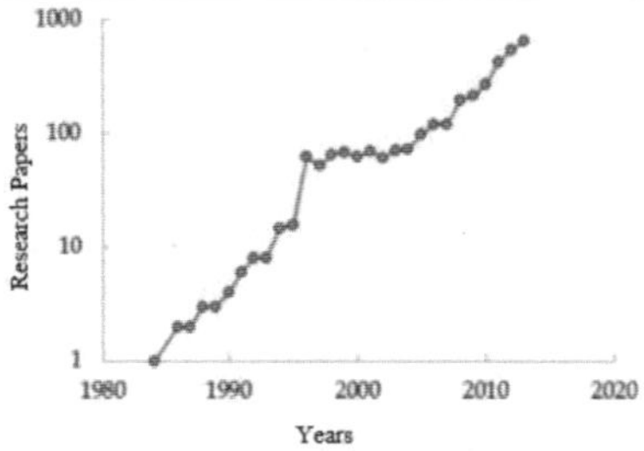


**Figure 2.** Research papers published on the topic of microalgal photobioreactors used in the mass cultivation to obtain biomass used in various applications as shown above [8].

## 2. Factors affecting the growth of microalgae

To attain the maximum production of microalgae via PBRs studies are going on to optimize the shapes of PBRs, regulating and controlling the environmental conditions, gas diffusion in PBRs and operational parameters like temperature, pH etc., (Table 1). The key elements which are responsible for algal growth are light source for the photosynthesis, medium containing proper nutrients and $CO_2$/air flow [1].

**Table 1.** Factors affecting growth of microalgae

| PBR Design and Installation | Environmental Parameters | Operational Parameters |
|---|---|---|
| Material | Light | pH |
| Geometric configuration | Temperature | Nutrients |
| Location and Orientation | $CO_2$/Air | Water and Salinity |
| Scaling | Predation or contamination | $O_2$ Removal |
| Surface area to volume ratio | | Cleaning |

### *2.1. pH*

pH defines the alkalinity or acidity of the water medium. The optimum range of pH for the growth of microalgae and photosynthetic ability is 6-8. The variation in pH can affect the growth of microalgae in many ways. During the daytime pH increases due to the photosynthetic assimilation of $CO_2$ by microalgae and during the night time pH decreases due to the respiratory process of the species [9]. Increase in $CO_2$ may lead to high biomass accumulation but causes decrease in pH which has an adverse effect on the physiology of microalgae [10].

### *2.2. Nutrients*

For the optimum growth of microalgae phosphorus and nitrogen are required as macro nutrients and Fe, Mg, B, Mo, K, Co, Zn, Mb are required as trace metals. The nutrients especially nitrogen and phosphorus together are of much importance in the production of lipids. Studies have reported that the growth medium which is deficient in nitrogen accelerates the algae to produce more lipids. Both the excess of nutrients as well as their deficiency can cause morphological and physiological changes in microalgae because they impede with the metabolic pathways [11]. Nowadays researchers are going for two phase algal growth, in the first phase micro algae are grown in the medium having excess nutrients and then transferred to the nutrient deficient medium.

### *2.3. Temperature*

Generally, algae can tolerate the temperature range of 15-40°C. Some studies revealed that microalgae can be best grown in the range of 25-30°C. The optimal temperature growth varies with the species of microalgae. It is difficult and expensive to control the temperature in the open cultivation systems because they are more susceptible to overheating, evaporation of the medium, cooling and composition of lipids. This is one of the reasons of opting the close systems i.e., the photobioreactors (PBRs), as it is comparatively easier to control the temperature in these systems [10].

### *2.4. Mixing*
Mixing is vital to prevent the problem of sedimentation and to promote the movement of algae between the dark and light regions of the cultivation systems. Otherwise, the algae at the surface will absorb all the light and can become photo inhibited while the algae which is in the inner or deeper part is deprived of the light. Furthermore, it aids in maintaining the homogeneous cell concentration in the medium. Mixing can be done by using mechanical stirrers like paddlewheels in the ponds or by giving bubbling in the PBRs [9].

### *2.5. Light intensity*
The intensity of light and the efficiency of photosynthesis plays a critical role in the productivity of algae as it affects the rate of growth, biomass production and lipid accumulation. The components which affect the light intensity are the density of algal biomass, depth of culture medium and the type of PBR. Increased light intensity is required if the cell concentration and depth of the culture is on the higher side but it should be kept in mind that sometimes direct sunlight and high-density artificial light may also lead to the photoinhibition. Overheating of the system should be avoided. Light intensity of 1000 Lux is appropriate for the culture in Erlenmeyer flasks while about 5000-10000 Lux is required for larger volumes. An important aspect is to provide the light-dark cycles to enhance the efficiency of photosynthesis. Light cycle is required for photochemical phase to produce ATP[1] and NADPH[2] and the dark cycle is provided for the biochemical phase to synthesize the essential biomolecules for the microbial growth [12].

### *2.6. Predators or invasive species*
They can be any type of living organisms, that are totally unwanted and unexpected in the microalgal culture system because they hinder the microalgae growth, cause nutrient deficiency, and pollute the medium. They can be fungi, insects, bacteria, or predator algae species. This is the reason behind the cultivation of extremophilic microalgae at industrial level because they can survive in extreme environment in which the predator species cannot. Open systems are more vulnerable to invasion by predator species while on the other hand close systems i.e., the PBRs helps in preventing the contamination by keeping the algae protected from the external environment [13].

## 3. Autotrophic vs Heterotrophic vs Mixotrophic Cultivation
Microalgae can go through different types of metabolisms like autotrophic, heterotrophic and mixotrophic [14]. Autotrophic culture is the most common mode of microalgae cultivation. In this cultivation process microalgae directly converts the inorganic carbon present into the organic matter through the photosynthetic process. As most of the microalgal cells are efficient in harnessing solar energy and utilizing $CO_2$ as the carbon source therefore contributing to the abatement of $CO_2$. Hence autotrophic mode is considered as environmentally sustainable and economic mode of microalgae cultivation [15].

In heterotrophic mode of cultivation microalgae use organic carbon for its growth and does not require light. Therefore acetate, glycerol, and volatile fatty acids (VFAs) may be utilized as sources of organic carbon for the synthesis of biomass. In heterotrophic cultivation the biomass yield is better than autotrophic cultivation, but the organic carbon is expensive as compared to $CO_2$. Therefore, to overcome this limitation organic carbon must be utilized from wastewater [16]. Mixotrophic cultivation is to counter the limitations of autotrophic and heterotrophic cultivation modes. So, in this mode of cultivation microalgae is utilizing inorganic carbon through photosynthesis as well as organic carbon.

[1] Adenosine triphosphate
[2] Nicotinamide adenine dinucleotide phosphate hydrogen

The mixotrophic cultivation mode is divided into two stages in the first stage the growth is heterotrophic, and the microalgae cells consume organic carbon present in the medium and when it is consumed up to a certain limit the autotrophic growth is induced and the assimilation of organic carbon start taking place [17].

## 4. Classification of photobioreactors (PBRs): The agencies for microalgae cultivation

The selection and design of PBRs for microalgae cultivation depends upon various factors such as mixing strategy, orientation, surface area to volume ratio (S/V), illumination, air/$CO_2$ supply, accumulation of oxygen and temperature. PBRs can be closed or open systems having various advantages and disadvantages but closed PBRs have an upper hand due to better control of the operational parameters. The prospects and limitations of both type of systems are summarized in the Table 2.

**Table 2.** Advantages and disadvantages of open and close PBRs. Adapted and modified [18, 19]

| No. | Advantages | Disadvantages |
|---|---|---|
| | **Open Systems** | |
| 1 | The construction, operation and maintenance are easy (easy to clean-up). | High risk of contamination by predators and heterotrophs. |
| 2 | Requirement of energy is low. | Poor biomass productivity (0.1-0.2 gm/l). |
| 3 | Economical compared to closed systems. | Harvesting is a tedious job as well as uneconomical. |
| 4 | Better for mass cultivation of microalgae. | $CO_2$ diffusion to atmosphere, evaporation losses and poor light utilization. |
| 5 | Does not compete with agricultural land. | Requirement of large area/space. |
| 6 | Is widely used to obtain protein, Beta carotene, astaxanthin. | Little control of culture and climatic dependence. |
| 7 | Solar trackers can be used for the cultivation in open systems. | Limited to few strains of microalgae and difficult to grow for long periods. |
| | **Closed Systems** | |
| 1 | Better control of culturing conditions and growth parameters. | The CAPEX and OPEX is high. |
| 2 | Little or no dependency on the climate. | Chances of overheating and biofouling. |
| 3 | Reduction in water and $CO_2$ losses. | Accumulation of oxygen can decrease the biomass yield. |
| 4 | Higher microalgal density, concentration and biomass productivity is achieved (2-8 mg/l). | The formation of algal biofilms causes fouling of PBR surfaces hindering the light penetration. |
| 5 | No or little risk of contamination or invasion by the predators | There is difficulty in scaling up of microalgae. |
| 6 | Higher biomass ratio is attained. | Shear stresses may damage the cells. |

### *4.1. Open systems*

Microalgae cultivation in open ponds using the energy obtained from sunlight has been regarded best for the past six decades and can be carried in covered or open places, artificial basins and natural water bodies like lakes and lagoons [20]. Some of the open systems are regarded as favorites due to their simple and economical construction and high production capacities like circular ponds and raceway ponds. Compressed earth and concrete are the building materials for the construction of open ponds.

While the closed ponds are generally constructed using acrylic or plexiglass which cost higher than the open ponds but comparatively less than the closed PBRs for the same cultivation conditions [21]. To fulfill the requirement of nutrients for the cultivation of microalgae nowadays the researchers are going for the integration of microalgae cultivation with wastewater treatment plants as a measure of phyco-remediation and to minimize the upstream processing cost. Open systems are significant in terms of the ease of scalability, economical perspectives, and massive microalgal biomass productivity. Albeit they have many advantages, but unmonitored light intensity and temperature can affect the tolerance of the microalgae thus reducing the biomass productivity. Poor light distribution is also an issue as the light can reach up to a certain depth thus the light distribution is uneven and open ponds can also not hold the optically dark zones [22]. Different configurations of open bioreactor systems are discussed in the following sections.

*4.1.1. Raceway ponds.* In this type of open systems microalgae culture is recirculated along the racetrack loop. They are made shallow with a depth of 15-50 cm to enhance the exposure to sunlight and subsequently good growth rate of microalgae cells [22]. The configuration is equipped with paddle wheels to provide recirculation and mixing as well as the baffles to direct the flow at the bends. The system is operated on a continuous basis with the help of paddle wheels to avoid any chances of sedimentation {Figure 4 and 5}. So that the algae cells are suspended within the culture medium during the cultivation cycle with simultaneous influent feed of microalgae broth and nutrients at the feeding port which is there at the front of paddle wheel. After the complete circulation in the loop microalgae containing water is removed at the harvesting port which is in the rearmost of the paddlewheel [23]. The turbulent mixing abets uniform density of microalgae and the formation of eddies will reduce the residence time of algae in the dark regimes. Microalgae cultivation gravitates towards open raceway ponds due to the advantages possessed by them. The world's largest raceway pond is being operated at Calipatria (USA) meant and specialized for the mass production of *spirulina* based products occupying a tremendous area of 4,44,000 $m^2$ [24]. The algae species reported in previous studies which are cultivated in this type of reactor system are *spirulina, chlorella, dunaliella, haemotococus*.

*4.1.2. Circular pond system.* The ponds have depth of 20-30 cm and diameter 40-50cm. It is the circular pond in which movement and mixing of the culture is assisted by a central pivot rotating agitator {Figure 4 and 6}, but it can only facilitate up to a size of 10,000 $m^2$ [25]. The CAPEX and OPEX is too high which is a barrier for circular ponds to be implemented for commercial scale. The other bottlenecks are the problems of temperature control and the vulnerability to contamination. In Asian countries it is cultivated commercially for *chlorella sp*. To overcome the limitations of open ponds and provide better control of microalgae culture conditions, researchers are going for closed ponds by just covering them with green house and reducing the problem of contamination associated with open ponds [20].

*4.1.3. Inclined surface system.* In this type of system, microalgae suspension flows over the planes having slope and given a cascade arrangement. The thickness of layer is < 1 cm and the arrangement facilitate the turbulence preventing the self-shading. Surface area to volume ratio of 130/m can be operated in these systems leading to high volumetric productivity (Figure 3). Due to short optical path high densities of 15-35 g/l can be operated. Productivity of 40g dry matter $m^{-2}d^{-1}$ can be achieved even in temperate zones [26].

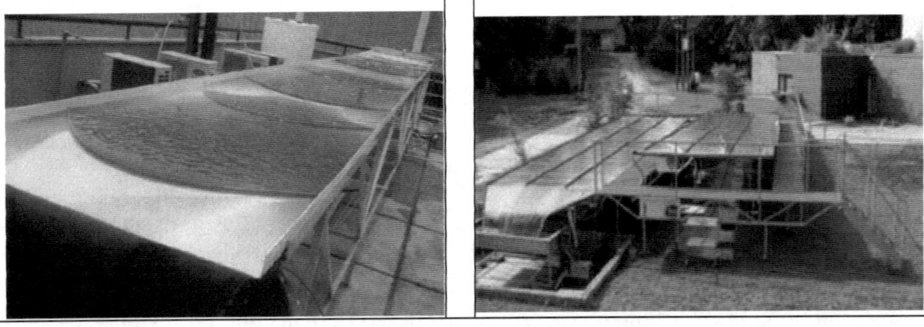

**Figure 3.** Inclined surface 'cascade' systems [26]

**Figure 4.** Open systems of algal cultivation [27]

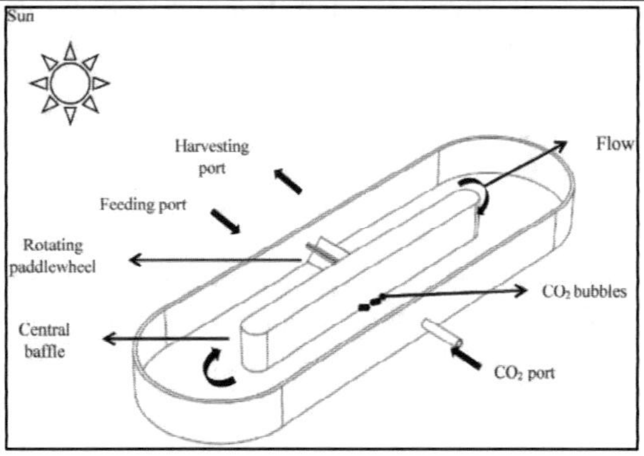


**Figure 5.** Schematic of raceway pond [20]

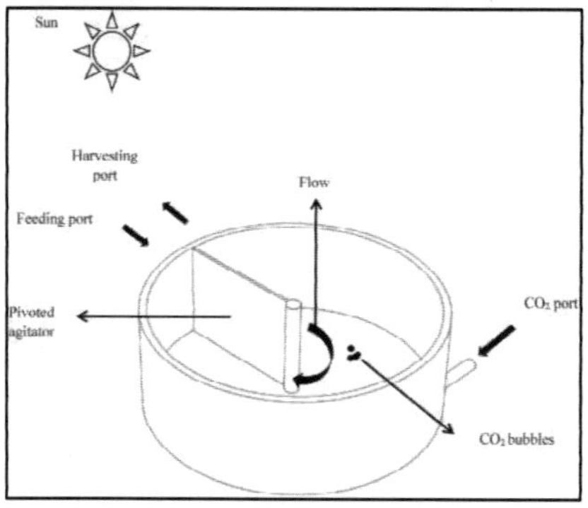


**Figure 6.** Schematic of circular pond [24]

### *4.2. Closed system*

Closed systems are studied and recommended to overcome the limitations of open systems. The growers are concerned to cultivate monoculture of microalgae in axenic enclosed PBRs which offer lower or no contamination chances in the production of biochemical and high value metabolites. It is a man-made closed vessel that helps microalgal cells to carry out photosynthesis in artificial illumination or sunlight as the source of energy [28]. Closed PBRs are generally characterized by the geometric configurations and the commonly manufactured closed PBRs are vertical and horizontal tubular PBRs, vertical column PBRs (bubble column and airlift systems), and flat panel PBRs. In the family of photobioreactors closed PBRs are 5 to 10 times more efficient than the open systems because they are having high surface to volume ratio but at the same time, they are uneconomic [29].

*4.2.1. Flat panel PBRs.* It is having a cuboidal shape of flat panels (plexiglass, polycarbonate, or plastic bags) from which dense microalgal culture is passed [30]. Since it has high surface area to volume ratio and cell densities, making it suitable for algae wastewater co-cultivation, as well as pure algae culture. The agitation is given by air bubbling or by the mechanical rotation of a motor through the perforated tube. Bubbling provides the mixing, prevents the oxygen accumulation, nutrients transfer and light distribution in an adequate manner [23].

Flat panel PBRs have many advantages for mass culture as the risk of accumulation of dissolved oxygen is less with high photosynthetic efficiency as compared to other closed PBRs. Self-shading of microalgae cells between the flat plates may disrupt the photosynthetic efficiency and microalgae growth. To overcome this problem LED lights can be used for the additional illumination [31]. Continuous research is going on to assess the microalgal biomass scale-up and lipid productivity in flat panel PBRs. Different views and photos are shown in Figure 7 and 8.

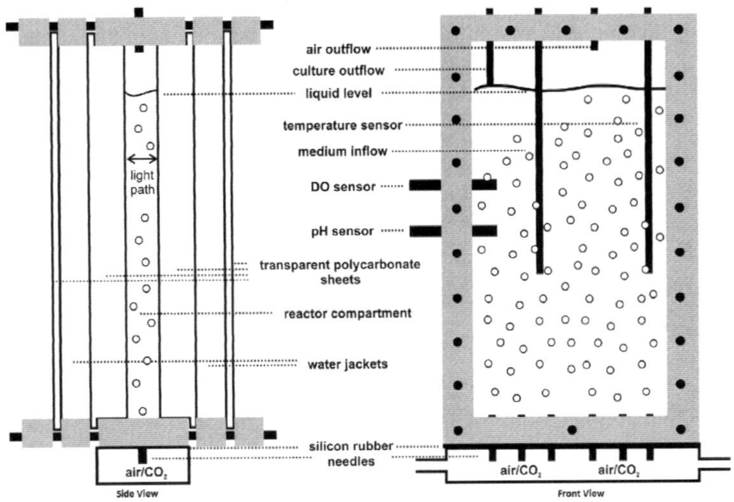


**Figure 7.** Side view and front view of flat panel PBR [32]

**Figure 8.** Pictures showing flat panel PBR [33]

*4.2.2. Tubular PBRs.* They are manufactured using plastic tubes, straight glass that are arrayed in vertical, inclined, or helical configuration, and fence like and horizontal [20]. The basic design principle of tubular PBRs can be divided into two parts. The first part is the tubular part or the reactor in which the microalgae are grown and the second one is the degasser unit which is used to remove the gas and the pump system which provides circulation and mixing of the microalgae culture (Figure 9). Optimum light penetration is the prime aspect leading to high biomass productivity by photosynthesis. Solar array is designed with small diameters up to 0.1 m to increase the light penetration through the dense microalgal culture.

Microalgae culture is circulated from the degasser unit to the solar array and return to the degasser unit in continuous operation mode. The configuration can be understood from the Figure 10. Usefulness of tubular PBR is proven for many existing operation modes such as semi-batch, batch, turbido-static and continuous thereby making it the most reliable type of PBR for scaling up due to better control [30].

*4.2.3. Vertical column PBRs.* These PBRs are economical, have easy operation and better control. Microalgae is cultivated in large volume of cylindrical shaped vessels. Vertical column PBRs can further be classified into bubble column or airlift reactors based on their flow pattern. A gas sparger system is the significant feature which is installed at the bottom of the reactor and converts the influent gas or air into tiny bubbles which are the driving force for the mixing, mass transfer of $CO_2$ and removing the $O_2$ produced during photosynthesis [13]. **Bubble column** PBRs are designed with height greater than twice their diameter. They have the features like low CAPEX, high surface area to volume ratio, satisfactory heat and mass transfer, relatively homogeneous culture density and no moving parts. Gas bubbling is provided by the spargers therefore the design of the sparger is critical relative to the area and volume of the cylinder. The illumination for autotrophic cultivation comes from outside the reactor but now days internal illumination is also getting acceptable because of higher light penetration capability and better light distribution. Photosynthetic efficiency and biomass yield greatly depends upon the gas flow rate and light/dark cycles are created when the liquid is circulated on regular basis from central dark zone to peripheral zone at a higher gas flow rate [34].

**Air lift** PBRs are designed on the principle that the cylindrical vessel is divided in interconnecting zones. The inner tube is called as gas riser or the up comer where the gas mixture flows upwards to the surface from the sparger. While in the outer portion called as the downcomer the medium flows downwards and circulates again into the riser. Further they can be subdivided into (a) draft tube airlift PBR and (b) Split cylinder PBR (Figure 11). In the former system a tube is inserted into the centre of the main cylindrical vessel two draft and the flow is forced in opposite direction. In the later one the culture flow in the cylinder is carried out with a rising region and the refluent region the design and flow directions can be understood from the Figure 11. Air lift PBRs are the systems in which mixing, gas transfer and air supply are carried simultaneously [19]. Table 4 shows pros and cons of various types of PBRs discussed in this paper.

The productivity of microalgae using different PBRs varies depending upon the type of microalgae species, type of PBR, light conditions and medium of nutrients used for the cultivation of microalgae. Performances of some of the PBRs with specific operating conditions with the species of cultured algae are shown in Table 4. It was found that flat plate PBRs have the highest biomass concentration and productivity followed by column and tubular type of PBRs.

**Figure 9.** Pictures showing tubular PBRs [35]

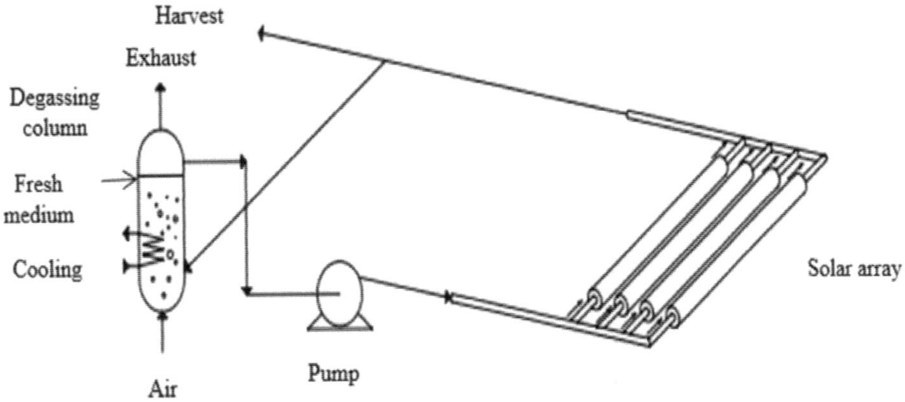


Figure 10. Mechanism in a tubular PBR [36]

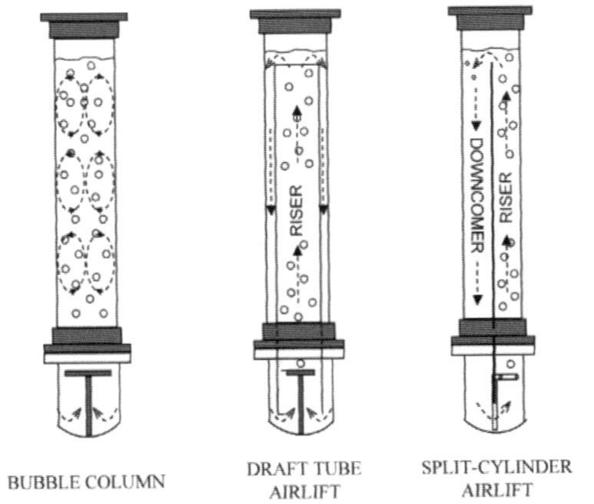


**Figure 11.** Configurations of Bubble column and airlift PBR [37]

**Figure 12.** Pictures showing vertical column PBRs [33]

**Table 3.** Salient features and limitations of various closed PBRs [12, 19, 38]

| Type of closed PBRs | Salient features | Limitations |
|---|---|---|
| Flat panel PBRs | • Large illumination surface area<br>• Low oxygen build-up<br>• High biomass productivity<br>• Economical installation and operation<br>• Good immobilization of microalgae<br>• Have good light path and easy to sterilize<br>• Suitable for outdoor cultures | • Difficulty in controlling culture<br>• Possibility of algal wall adhesion<br>• Scaling-up requires support materials<br>• Small degree of hydrodynamic stress |
| Tubular PBRs | • Temperature control can be monitored effectively<br>• Mass transfer can be improved by installing internal mixers<br>• Good biomass productivities<br>• Economical and suitable for outdoor cultivation | • Difficulty in scaling-up<br>• Accumulation of O2<br>• Acidification may occur due to inadequate fixation of $CO_2$<br>• Requires large land space<br>• High microalgae density can shade light penetration in the tubes<br>• Some degree of wall growth and fouling |
| Bubble column PBRs | • High potential of scalability<br>• Reduced photoinhibition and photooxidation<br>• High mass transfer due to sparging<br>• Bubbles promote light scattering and penetration<br>• Reduced physical stress to microalgal cells | • Sedimentation of microalgal cells<br>• Random flow pattern with the microalgae cultivation |
| Airlift PBRs | • High productivity and biomass concentration<br>• Efficient light utilization<br>• Comparatively economical<br>• High $CO_2$ utilization efficiency<br>• Good for immobilization of microalgae | • Scale up process is difficult<br>• Insufficient turbulence creation by the airlift mechanism<br>• Restricted working volume due to the insertion of additional shaft |

**Table 4.** Performance of different types of PBRs with operating conditions

| Type of PBR | Algae Species | Light Intensity (μmol/ $m^2$ s) | Source of light/Light-Dark cycle(h:h) | Biomass concentration (g/l) | Biomass production | Ref. |
|---|---|---|---|---|---|---|
| Flat plate | *Botryococcus braunii* | 100 | Fluorescent/24:0 | 96.4 | 0.71 g/$m^2$ | [39] |
| Flat plate | *Chlamydomonas reinhardtii* | 1500 | LED/24:0 | 4.5 | 29–54 g/$m^2$·d | [40] |
| Column | Green algae (genus *Chlorella* and *Stigeoclonium*) | 204 | Metal Halide/12:12 | 0.49-0.84 | 0.039-0.084 g/L. d | [41] |
| Column | *Dunaliella tertiolecta* | 1000 | LED/Continuous | 19.78 | 10.18 g/L. d | [42] |
| Tubular | *Phaeodactylum tricornutum* | 50–700 W/$m^2$ | Sunlight/Natural cycle | 0.2–12 | 1.4–3.29g/L. d | [43] |
| Tubular | *Nannochloropsis salina* | 500 | LED/24:0 | 7.2 | 0.888g/L. d | [44] |

*4.3. Membrane Photobioreactors (MPBR)*

In a nutshell the open systems have the limitation of susceptibility to contamination and inadequate control of the operational parameters. While in closed PBRs the problem of washout persists, which causes reduction in concentration of obtained biomass. Since microalgae are having a smaller size and negligible relative density as compared to water, they are difficult to settle, rather they form homogeneous suspension. Therefore, it is almost impossible to decouple the retention time of microalgae (MRT) from the dilution rate (D) in an ordinary PBR merely using settling but this can be achieved by using membrane photobioreactor (MPBR)[45]. Using MPBR serves two purposes i.e., they help to hold the biomass inside the reactor (preventing washout) and assisting $CO_2$ transfer. MPBR provides complete retention of microalgae cells while the medium passes through (as permeate) thereby enhancing the concentration of biomass in the bioreactor. The biomass concentration can be better controlled by partly returning the retentate. As the biomass is concentrated in the retentate stream, the permeate can be reused as feed medium. PBRs coupled with membrane for harvesting algal biomass and wastewater treatment are shown in Figures 13 and 14.

As discussed in earlier sections of review that microalgae can be used to treat wastewater and simultaneously produce biofuels, compensating the cost in an ecologically sustainable manner. To make this technology feasible it is again necessary to separate and harvest the suspended algal biomass from water both to produce biofuels and to get better effluent after wastewater treatment. Although this has proven too difficult and uneconomical hence, to overcome these problems researchers are studying and employing the membranes in PBRs. The membrane will act to separate solid and liquid, thereby helping to isolate the microalgal cells from the effluent[46]. Furthermore, the hydraulic retention time (HRT) and solids retention time (SRT) of the bioreactor can be controlled independently during the culture and the microalgal biomass concentration is not affected by the hydraulic loading and the growth of algal cells. Consequently, improved performance in terms of nutrients/pollutants removal and algal biomass productivity can be achieved in MPBR [47].

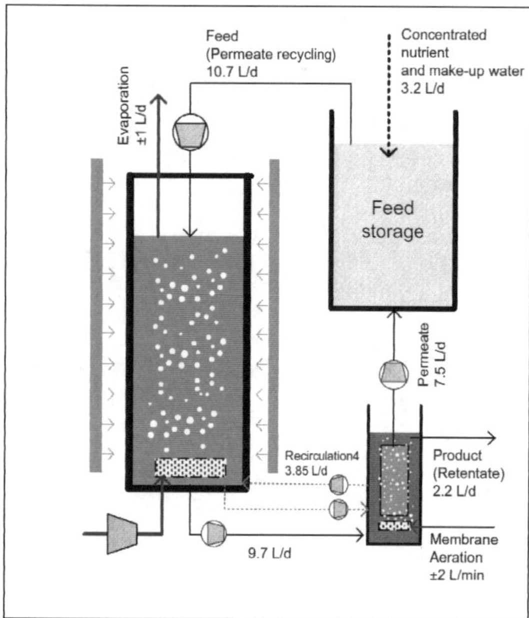


**Figure 13.** MBPR setup for harvesting microalgae [47]

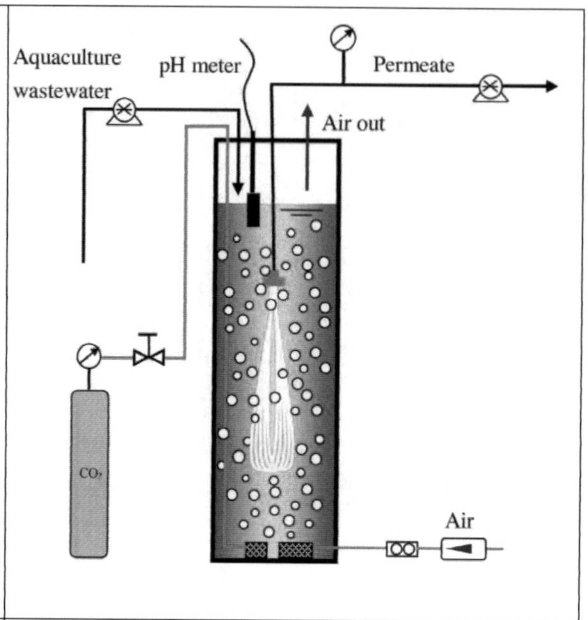


**Figure 14.** MPBR setup for wastewater treatment [48]

## Conclusion

The review concludes that PBRs have better control of cultivation conditions as well as higher productivity as compared to open systems and natural water bodies. PBRs are expensive as compared to open systems but this bottle neck can be resolved by enhancing the yield of microalgal biomass and deriving various other useful metabolites. The review provide an insight to the factors responsible for the optimum growth of microalgae with different modes of growth. The details of different types of open systems and PBRs are elaborated with the figures to understand the mechanism of their operation and their prospects. Since all the PBRs are having advantages and disadvantages therefore having a clear understanding of the features will help the phycologists to choose the suitable PBR for their research.

## Acknowledgement

The authors are thankful for Malaysia-Japan International Instituteof Technology for providing the Algae and Biomass Research laboratory to facilitate the smooth conduct of the study.

## References

[1] Huang, G., F. Chen, D. Wei, X. Zhang, and G. Chen, Biodiesel production by microalgal biotechnology. *Applied energy*, 2010. **87**(1): p. 38-46.
[2] Hamed, I., The evolution and versatility of microalgal biotechnology: a review. *Comprehensive Reviews in Food Science and Food Safety*, 2016. **15**(6): p. 1104-1123.
[3] Pacheco-Torgal, F., C. Buratti, S. Kalaiselvam, C.G. Granqvist, and V. Ivanov, Nano and Biotech Based Materials for Energy Building Efficiency. 2016, Springer International Publishing Switzerland.
[4] Huang, Q., F. Jiang, L. Wang, and C. Yang, Design of photobioreactors for mass cultivation of photosynthetic organisms. *Engineering*, 2017. **3**(3): p. 318-329.

[5] Yadala, S. and S. Cremaschi, Design and optimization of artificial cultivation units for algae production. *Energy*, 2014. **78**: p. 23-39.

[6] Sheng, A., M. Bilad, N. Osman, and N. Arahman, Sequencing batch membrane photobioreactor for real secondary effluent polishing using native microalgae: process performance and full-scale projection. *Journal of Cleaner Production*, 2017. **168**: p. 708-715.

[7] Elrayies, G.M., Microalgae: prospects for greener future buildings. *Renewable and Sustainable Energy Reviews*, 2018. **81**: p. 1175-1191.

[8] Olivieri, G., P. Salatino, and A. Marzocchella, Advances in photobioreactors for intensive microalgal production: configurations, operating strategies and applications. *Journal of Chemical Technology & Biotechnology*, 2014. **89**(2): p. 178-195.

[9] Johnson, M.B., Microalgal biodiesel production through a novel attached culture system and conversion parameters. 2009, Virginia Tech.

[10] Pragya, N., K.K. Pandey, and P. Sahoo, A review on harvesting, oil extraction and biofuels production technologies from microalgae. *Renewable and Sustainable Energy Reviews*, 2013. **24**: p. 159-171.

[11] Mercer, P. and R.E. Armenta, Developments in oil extraction from microalgae. *European journal of lipid science and technology*, 2011. **113**(5): p. 539-547.

[12] Yousuf, A., Fundamentals of Microalgae Cultivation, in *Microalgae Cultivation for Biofuels Production*. 2020, Elsevier. p. 1-9.

[13] Ugwu, C., H. Aoyagi, and H. Uchiyama, Photobioreactors for mass cultivation of algae. *Bioresource technology*, 2008. **99**(10): p. 4021-4028.

[14] Zhu, L., Microalgal culture strategies for biofuel production: a review. *Biofuels, Bioproducts and Biorefining*, 2015. **9**(6): p. 801-814.

[15] Wahidin, S., A. Idris, and S.R.M. Shaleh, The influence of light intensity and photoperiod on the growth and lipid content of microalgae Nannochloropsis sp. *Bioresource technology*, 2013. **129**: p. 7-11.

[16] Cai, T., S.Y. Park, and Y. Li, Nutrient recovery from wastewater streams by microalgae: status and prospects. *Renewable and Sustainable Energy Reviews*, 2013. **19**: p. 360-369.

[17] Mohan, S.V. and M.P. Devi, Salinity stress induced lipid synthesis to harness biodiesel during dual mode cultivation of mixotrophic microalgae. *Bioresource technology*, 2014. **165**: p. 288-294.

[18] Yin, Z., L. Zhu, S. Li, T. Hu, R. Chu, F. Mo, D. Hu, C. Liu, and B. Li, A comprehensive review on cultivation and harvesting of microalgae for biodiesel production: Environmental pollution control and future directions. *Bioresource Technology*, 2020. **301**: p. 122804.

[19] Yen, H.-W., I.-C. Hu, C.-Y. Chen, D. Nagarajan, and J.-S. Chang, Design of photobioreactors for algal cultivation, in *Biofuels from algae*. 2019, Elsevier. p. 225-256.

[20] Tan, X.B., M.K. Lam, Y. Uemura, J.W. Lim, C.Y. Wong, and K.T. Lee, Cultivation of microalgae for biodiesel production: a review on upstream and downstream processing. *Chinese Journal of Chemical Engineering*, 2018. **26**(1): p. 17-30.

[21] Singh, R. and S. Sharma, Development of suitable photobioreactor for algae production–A review. *Renewable and Sustainable Energy Reviews*, 2012. **16**(4): p. 2347-2353.

[22] Lam, T.P., T.-M. Lee, C.-Y. Chen, and J.-S. Chang, Strategies to control biological contaminants during microalgal cultivation in open ponds. *Bioresource technology*, 2018. **252**: p. 180-187.

[23] Faried, M., M. Samer, E. Abdelsalam, R. Yousef, Y. Attia, and A. Ali, Biodiesel production from microalgae: Processes, technologies and recent advancements. *Renewable and sustainable energy reviews*, 2017. **79**: p. 893-913.

[24] Suparmaniam, U., M.K. Lam, Y. Uemura, J.W. Lim, K.T. Lee, and S.H. Shuit, Insights into the microalgae cultivation technology and harvesting process for biofuel production: a review. *Renewable and Sustainable Energy Reviews*, 2019. **115**: p. 109361.

[25] Lee, J., K. Lee, D. Sohn, Y.M. Kim, and K.Y. Park, Hydrothermal carbonization of lipid extracted algae for hydrochar production and feasibility of using hydrochar as a solid fuel. *Energy*, 2018. **153**: p. 913-920.

[26] Masojídek, J. and G. Torzillo, Mass cultivation of freshwater microalgae. 2014.

[27] Shen, Y., W. Yuan, Z. Pei, Q. Wu, and E. Mao, Microalgae mass production methods. *Transactions of the ASABE*, 2009. **52**(4): p. 1275-1287.

[28] Fulbright, S.P., A. Robbins-Pianka, D. Berg-Lyons, R. Knight, K.F. Reardon, and S.T. Chisholm, Bacterial community changes in an industrial algae production system. *Algal research*, 2018. **31**: p. 147-156.

[29] Öncel, S., A. Köse, and D. Öncel, Façade integrated photobioreactors for building energy efficiency, in *Start-Up Creation*. 2016, Elsevier. p. 237-299.

[30] Ting, H., L. Haifeng, M. Shanshan, Y. Zhang, L. Zhidan, and D. Na, Progress in microalgae cultivation photobioreactors and applications in wastewater treatment: A review. *International Journal of Agricultural and Biological Engineering*, 2017. **10**(1): p. 1-29.

[31] Xu, L., P.J. Weathers, X.R. Xiong, and C.Z. Liu, Microalgal bioreactors: challenges and opportunities. *Engineering in Life Sciences*, 2009. **9**(3): p. 178-189.

[32] Bahadar, A. and M.B. Khan, Progress in energy from microalgae: a review. *Renewable and Sustainable Energy Reviews*, 2013. **27**: p. 128-148.

[33] Bitog, J.P., I.-B. Lee, C.-G. Lee, K.-S. Kim, H.-S. Hwang, S.-W. Hong, I.-H. Seo, K.-S. Kwon, and E. Mostafa, Application of computational fluid dynamics for modeling and designing photobioreactors for microalgae production: a review. *Computers and Electronics in Agriculture*, 2011. **76**(2): p. 131-147.

[34] Loubière, K., E. Olivo, G. Bougaran, J. Pruvost, R. Robert, and J. Legrand, A new photobioreactor for continuous microalgal production in hatcheries based on external-loop airlift and swirling flow. *Biotechnology and bioengineering*, 2009. **102**(1): p. 132-147.

[35] Qiu, F., Algae architecture [Master]. *TU Delft: Delft University of Technology*, 2014.

[36] Andersen, R.A., Algal culturing techniques. 2005: Elsevier.

[37] Sevilla, J.F., M. Cerón García, A. Sánchez Mirón, E.H. Belarbi, F.G. Camacho, and E.M. Grima, Pilot-plant-scale outdoor mixotrophic cultures of Phaeodactylum tricornutum using glycerol in vertical bubble column and airlift photobioreactors: studies in fed-batch mode. *Biotechnology progress*, 2004. **20**(3): p. 728-736.

[38] Chew, K.W., S.R. Chia, P.L. Show, Y.J. Yap, T.C. Ling, and J.-S. Chang, Effects of water culture medium, cultivation systems and growth modes for microalgae cultivation: a review. *Journal of the Taiwan Institute of Chemical Engineers*, 2018. **91**: p. 332-344.

[39] Ozkan, A., K. Kinney, L. Katz, and H. Berberoglu, Reduction of water and energy requirement of algae cultivation using an algae biofilm photobioreactor. *Bioresource technology*, 2012. **114**: p. 542-548.

[40] de Mooij, T., G. de Vries, C. Latsos, R.H. Wijffels, and M. Janssen, Impact of light color on photobioreactor productivity. *Algal research*, 2016. **15**: p. 32-42.

[41] Arias, D.M., E. Uggetti, M.J. García-Galán, and J. García, Cultivation and selection of cyanobacteria in a closed photobioreactor used for secondary effluent and digestate treatment. *Science of the total environment*, 2017. **587**: p. 157-167.

[42] Hu, J.-Y. and T. Sato, A photobioreactor for microalgae cultivation with internal illumination considering flashing light effect and optimized light-source arrangement. *Energy Conversion and Management*, 2017. **133**: p. 558-565.

[43] Slegers, P., P. Van Beveren, R. Wijffels, G. Van Straten, and A. Van Boxtel, Scenario analysis of large scale algae production in tubular photobioreactors. *Applied energy*, 2013. **105**: p. 395-406.

[44] Iluz, D. and S. Abu-Ghosh, A novel photobioreactor creating fluctuating light from solar energy for a higher light-to-biomass conversion efficiency. *Energy Conversion and Management*, 2016. **126**: p. 767-773.

[45] Acién, F., J. Fernández, J. Magán, and E. Molina, Production cost of a real microalgae production plant and strategies to reduce it. *Biotechnology advances*, 2012. **30**(6): p. 1344-1353.
[46] Bilad, M., H.A. Arafat, and I.F. Vankelecom, Membrane technology in microalgae cultivation and harvesting: a review. *Biotechnology advances*, 2014. **32**(7): p. 1283-1300.
[47] Bilad, M., V. Discart, D. Vandamme, I. Foubert, K. Muylaert, and I.F. Vankelecom, Coupled cultivation and pre-harvesting of microalgae in a membrane photobioreactor (MPBR). *Bioresource technology*, 2014. **155**: p. 410-417.
[48] Gao, F., C. Li, Z.-H. Yang, G.-M. Zeng, L.-J. Feng, J.-z. Liu, M. Liu, and H.-w. Cai, Continuous microalgae cultivation in aquaculture wastewater by a membrane photobioreactor for biomass production and nutrients removal. *Ecological engineering*, 2016. **92**: p. 55-61.

# Review on recent advance biosynthesis of $TiO_2$ nanoparticles from plant-mediated materials: characterization, mechanism and application

**Syamsutajri Syamsol Bahri[a], Zawati Harun[a*], Siti Khadijah Hubadillah[b], Wan Norhayati Wan Salleh[c], Nurafiqah Rosman[c], Noor Hasliza Kamaruddin[a], Faiz Hafeez Azhar[a], Norsuhailizah Sazali[a], Raja Adiba Raja Ahmad[a], Hatijah Basri[d]**

[a]Advanced Manufacturing and Materials Centre (AMMC), Faculty of Mechanical and Manufacturing Engineering, Universiti Tun Hussein Onn Malaysia, 86400 Parit Raja, Batu Pahat, Johor.
[b]Faculty of Engineering Technology, University Malaysia Perlis (UniMAP), Kampus, UniCITI Alam, Sungai Chuchuh, Padang Besar 02100, Perlis, Malaysia
[c]Advanced Membrane Technology Research Centre (AMTEC), School of Chemical and Energy Engineering, Faculty of Engineering, Universiti Teknologi Malaysia, 81310 Skudai, Johor Bahru, Malaysia
[d]Faculty of Applied Sciences and Technology, Universiti Tun Hussein Onn Malaysia, Pagoh Education Hub, Muar, Johor, Malaysia

**zawati@uthm.edu.my*

**Abstract.** Titanium dioxide, $TiO_2$ nanoparticles are being utilized in various application of science and technology including membrane, medical, electrical and chemical field for their respective worth which generally attributed to the self-cleaning and photocatalytic ability, good band gap, an antibacterial as well as physical and chemical stability. As commonly known conventional $TiO_2$ nanoparticles synthesized using chemicals as reducing agents has become accountable for various biological risks due to their general toxicity, thus engendering the serious concern in developing environment friendly processes. Naturally derived products, such as extracts of plants that composed of biomolecules, have been used intensively recently as a reductant agent, that also sometimes can be acting as capping agents after synthesis process. These natural biomolecules mostly consisted of polyphenols have been identified to be actively play a role in this biosynthesis of nanoparticles from any plants extract that able to form different shapes and sizes of nanoparticles with better surface reactive area, characteristic and properties. Therefore, biosynthesis can be considered as a driving force for the greener, safe and environmentally friendly for many applications that have used $TiO_2$ particles either used as additive, purely or in composite form. The present review targets on the 'greener' routes of synthesis $TiO_2$ nanoparticles with an emphasis on experimental conditions based on sustainable methodologies and also explores the huge plant diversity to be utilized. The use of 'greener' not only reduces the cost of synthesis but also minimizes the need of using hazardous chemicals and stimulates green synthesis. This review also focuses on aspects characteristic and properties that generated from the output of this green process that make it strongly applicable to certain applications as for binding of biomolecules, to the biosynthesized is significantly benefit to biomedical fields. It is expected that these outstanding findings will encourage researchers and attract newcomers to continue and extend the exploration of possibilities offered by nature and the design of innovative and safer methodologies towards the synthesis of nanomaterials, possessing desired features and exhibiting valuable properties that can be exploited in a profusion of fields.

Keywords: Titanium dioxide; $TiO_2$; Biosynthesis; Plant extract; Green synthesis

## 1. Introduction

Nanoparticles has been integrated in membrane for various application such as water treatment [1–4], gas separation [5], wound dressing [6,7] etc. The use of nanoparticles as an additive in nanoscale range is considered important as it can provide larger surface area resulting in good characteristics and membrane performance such as anti-fouling, photocatalytic and anti-bacterial activity. However, there is an issue that need to pay attention regarding to nanoparticles potential to human health and environmental risk of nanoparticles [4]. Recently, biosynthesized nanoparticles have drawn great attention resulting in exploration of green technology through biosynthesis method with added functions such as antibiotic resistance [8,9] and biocompatibility [10] compared to the chemical and physical method. Incorporation of these biosynthesized nanoparticles as an additive with well-defined nanostructure bring significant effect in their properties and characteristics also can produce a good membrane performance. Some reviews have already described the overview particular biosynthesized nanoparticles development with various application [11–14].

There is various synthesis approach have been used in nanoparticles production include chemical, physical and green synthesis methods. These methods are still experiences problems such as aggregation and stability, crystal growth control, size, morphology and distribution because they are still in the development stage. Besides, the separation of the nanoparticles produced become an important consideration for further applications [15]. The approach of green technology through biosynthesis using bacteria, actinomycetas, yeasts, fungi, algae and plant extract provides advancement over chemical and physical method. It also in fact able to eliminate the use of hazardous chemicals, cost effective, eco-friendly, easily scaled up for large scale synthesis [15] and more energy efficient without using high pressure and temperature [16]. Although chemical and physical methods are extensively used for nanoparticles production but they require the use of reactive and toxic reducing agents such as sodium borohydride ($NaBH_4$), N, N-dimethyl formamide (DMF), Tollen's reagent and hydrazine hydrate [17]. This will cause undesired detrimental impacts on the environment, plant and animal life it supports [11].

The development of facile, reliable and effective green synthesis process is continuously in progress for the nanoparticles production. The efforts made by all researchers is vitally important to produce nanoparticles that is stable, more reliable and well-functionalized. In application such as biomedical, biosynthesis of nanoparticles is still under exploitation to avoid adverse effects [13]. Thus, there is a need to explore sustainable process for nanoparticles synthesis. The concern may include the viability of economy, sustainability of environment, adaptability of social and the local resource availability (see Fig. 1.) [11,18–20]. Generally, nanoparticles involved two main synthesis approaches which is either from "top to bottom" or "bottom to up" (see Fig. 2.). In top to bottom approach, lithographic technique such as milling, sputtering thermal and grinding are used to break down bulk material into fine particles by size reduction while in bottom to up approach, atoms are self-assemble to form new nuclei which grow into nanoscale particles as shown in Fig. 2. Certain limitation and drawback from these techniques mentioned before leads to the green synthesis technology via biosynthesis that does not required toxic chemicals thus develop environment friendly processes. Moreover, limitation to biomedical applications using the resulting nanoparticles could be reduced because biosynthesis routes mostly are free from contamination of toxic by-products produced by physicochemical route that normally will attached to the nanoparticles during synthesis [21].

Recent studies have point out the potential of biosynthesized $TiO_2$ nanoparticles for various application including remediation of environmental. The interest is increasing due to its self-cleaning and photocatalytic ability, good band gap, an antibacterial agent as well as physical and chemical stability. Thus, the possibilities offered by nature and the design of innovative and safer methodologies towards the biosynthesis of $TiO_2$ nanoparticles is a great step forward in the development of nanomaterials, possessing desired features and exhibiting valuable properties that can be exploited in a profusion of fields. The current review intends to discuss on the important of biosynthesis of titanium dioxide nanoparticles as an additive by focusing on the characteristic and properties that applicable to certain applications with an emphasis on experimental conditions based on sustainable methodologies and also explores the huge plant diversity to be utilized.

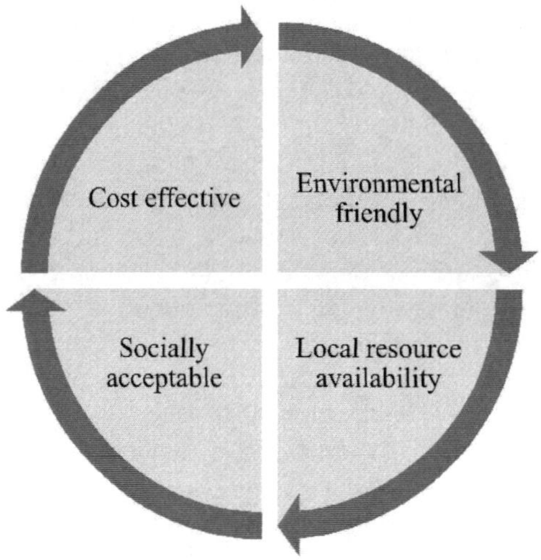


Fig. 1. Sustainable green technology.

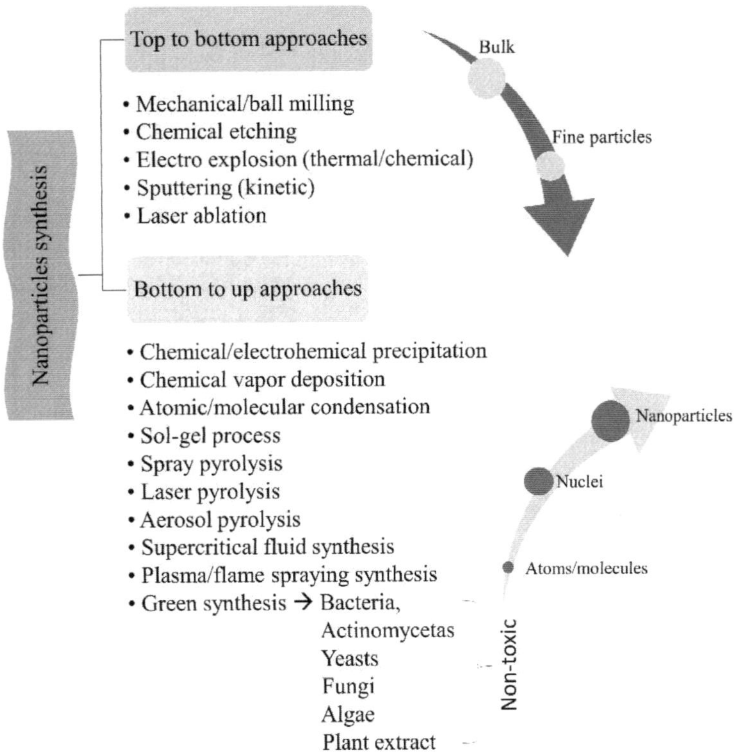


Fig. 2. Important approaches of nanoparticles synthesis.

## 2. Biosynthesis of $TiO_2$ nanoparticles by plant extracts – the important

Many researchers from literature mentioned that $TiO_2$ nanoparticles have potential oxidation strength, non-toxicity, high photocatalytic activity, interesting optical and chemical stability. They also possess antimicrobial property and antibacterial catalytic which leads to various industrial applications such as photocatalyst, catalyst supports and pigments [13,22]. Moreover, biosynthesized nanoparticles may have different properties prepared by other techniques and also can have enhanced biocompatibility and stability probably due to the active biomolecules or capping agent coated on the surface. Mainly, the stabilization of nanoparticles in biosynthesis perspective depend much on the choice of the reducing agent, solvent medium and non-toxic material. A green and facile biosynthesis using plant materials was achieved in continuation of the efforts of recent biosynthesis by bacteria, actinomycetes and fungi [23].

To date, $TiO_2$ nanoparticles have been synthesized mostly using different type of plant extracts. Extracts of plants commonly obtained by using leaves, flowers, roots and stems. Most of them are categorized as herbs in order to apply in biomedical application. Common herbs are not only used as fragrance or flavour in the food industry but most of them sometimes can exhibit antimicrobial properties [24]. The use of these extracts in the nanoparticle production assembly was shown more stable because their reaction rate as reducing agent (able to reduce metal or metal oxide ions) as well as stabilizing agent is much faster than fungi, bacteria or other organisms. Furthermore, plant extract provides advanced advantage as the organism mostly required stringent control on cell culture. The green method is also easy and safe for industrial production scale-up of well-dispersed nanoparticles as it does not require harmful and toxic chemicals [15] (see Fig. 3.). Commonly, water or organic solvents are used to extract phytochemicals based on the group or type and the process is done with less temperature which is not greater than 100°C (see Table 1.). Moreover, there is a significance effect of plant extract concentration on the size, shape as well as distribution of the nanoparticles revealed by previous studies [25].

The combination of functionalized surface that contains biological molecules compounds in plant extracts such as polysaccharides, alkaloids, tannins, phenolics, saponins, carbonyl and hydroxyl are responsible for reduction and stabilization of $TiO_2$ ions and already established in several reported journals [26,27]. The plant extracts are employed for focusing on the principles of green chemistry such as aloe vera, neem, lemon grass, Indian gooseberry, tulsi and corriandrum [13,28]. They also have medicinal values, environmental benign with substantial health benefits yet having a complex chemical structure which are not present in physical and chemical synthesized nanoparticles [12,19]. Moreover, in order to optimize the extraction of these biological molecules compounds, various studies have been carried out including extraction temperature, type of solvent as well as plant drying technique [19,29,30].

Several experimental parameters were screened, namely extraction temperature, solvents used, precursor type, plant extract volume, drying temperature, reaction time and calcination step as shown in Table 1 with pictures of the extract plants illustrated in Fig. 4. Generally, different experimental sets generate similar nanoparticles mostly spherical in shape. To summarized the screened parameters, the biosynthesized $TiO_2$ nanoparticles composed as result from the mixing of plant extract from 5 to 80 ml with titanium dioxide precursor. Then, they were dried at either room temperature or using temperature not higher than 100 °C with reaction time from two hours or until overnight. Some experiment involved calcination step ranging from 450 °C to 600 °C with a maximum 4 h reaction time to remove organic groups. Most experiment that using $TiO_2$ as precursor specifically $TiO_2$ bulk particles may not require calcination step as in experiment conducted by Abdul Jalill et al. [31] and others shown in Table 1. A study by Kharissova et al. [18] added some common assumptions about green chemistry that at least one role in biomolecules supposedly present in plant either the hydrolysis of the Ti(IV) antecedent, polymerization of the different intermediates, solubilizing or most importantly the reduction of the metal salts roles.

The first step of preparation of plant extract normally involved collection of the plant of interest from local garden or at common available place. The plants were cleaned and washed thoroughly with either distilled water or tap water to remove dust. Some were shade dried for about 2 weeks at room temperature under dust free condition and cut before further sieved, grinded or blended to obtain the finest powder such as experiment conducted by Sundrarajan and Gowri [28]. Then the dried powder was boiled at certain temperatures with a known gram of the dried powder. This extraction technique namely infusion filtered the resulted infusion extract containing all the biochemical that will be used as a reducing and stabilizing agent in the biosynthesis. Usually, after biosynthesis, the obtained precipitate of nanoparticles was filtrated to separate them from the solution then the ionic impurities were removed by ethyl alcohol through washing. Lastly, they were dried or calcined depends on the parameter used. The leaf extract of *Nyctanthes arbortristis* as reported by Sundrarajan and Gowri [28] promotes the production of higher crystallinity and purity of $TiO_2$ nanoparticles with spherical in shape and 100 nm in size where the experimental conditions were as follows: 0.4M of titanium tetraisoproxide (TTIP) in ethanolic leaf extract (50ml of ethanol mixed with 1g dried leaves) stirred at 50°C and calcined at 500°C at 3 hours. Patidar et al. [32] also used shade dried leaves of Moringa Oleifera by mixing 0.5M of TTIP in ethanolic leaf extract (10g dried leaves was mixed with 100 ml ethanol), stirred at 50°C and calcined at 500°C for 5 hours producing tetragonal body-centred titanium dioxide with average size of 12 nm. The color change during stirring or autoclave might indicate the proof of nanoparticles formation due to surface plasmon resonance excitation which is dependent on the size distribution and shape of the nanoparticles [33]. But technically, some reacton didn't show any color change and there was no proof for the formation of nanoparticles such as experiment conducted by Rajakumar et al. [34] and Santhoskumar et al. [35] where the pure $TiO(OH)_2$ and aqueous leaf extract of P. guajava. Usually, those pure $TiO_2$ and leaves extract didn't show any color change unless if they were mixed under certain conditions.

The biomolecules extraction from plant is an important step in the biosynthesis. Here, researcher need to choose which extraction method to be used as type of plant, temperature control, stirring rate, plant volume/concentration and types of solvents used such as water or organic solvent (hexane, ethanol, methanol, ethyl alcohol etc.) that have difference in polarity can bring impact in term of accuracy of type of bioactive molecules present and the total amount of the bioactive compounds obtained [36]. Generally, most researchers used conventional extraction methods due to its simplicity to facilitate the extraction of biomolecules with high concentration with systems that consume safe solvents such as pure ethanol [37] as illustrated in Fig. 5. In selection of solvents, it is important to consider solubility, cost and safety as it is crucial part for solvent extraction [30]. Besides, solvent used are chosen based on the solute (plant extract) polarity and usually researcher eliminate the water interference by employed a dry powder of plant material. The solute will properly dissolve if the solvent has a similar polarity. Common solvents, from high polarity to less polarity is listed as follows: Water > Methanol > Acetone > Ethylacetate > Chloroform > Hexane. There are many research study suggested that high polarity solvents can extract antioxidant effectively [29]. In addition, Koffi et al. [38] reported that higher concentrations of phenolics compound was extracted from ethanolic extracts of Ivorian plants compared to methanol, water and acetone.

On the other hand, the size of nanoparticles can be affected by the pH value of the medium under formation [21]. In this review, it was suggested that alkaline pH values (pH 7) using various extracts of plants can change the charge of natural biochemical that further affects their capability to bind and reduce $TiO_2$ ions efficiently thus producing variety of yield and morphology includes shape and size of nanoparticles. For example, in the case of biosynthesis of $TiO_2$, the extracts peels of *Citrus sinensis (L.)* was added dropwise until pH 7 producing a rutile phase, tetragonal in shape with size of 19nm. A spherical $TiO_2$ nanoparticles with distinct edges without aggregation and size 7 nm was obtained using extract of *Hibiscus rosa sinensis.* Rajkumari et al. [39] obtained a spherical, polydispersed biosynthesized $TiO_2$ nanoparticles with size ranging from 20 to 50 nm using the leaf extracts of *Aloe barbadensis mill.* Another example of size and shape controlled biosynthesized nanoparticles by pH 7 is Rao et al. [40] which reported that by using Aloe vera leaf extract, an irregular shape with 20 (tetragonal), 32 and 60-80 nm in size was obtained using XRD, PSA and SEM analysis technique. In addition, a finding by Pratap et al. [41] has reported that the low pH of biosynthesized $TiO_2$ nanoparticles solution using leaf extracts of *Jatropha curcas. L* might promote agglomeration of the nanoparticles however biochemical present perhaps can enhance the dispersion of the nanoparticles thus reduce the agglomeration.

Most experimental procedure that involved calcination process obtained $TiO_2$ biosynthesized nanoparticles with tetragonal structure as reported by Patidar et al. [32], Rao et al. [42], Rao et al. [40], Pratap et al. [41] and Sundrarajan et al. [43] except for Sundrarajan and Gowri [28] who obtained cubic structure. Specifically, the biosynthesized $TiO_2$ nanoparticles are produced when calcined by 400°C to 600°C from 3 to 5 hours producing those two structure. Thus, it is possible to obtain biosynthesized $TiO_2$ nanoparticles with significant output (type of biochemical content, extract volume, yield etc) depending by experimental conditions/parameter such as listed in Table 1. In addition, a study by Singh et al. [23] has reported that biosynthesized nanoparticles require time shorter than physiochemical approaches by utilizing various sources of plant.

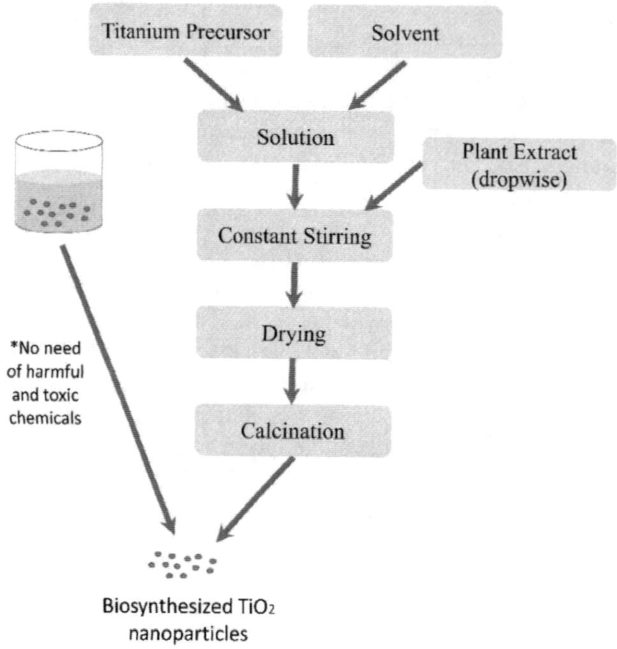


Fig. 3. Procedure of biosynthesis of $TiO_2$. Adapted from Nabi et al. [44] with some modification.

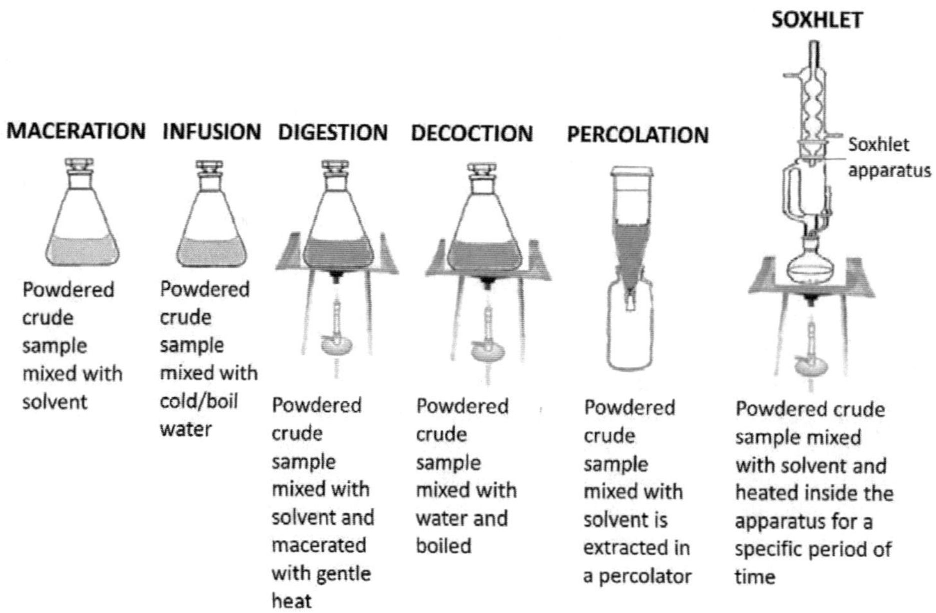


Fig. 4. Plant-mediated materials used for biosynthesis of $TiO_2$ nanoparticles (sort according to name)

Table 1. Parameter of biosynthesized $TiO_2$ nanoparticles using plant extract (sort according to $TiO_2$ precursor).

| No | Name and part of plant | Extract method | Titanium precursor | Extract volume, ml | Stirring temp, °C / Time, h | Drying temp, °C / Time, h | Calcined temp, °C / Time, h | Ref |
|---|---|---|---|---|---|---|---|---|
| 1. | *Nyctanthes arbortristis*, leaf | Ethanol, 50°C | TTIP | 1 g dried leaves + 50 ml ethanol | 50 / 4 | - / 4 | 500 / 3 | [28] |
| 2. | *Cynodon dactylon, leaf* | Water, Boil | TTIP | 100 | 50 / 1 | 80 / 2 | 500 / 5 | [45] |
| 3. | *Moringa oleifera leaf* | Ethanol, 50°C | TTIP | 10 g extract + 100 ml ethanol | 50 / 4 | Room temperature / - | 500 / 5 | [32] |
| 4. | *Citrus sinensis (L.) (orange waste)*, peel | Water, 90°C | TTIP | Dropwise (until pH 7) | Room temperature / 3 | 80 / overnight | 600 / 3 | [42] |
| 5. | *Azadirachta indica*, leaf | Water, 60°C | TTIP | 10 | 50 / 5 (dark brown color) | - | - | [46] |
| 6. | *Hibiscus rosa sinensis* (shoe black plant), petals | Water, 70°C | $TiOSO_4$ | 5 (dropwise until pH 7) | - / 3 | 100 / 3 | - | [47] |
| 7. | *Catharanthus roseus*, leaf | Water, Boil | $TiO_2$ (powder) | 20 | 50 / 4 (light green) | - | - | [48] |
| 8. | *Curcuma longa* (turmeric), leaf | Water, 40°C | $TiO_2$ (bulk particles) | 50 | | Room temperature / - | - | [31] |
| 9. | *Echinacea purpurea, leaf* | Water, 90°C | $TiO_2$ (aqueous solution) | 10 | | - | - | [49] |
| 10. | *Vigna radiate,* seed | Water, Boil | $TiO_2$ (aqueous solution) | 20 | | - | - | 45] |
| 11. | *Cassia fistula, leaf* | Water, Boil | $TiO_2$ | 20 | - / 24 (milky color) | - | - | [50] |
| 12. | *Citrus limon,* leaf | Water, Boil | $TiO_2$ | 20 | | - | - | [51] |
| 13. | *Aloe barbadensis mill, leaf* | Water, 70°C | $TiCl_4$ | Dropwise (until pH 7) | - / 3 (brown) | 100 / 7 | - | [39] |
| 14. | *Aloe vera, leaf* | Water, 90°C | $TiCl_4$ | Dropwise (until pH 7) | - / 4 | 100 / overnight | 500 / 4 | [40] |
| 15. | Jatropha curcas, leaf | Water, 80°C | $TiCl_4$ | 80 | Room temperature / | Air dried / - | 450 / 3 | [41] |

| | | | | | | | | |
|---|---|---|---|---|---|---|---|---|
| | | | | | 20 (whitish-brown) | | | |
| 16. | *Tinospora cordifolia*, stem | Organic solvent, water | $TiCl_4$ | 500 mg of extract + 20 ml ethanol | | 70 / - | - | [52] |
| 17. | *Morinda citrifolia (M. citrifolia)*, leaf | Ethanol, 50°C | $TiCl_4$ | 50 | | 100 / 5 | 400 / 4 | [43] |
| 18. | *Ageratina altissima, leaf* | Water, 60°C | $TiO(OH)_4$ | 25 | | - | - | [53] |
| 19. | *Calotropis gigantean* (Crown flower), flower | Water, 60°C | $TiO(OH)_2$ | 50 mg | | 90 / 2 | - | [54] |
| 20. | *Solanum trilobatum (purple fruited pea eggplant), leaf* | Water, 25°C | $TiO(OH)_2$ | 20 | | - | - | [55] |
| 21. | *Euphorbia heteradena Jaub*, root | Water, 70°C | $TiO(OH)_2$ | 15 | | Room temperature / 2 | - | [56] |
| 22. | *Eclipta prostrate L.* (False daisy), leaf | Water, 60°C / 10 | $TiO(OH)_2$ | 15 | | - | - | [34] |
| 23. | *Euphorbia prostrata*, leaf | Water, 60°C | $TiO(OH)_2$ | 20 | | - | - | [57] |
| 24. | *Jatropha curcas*, latex | 0.3% latex | $TiO(OH)_2$ | 1 ml crude latex | | Room temperature / - | - | [58] |
| 25. | *Mangifera indica L*, leaf | Water, 60°C | $TiO(OH)_2$ | 15 | | - | - | [59] |
| 26. | *Psidium guajava*, leaf | Water, 60°C | $TiO(OH)_2$ | 20 | | - | - | [35] |
| 27. | *Glycosmis cochinchinensis*, leaf | Water, 85°C | $TiO(OH)_2$ | 25 | | - | - | [60] |
| 28. | *Annona squamosa (sugar-apple), peel* | Water, 25°C | $TiO(OH)_2$ | 20 | | - | - | [33] |
| 29. | *Aloe vera gel* | Water, boil | $TiO(OH)_2$ | 20 | - / 24 (light green) | 120 / 1 | - | [61] |

Fig. 5. Conventional methods used for plant extraction.

## 3. Characterization of biosynthesized $TiO_2$ nanoparticles – control of size and shape

There are many techniques have been applied to characterize biosynthesized nanoparticles and the characterization field is still in trend. Many discussions regarding which suitable techniques should be performed to provide information data and researchers currently try to find standard regulatory methodologies or protocols for the biosynthesized characterization in various field includes environmental and biomedical field [44]. Understanding of these techniques will really help readers to get appropriate techniques for their research needs. Given here is some important techniques and relevant references for details.

The biosynthesized $TiO_2$ nanoparticles are characterized by certain attributes such as surface area [62], morphology [28], size [63], thermal stability [64], optical activity [60] etc. These characteristic parameters can bring direct effects on the nanoparticles behaviour especially in downstream application. The significant characterization techniques for $TiO_2$ nanoparticles that often used are Fourier transform infrared spectroscopy (FTIR) [34], X-ray diffraction (XRD) [32], field emission scanning electron microscopy (FE-SEM) [51], scanning electron microscopy (SEM) [28], energy dispersive analysis of X-rays (EDX)/(EDAX) spectroscopy [35], ultraviolet visible (UV–Vis) spectroscopy [31], [49], atomic force microscopy (AFM) [65], particle size analyzer (PSA) [40], dynamic light scattering (DLS) [66], transmission electron microscopy (TEM) [60] and zeta potential [67].

The most necessary parameters sought for characterization of biosynthesized nanoparticles are size and surface functional group. Regarding this need, XRD, DLS, TEM, PSA and AFM can provide information of the size. XRD can clarify a crystal structure, crystalline phase as well as diffraction pattern of the nanoparticles. The structural information and the rendered pattern obtained then will be compare with the standards. Here, Debye-Scherer's formula is used to calculate the average crystallite size. In PSA technique, ultrasonicator is used to dispersed the nanoparticles in the distilled water in order to obtain a mean value from histogram of the dispersed nanoparticles. DLS is also a simple technique for average size determination in the liquid solution phase. Besides size, AFM is preferable technique to characterize topography, roughness, morphology, surface texture and granularity volume of the biosynthesized nanoparticles.

Meanwhile FTIR is used to identify all the functional groups of the compounds for chemical bond analysis. The stability of the colloidal suspensions is the main challenge in the biosynthesis. Usually, biosynthesized nanoparticles tend to aggregate in order to reduce surface tension because the surface are coated with other compounds present [68,69]. To stabilize it, capping agents are commonly needed to keep the nanoparticles from aggregating during the procedure. That is why those compounds (capping agent) as well as compounds present in the biosynthesis solution need to be identified by FTIR. In fact, if difference bonds (C–C, C=C, C≡C, C–O, C=O, O–H, and N–H) are present, they can be identified because they have diverse vibrational frequencies [35]. Thus, the surface of nanoparticles can be well characterize using this technique through the determination of organic functional groups, surface chemical residues and surface chemistry by detecting their frequency absorption band characteristics in the infrared spectrum.

Morphology information such as shape of the nanoparticles can be obtained by using SEM, FESEM, TEM and AFM. In fact, FESEM is used in order to get a more accurate data of surface properties. In comparison, TEM provide better resolution and internal structure information of the nanoparticles. All these techniques also can be used to estimate the average size of the biosynthesized nanoparticles. EDX and EDS basically has same function as they are used to proves and estimate the chemical and elemental composition, type as well as concentration of the nanoparticles.

Other techniques may be needed depending on the applications. For example, the light absorption properties of nanoparticles and their interaction information are needed for environmental remediation. Common $TiO_2$ only have absorbance not more than 380 nm [65] but can extend into visible range if they have ligands attached on their surfaces or by doping [70]. Regarding this need, UV-VIS technique is used for qualitative and quantitative analysis to obtain the reflectance and absorbance spectra in order to identify their optical properties and certain classes of compounds (pure and biological mixtures) in the ultraviolet and the visible light region. UV-Vis spectroscopy can detect natural compounds such as phenolic including polymer dyes, tannins and phenols that formed complexes with iron in the UV range [71]. This technique was found to reduce cost, less selective, can give information of the total content of phenolic extract (280 nm), phenolic acids (360 nm), flavones (320 nm), and also less time consuming compared to other techniques [72]. In fact, various nanoparticles ranging size from 2 to 10 nm can be characterized using light wavelength within the 300-800 nm [73]. Moreover, photocatalytic activity could be enhanced with the help of stronger UV absorption intensity where higher separate electrons or holes produced and higher electrons could be promoted from the valence band into the conduction band [52].

Besides, other techniques that useful in the characterization is the TG/DTA and zeta potential. The thermal and thermal stability of the nanoparticles can be expressed by using TG/DTA technique. Zeta potential is also important because most nanoparticles have a surface charge. Some studies mentioned that zeta potentials either positive or negative that have an absolute value of more than 30 mv can maintain colloidal stability by electrostatic repulsion [74,75]. Generally, nanoparticles suspension contains ions that affect the measurement of zeta potential through its strength and valency of ions. Increased ionic strength and valency can reduce the zeta potential by compressing the electric double layer [76]. Besides, the concentration of hydrogen ions in medium and the pH can also influence the zeta potential because the nanoparticles acquire more positive charge when the suspension is acidic. So, it is important to give information of the nanoparticles suspension with indication of solution pH including the composition of the medium and ionic strength [74]. For example, Sanker et al. [46] obtained biosynthesized $TiO_2$ nanoparticles with interconnected spherical in shape having a zeta potential of -24 mV in 124 nm of nanoparticle size. 5 mM titanium isopropoxide solution was reduced at 50 °C with pH 1.5 indicating that the biosynthesized TiO2 nanoparticles using *Azadirachta indica* leaf extract were synthesized successfully. It is said that the electrical charge on the nanoparticles surfaces is high resulted from the high zeta potential's absolute value.

Taking advantage of plant resources in Table 2, the biosynthesis of $TiO_2$ produce nanoparticles with size ranging from 7 to 150 nm. Mostly they are spherical in shape. Some of them are oval, aggregated, poorly dispersed, agglomerated and even have irregular shape. This might due to the biomolecules compounds coated at the surface that responsible as capping agent during the biosynthesis. Sahaya et al. [47] has compared their biosynthesized $TiO_2$ nanoparticles using dried petals identified as *Hibiscus-rosa-sinensis L.* with chemically synthesized nanoparticles. They found that their functionalized biosynthesized $TiO_2$ nanoparticles are spherical with distinct edges without aggregation and stable with good dispersibility compare to chemically synthesized $TiO_2$. Besides, in chemically synthesized $TiO_2$, there was a sharp diffraction peak observed whereas there is slight broadening and less intensity observed in biosynthesized $TiO_2$ nanoparticles with calculated crystallite size were 24 nm and 7nm, respectively. The broadening of the peak of biosynthesized TiO2 nanoparticles confirms the size reduction which indicates that their size is smaller than the chemically synthesized $TiO_2$ nanoparticles supported by green synthesis protocol [77]. In addition, Sahaya et al. [47] has suggested that the surface of biosynthesized $TiO_2$ nanoparticles was coated by the phytochemicals present in the petals extract thus reducing the XRD peak intensity probably due to an internal strain in the particles that make signal:noise ratio to decrease. This result also may enhance the dispersibility, stability as well as their bioavailability. On the other hand, the chemically synthesized $TiO_2$ nanoparticles are bare and uncapped indicated by its high intense peaks. Moreover, the effect of toxic related on size is expected none due to the reduced capability of nanoparticles to penetrate the cellular structure caused by the agglomeration of nanoparticles [52].

Another comparative study was reported by Hariharan et al. [45]. They have compared powder between biosynthesized $TiO_2$ and leaf extract powder of *Cynodon dactylon*, namely powder A and powder B, respectively. Both were calcinated at 500 for 5 hours. No peaks were observed for the powder B whereas there was broad XRD diffraction peaks were obtained for powder A indicating anatase $TiO_2$ and powder A has small crystalline size. There were no clear peaks observed for the powder B might be due to the organic compounds present in the *Cynodon dactylon* leaf powder. In this part, it could conclude that by altering the bioactive functional group and environmental growth can control of the shape and size of biosynthesized TiO2 nanoparticles including by improving the reaction conditions/parameters as supported by Singh et al. [78]. An interesting finding by Hudlikar et al. [58] mentioned that the integrity and stability of their nanoparticles contributed from the surface bound capping proteins can be check by treating them with 1% sodium dodecyl sulfate (SDS) detergent and heated at 85 °C for 30 min. This treatment lead to an aggregation/clumping of nanoparticles at a short period of time due to protein denature supported by FTIR spectroscopy analysis where most of the bands disappear including C-H stretch, N-H stretch and carbonyl stretch (C-O-C or C-O). From this finding we can note that the surface coating (peptide/protein) can be removed rather than utilized as capping agent depend on applications where surface coating is not needed such as in thin film formation and particle annealing.

Table 2. Characterization technique, size, morphology and application of biosynthesized $TiO_2$.

| No | Name and part of plant | Characterization | Size (nm) | Morphology | Application | Ref |
|---|---|---|---|---|---|---|
| *1.* | *Nyctanthes arbortristis*, leaf | XRD, SEM, PSA | 100 (XRD)<br>100-150 (SEM) | Spherical | Biomedical systems | [28] |
| *2.* | *Cynodon dactylon*, leaf | XRD, FTIR, Laser Raman spectroscopy, SEM | - | Irregular shape | Antibacterial and anticancer activity | [45] |
| *3.* | *Moringa oleifera leaf* | XRD, UV-VIS | 12 | - | - | [32] |
| *4.* | *Citrus sinensis (L.) (orange waste), peel extract* | XRD, PSA, FTIR, TG/DTA | 19 (XRD – rutile phase, tetragonal)<br>24 (PSA) | - | - | [42] |
| *5.* | *Azadirachta indica (Neem), leaves* | FESEM, EDAX, XRD, UV-Vis, FTIR, Zeta potential | 124 (DLS) | Spherical, interconnected | Photocatalytic degradation activity | [46] |
| *6.* | *Hibiscus rosa* sinensis (shoe black plant), flower | XRD, SEM, FTIR | 7 (XRD) | Spherical, dispersed, aggregated | Antibacterial activity | [47] |
| *7.* | *Catharanthus roseus*, leaf | XRD, FTIR, SEM, AFM | 100 (XRD - cubic)<br>25-110 (SEM) | Irregular, uneven and clustered | Antiparasitic activity | [48] |
| *8.* | *Curcuma longa* (turmeric), leaf | AFM, UV-VIS, XRD, SEM | 43.08-92.6 | Spherical | Reduced fungal growth, pathogenicity and spores | [31] |
| *9.* | *Echinacea purpurea, leaf* | UV-VIS, SEM, TXRF, FTIR | 120 | Spherical, agglomerated, poorly dispersed | - | [49] |
| *10.* | *Vigna radiate*, seed | FTIR, SEM | - | Oval | Antibacterial activity | [45] |
| *11.* | *Cassia fistula, leaf* | UV-Vis, XRD, FTIR, AFM, SEM, TGA | crystalline, cubic | Nearly spherical, agglomerated | Antibacterial activity | [50] |
| *12.* | *Citrus limon*, leaf | FTIR, FESEM-EDS | - | Irregular shape | Antibacterial activity | [51] |
| *13.* | *Aloe barbadensis mill, leaf* | UV-Vis, SEM, HRTEM, FESEM, EDS, FTIR, XRD, Raman spectra | 20-50 (SEM)<br>20 (HRTEM, FESEM) | Spherical, polydispersed | Biofilm | [39] |
| *14.* | *Aloe vera, leaf* | TEM, XRD, SEM, PSA, TG/DTA | 20 (XRD – tetragonal)<br>32 (PSA)<br>60-80 (SEM) | Irregular shape | - | [40] |

| | | | | | | |
|---|---|---|---|---|---|---|
| *15.* | *Jatropha curcas L*, leaf | UV-Vis, FESEM, EDS, FTIR, XRD, DLS, BET, BJH | 13 (XRD- tetragonal)<br>75 (DLS)<br>10-20 (FESEM) | Spherical | Wastewater treatment<br>Photocatalytic activity | [41] |
| *16.* | *Tinospora cordifolia*, stem | XRD, SEM, UV-Vis | 8-10 (XRD) | Irregular shape (SEM) | Photocatalytic activity<br>Antibacterial activity | [52] |
| *17.* | *Morinda citrifolia (M. citrifolia)*, leaf | XRD, FTIR, UV-Vis DRS, UV-Vis, Raman spectroscopy, SEM, EDX | 10 (XRD) –tetragonal rutile<br>15-19 (SEM) | Spherical (SEM) | Antimicrobial activity | [43] |
| *18.* | *Ageratina altissima, leaf* | UV-vis, FTIR, XRD, FESEM, EDX | 60-100 (polycrystalline) | Spherical | Photocatalytic activity | [53] |
| *19.* | *Calotropis gigantean* (Crown flower), flower | XRD, FTIR, SEM, EDX | 10.52 (XRD)<br>160-220 (SEM) | Spherical, oval, aggregated | Acaricidal activity | [54] |
| *20.* | *Solanum trilobatum (purple fruited pea eggplant), leaf* | XRD. FTIR, SEM, EDX, AFM | 70 (SEM) | Spherical, oval, aggregated, uneven | Antiparasitic efficacies, Larvacidal and pediculocidal activities | [55] |
| *21.* | *Euphorbia heteradena Jaub*, root | UV-Vis, XRD, TEM | 20 ± 3 | Spherical, rutile | - | [56] |
| *22.* | *Eclipta prostrate L.* (False daisy), leaf | FESEM, FTIR, XRD, AFM | 36-68 (XRD) | Spherical, quite polydispersed | - | [34] |
| *23.* | *Euphorbia prostrata*, leaf | TEM, XRD, FTIR, GC-MS | 83.22 (XRD) | Spherical, quite polydisperse | Antileishmanial activity | [57] |
| *24.* | *Jatropha curcas*, latex | TEM, SAED, XRD, FTIR, EDAX | 25-50 (TEM)<br>>50 (irregular shape) | Spherical and uneven | Biotechnology, environmental, biomedical and electronic systems | [58] |
| *25.* | *Mangifera indica L*, leaf | UV-Vis, FTIR, XRD, AFM, SEM, TEM | 30 ± 5 (TEM) | Spherical, oval | Larvicidal activity | [59] |
| *26.* | *Psidium guajava*, leaf | XRD, FTIR FESEM, EDX | 32.58 (XRD) | Sperical, cluster (FESEM) | Antibacterial and antioxidant activity | [35] |
| *27.* | *Glycosmis cochinchinensis,* leaf | XRD, UV-Vis, FTIR, SEM–EDS, TEM | 40 ± 5 | Spherical | Antibacterial activity | [60] |

| | | | | | | |
|---|---|---|---|---|---|---|
| *28.* | *Annona squamosa (sugar-apple), peel* | EDS, UV-Vis, TEM, SEM, XRD | 23 ± 2 | Polydispersed and spherical clusters | Bio therapeutic , bioengineering and electronics | [33] |
| *29.* | *Aloe vera gel* | XRD, FTIR, UV-Vis, DRS, AFM | 11 (XRD – tetragonal) 80-90 (PSA) | Spherical | Photocatalytic activity | [61] |

## 4. Mechanism of biosynthesized $TiO_2$ nanoparticles

Researchers have given attention on the detection, characterization of biomolecules as well as biomolecules mechanisms involved in the biosynthesis. But yet at the present time, the actual mechanism of biosynthesized nanoparticles by plant extract is still not clear however many studies proved that biomolecules compounds such as alkaloids, polysaccharides, alcoholic and phenolic compounds in plant materials could be involved in formation, stabilization and bioreduction of the nanoparticles [11], [20]. Many research study also suggests that in synthesizing nanoparticles, different plant species exhibit different mechanisms [21]. Reducing capacity of plants and reduction potential of ions can affect the production of nanoparticles which depend on the biomolecules presence in plants. For example, Nasrollahzadeh and Sajadi [56] have determined a possible mechanism of bioreduction of titanyl hydroxide illustrated in Figure 5. At first, they carried out FTIR analysis to identify the biomolecules present. As a result, they found, it was confirmed that the hydroxyl groups of phenolics in E. *heteradena Jaub* root extract is responsible for $TiO(OH)_2$ reduction and also as capping ligands to the surfaces of the nanoparticles. Furthermore, the FT-IR data shows demonstrative differences in the location and shape of signals indicate that there is interaction with involved sites of phytochemicals in the biosynthesis process thus revealed that polyphenolics might be adsorbed on the surface of the nanoparticles.

It has been reported that biomolecules components such as flavonoids, alkaloids, polyphenols, terpenoids, heterocyclic compounds, and polysaccharides have significant roles in reduction, stabilization as well as capping agent of highly mono disperse nanoparticles as well as preventing their aggregation [47]. It has been reported that these biomolecules could coat and cap the nanoparticles thus lead to a synergistic effect in biomedical applications [79]. For instance, Rajkumari et al. [39] showed that carboxylic acids of the major phytocompound of Aloe vera leaf extract represented by the broad OH stretch in FTIR analysis which also indicate the presence of terpenoids, flavonoids and proteins could assist in forming and fabrication of the biosynthesized $TiO_2$ nanoparticles thus acts as capping and reducing agent in the biosynthesis process. The same result was obtained by Sankar et al. [46] using *Azadirachta indica.* A study research by Pratap et al. [41] showed the presence of hydroxyl group in *Jatropha curcas L.* indicating that the surface of the biosynthesized $TiO_2$ nanoparticles was capped by the presence of phenols or polyphenolic tannins. They also proved the evidence of biosynthesized process by revealed a significant absorption peak at 336 nm in the range 200 to 800 nm. The proposed reaction mechanism can be seen in Figure 6. Moreover, Nithya et al. [61] have reported that stretching vibration of hydroxyl group and amino groups present in alcohols, phenols and amines was observed in FTIR spectrum thereby confirming the existence of $TiO_2$ in biosynthesis by biomolecule attachment using aloe vera extract.

Roopan et al. [33] has showed possible mechanism of $TiO_2$ nanoparticles formation where hydroxyl group from the *A. squamosa* aqueous peel extract serves as catalyst in the reaction. Here, titanyl hydroxide, $TiO(OH)_2$ was hydrated to produce $TiO_2$ nanoparticles by heating it with the extract at 60 °C. They also reported that hydroxyl functional group from water soluble compounds are responsible for the $TiO_2$ nanoparticles stabilization as shown in the pathway. Future investigations might focus on the optimization of reaction conditions for high amounts of biomolecules involved in the production process as well as the stabilization of biosynthesized nanoparticles. Although there is no specific information/ explanation (biochemically elucidated) on the biochemically mechanism, there is no doubt that those biochemical molecules significantly involved in the reduction, stabilization process reaction as well as great capping agent. Thus, by understanding the processes will help improvement of the nanoparticles productivity and accumulation capacity.

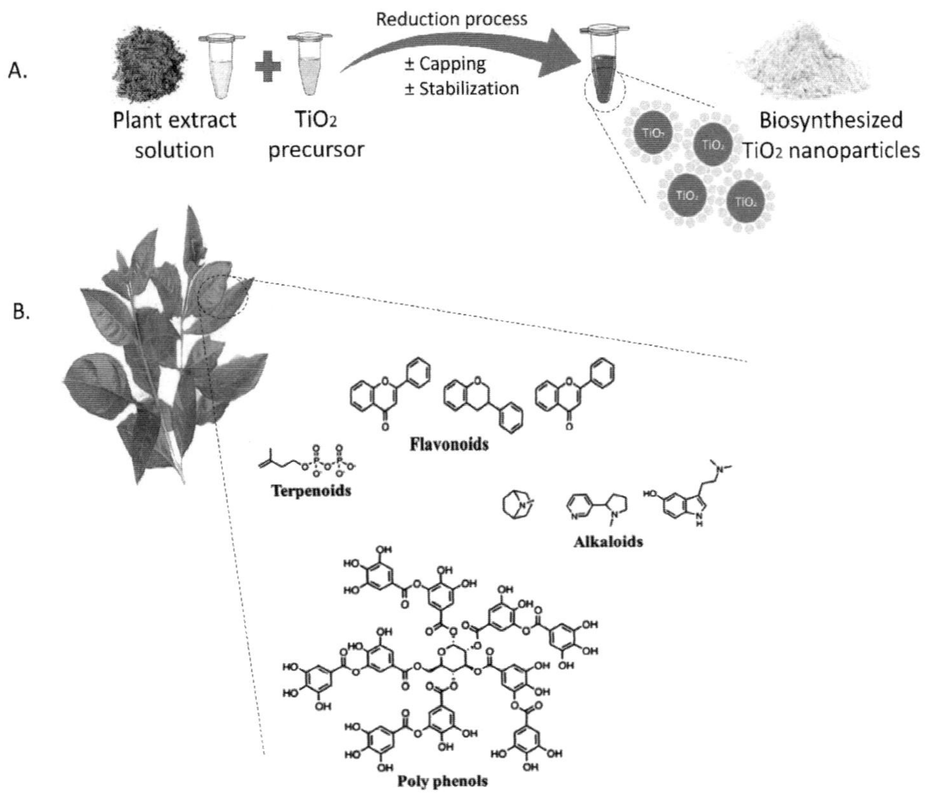


Fig. 6. (A) General mechanism for biosynthesis of $TiO_2$ nanoparticles; (B) Chemical structures of some compounds of plant extracts

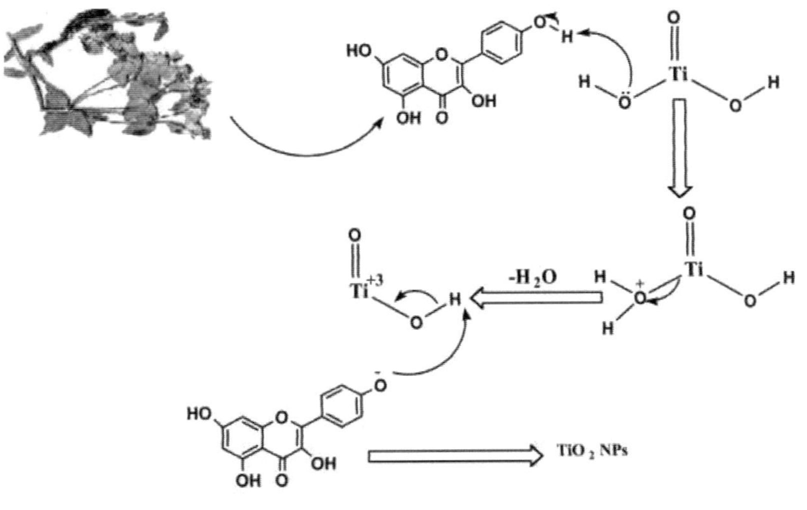


Fig. 7. Possible reaction mechanism of bioreduction of titanyl hydroxide to $TiO_2$ nanoparticles. Adapted from Nasrollahzadeh and Sajadi [56].

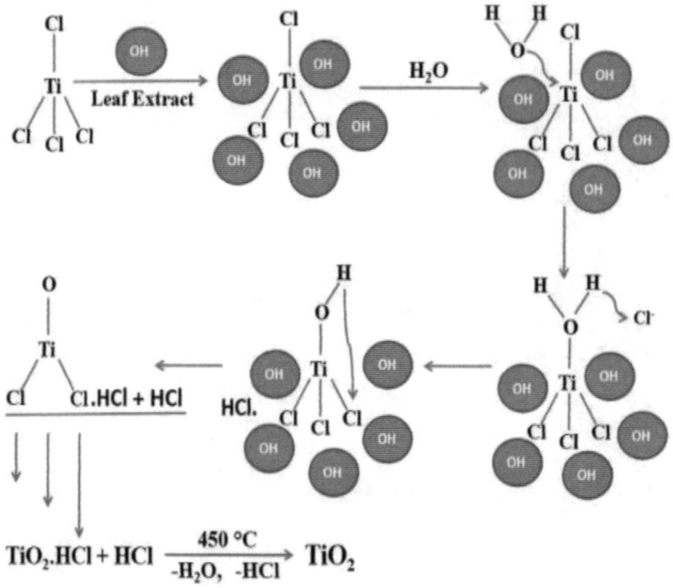


Fig. 8. Possible reaction mechanism for the formation of biosynthesized $TiO_2$ nanoparticles using leaf extract of *Jatropha curcas L.* Adapted from Pratap et al. [41].

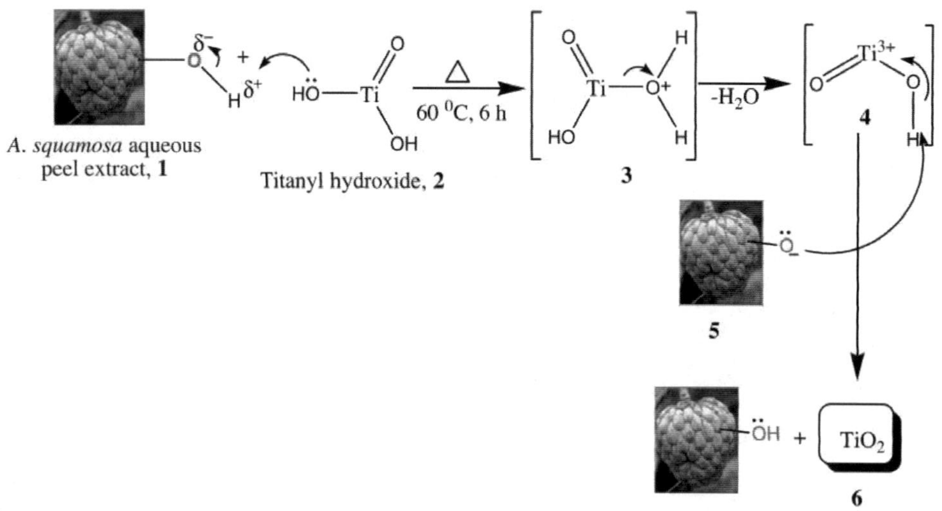


Fig. 9. Possible reaction mechanism for the dehydration of titanyl hydroxide into $TiO_2$ nanoparticles using *A. squamosa* aqueous peel extract. Adapted from Roopan et al. [33].

## 5. Applications of biosynthesized $TiO_2$ nanoparticles using plant extract

The use of materials in biosynthesis of $TiO_2$ nanoparticles such as plant extract (flower, seed, leaf, peel etc.) provide compatibility and eco-friendliness benefits for pharmaceutical, biomedical, as well as other applications as they can minimize the use of toxic and harmful chemicals. They also have superior photocatalytic oxidation properties that attracted great interest for water and waste water treatment. Some stabilized biosynthesized $TiO_2$ nanoparticles discussed before also can be useful for remediation of pollutant [11]. The reactive superoxide and hydroxyl radicals generated by $TiO_2$ under UV light help to degrade toxic as well as organic pollutants present in the water such as bacteria, viruses and other inactive microorganisms [80]. In this review, we try to discuss some potential applications that use biosynthesized $TiO_2$ nanoparticles using plant-mediated materials based on the applications listed in Table 2.

*5.1. Biomedical application*

There are several published reports on antibacterial activity effects of biosynthesized $TiO_2$ nanoparticles using various plant extract such as leaf, seed, stem and flower part. For example, biosynthesized $TiO_2$ using *Cynodon dactylon* leaf extract exhibited a good antibacterial activity against gram–negative bacteria as reported by Hariharan et al. [45]. using agar well diffusion method. They found a trend where activity of bacteria increased when $TiO_2$ concentration increased. Nonetheless, only when the inhibition concentration was 10μm, the effect of antibacterial activity was minimum. Stabilized $TiO_2$ nanoparticles by using Hibiscus flower extract prepared by Sahaya et al. [47] is considered to have antimicrobial property against pathogenic bacteria. In comparable with standard antibiotic they also claim that their nanoparticles are efficient for antibacterial activity thus have potential in biomedical applications due to its enhanced stability, surface coatings and dispersibility compare to chemically synthesized $TiO_2$ nanoparticles. The antibacterial activity tested by Swati et al. [50] against pathogenic strains of *Escherichia coli* and *Staphylococcus aureus* shows effective effect through biosynthesized $TiO_2$ using *Cassia fistula*. Besides, an eco-friendly biosynthesis of Titanium nanoparticles using leaf extract of Citrus limon investigated by Farook et al. [51] show interesting finding where they reveal the irregular shape nanoparticles can inhibit the growth of *Escherichia coli* even in low concentration while the growth - inhibitory effects of *Pseudomonas aeruginosa* were mild with Ampicillin as positive control. The obtained result was considered to be effective against clinical pathogen due to the presence of hydroxyl bonds and can be further used in antibiotic drugs.

Biosynthesized $TiO_2$ nanoparticles using plant mediated namely *Cynodon dactylon* are demonstrated not only can show the inhibitory effect on *E. coli* growth, they also can enhance anticancer activity against lung cancer (A549). Cancer has universal properties as tumor cells where they cannot control their own cell proliferation with changes of enzymatic and biochemical parameters Akhtar et al. [81]. So by using controlling agents such as bio-based nanoparticles perhaps the growth of cancerous cell can be regulated with systematic mechanism of cell cycle. Another study by Suman et al. [82] reported that the biosynthesized silver nanoparticles exhibited a strong cytotoxic effect in HeLa cell lines compared to other conventional synthetic drugs. In continuation, recent studies using other biosynthesized nanoparticles shows that nanoparticles derived from plant can potentially control cell growth tumor as free radical formation can be control relatively from the cell. These common free radicals usually damage the function of a normal cell and induce cell proliferation.

Antiparasitic activity of biosynthesized $TiO_2$ nanoparticles has been reported by Velayutham et al. [48] and Rajakumar et al. [55] by using *Catharanthus roseus* and *Solanum trilobatum* (purple fruited pea eggplant) leaf, respectively. Velayutham et al. [48] has utilized *Catharanthus roseus* against the adults of hematophagous fly, Hippobosca maculata Leach (Diptera: Hippoboscidae), and sheep-biting louse, Bovicola ovis Schrank (Phthiraptera: Trichodectidae). They prepared concentration ranging from 6.12 to 100 mg/L and 5.0 to 25 mg/L for aqueous plant extracts ($TiO_2$) and synthesized $TiO_2$ NPs, respectively. Within 24h exposure, the dead parasites numbers were counted and mortality percentage from five replicates in average were calculated. As reported by Rajakumar et al. [55] the adulticidal activity of biosynthesized $TiO_2$ NPs using S*olanum*. trilobatum showed maximum activity against P. h. capitis and H. a. anatolicum and larvae of A. subpictus. Base on the result it was conclude that bio synthesized can controls the adult head louse P. h. capitis, larvae of cattle tick H. a. anatolicum,and fourth instar larvae of malaria vector A. subpictus effectively. Other acaricidal and larvicidal activity is reported by Rajakumar et al. [59] using extract leaf of *Mangifera indica L.* against the larvae of Rhipicephalus (Boophilus) microplus, Hyalomma anatolicum anatolicum and Haemaphysalis bispinosa (Acari: Ixodidae), fourth instar larvae of Anopheles subpictus,and Culex quinquefasciatus (Diptera:

Culicidae). All the published report can provide a great excellent by helping in parasite control and widen the biosynthesized nanoparticle based technologies. Lastly, Sundrarajan et al. [43] also demonstrated that their green synthesized $TiO_2$ nanoparticles has superior antimicrobial activity against pathogen namely *Staphylococcus aureus, Escherichia coli, Bacillus subtilis, Pseudomonas aeruginosa, Candida albicans, and Aspergillus niger* (gram-positive bacteria) by testing their inhibitory activity using agar well-diffusion method.

### *5.2. Water and wastewater treatment application*

Recent eco-friendly nanoproducts available in commercial market with high demand is water purifier, water filter and other home-made products [83]. The material selection using titanium dioxide nanoparticles is preferable in photocatalytic reactions due to their abundance, photostability property as well as affordable in price. They also have excellent photocatalytic ability because of their strong oxidizing power, long-term stability and non-toxic [84]. However, the enriched photocatalytic degradation activity might be influenced by the charge, morphology and size of the synthesized nanoparticles. Organic compounds have been reduced effectively using biosynthesized TiO2 nanoparticles. For example, a study reported by Sankar et al. [46] shows that the colloidal biosynthesized $TiO_2$ nanoparticles using *Azadirachta indica (Neem)* leaves extract with a mean particle size of 124 nm and a zeta potential of -24 mV have been used to degrade industrially harmful methyl red dye from aqueous solutions under bright sunlight catalytically. Their result also shows that the dye degradation regulated efficiently when the incubation time of titanium dioxide nanoparticles with methyl red solution increased.

Pratap et al. [41] demonstrated a remarkable potential for wastewater treatment where biosynthesized $TiO_2$ nanoparticles using leaf extract of *Jatropha curcas* was applied in removal of chemical oxygen demand (COD) and chromium (Cr) simultaneous from secondary treated tannery wastewater (TWW). It was shown that the treatment can remove 82.26% and 76.48% of COD and Cr, respectively in a self-designed and fabricated Parabolic Trough Reactor (PTR) during the photocatalytic treatment of wastewater. The adsorption that related to the surface area increased of the biosynthesized $TiO_2$ nanoparticles might be the main reason for Cr removal from TWW. This achievement can provide an alternative to a clean-green treatment solution. Moreover, this study revealed that nanoparticles with pure crystalline anatase phase, high surface area, size in nano and large amount surface hydroxyl groups might increase solar photocatalytic activity. Normally, *Jatropha curcas* L. leaves is considered as waste material in many Asian countries but they can be find easily, and environmental friendly. Thus, the utilization of the nanoparticles is a very promising approach in wastewater to overcome environmental threats. Maurya et al. [52] used *Tinospora cordifolia* extract to biosynthesize TiO2 nanoparticles to enhance photocatalytic activity of plant extract/TiO2 composites. The authors also revealed an enhanced absorption capability by using UV-vis measurements.

*Ageratina altissima* leaf extract that been used in biosynthesis of $TiO_2$ nanoparticles shows enhanced percentage of decoloring of methylene blue, alizarin red, crystal violet, and methyl orange dyes with 86.79%, 76.32 %, 77.59 % and 69.06 % in percentage, respectively. The results present by Ganesan et al. [53] has indicated clearly that the biosynthesized nanoparticles can be used as an alternative to degrade and remove pollutants released from dyeing and textile industries as well as act as photocatalytic agent for water remediation in future. Nithya et al. [61] reported photocatalytic activity for Rhodamine B dye and the % of decolourization was effective for biosynthesized $TiO_2$ nanoparticles as compared to $TiO_2$ under visible light irradiation. The authors has been used *Aloe vera gel* extracts in the biosynthesis of $TiO_2$ nanoparticles.

### *5.3. Other applications*

There has been a great interest towards plant latex based strategies for nanoparticles fabrication in emphasizes to develop an economical and simple synthesis routes. Candidate such as *Jatropha curcas* can be considered as alternative facile approach for biosynthesis of $TiO_2$ nanoparticles in the electronic production of photo electrochemical energy as reported by Hudlikar et al. [58]. A research study by

Abdul Jalill et al. [31] discussed that their biosynthesized TiO2 nanoparticles can give agronomical impact where they found out that fungal and spores were reduced more effectively by biosynthetic $TiO_2$ nanoparticles using *Curcuma longa* aquatic extract (turmeric) compare with industrial synthetic nanoparticles. Their purpose is to investigated the effect of biosynthesized $TiO_2$ nanoparticles using *Curcuma longa* aquatic extract on the growth, sporulation, pathogenicity of *Fusarium graminearum* and some wheat plants parameters and compare them with standards industrials nanoparticles. Their study is in agreement with the results of Abdul Jalil and Yousef [85] who agreed that the use of industrial $TiO_2$ nanoparticles (50 nm of anatase shape) can increase the number of total chemical compounds in leaves plant thus give negative effect on the percentage of germination, rate of germination and the length of root and shoot. In addition, wheatgrass seeds that exposed to high concentration of $TiO_2$ nanoparticles (80 ppm) can lead to diminished germination rate [86]. It is well mentioned that the effects of nanoparticles on plant growth strongly depends on the synthesis methods of nanoparticles where concentration, shape, size, chemical and physical properties of the nanoparticles can be altered. [87,88].

## 6. Concluding remarks

This review highlights the characterization and mechanism of $TiO_2$ nanoparticles via biosynthesis method and their potential in biomedical, water and wastewater treatment as well as in other application. This review also discusses the characteristic such as size, morphology of the nanoparticles that related to the procedures and properties of the materials. Plant related materials has been proved to be a good reducing, stabilizing as well as capping agent for facile biosynthesis of $TiO_2$ nanoparticles. In fact, they contain active compound that can enhanced properties of the nanoparticles with controlled shapes and sizes. By comparing to conventional methods, several studies confirm that nanoparticles produced via biosynthesis route using natural resources has great advantages such as are less toxic contaminants, less of subsequent complex chemical synthesis, environmental friendly, cost effective and stable nanoparticles production with controlled shapes and sizes. Therefore, much effort is being made to implement biological synthesis methods with these proven advantages.

To achieve the sustainability of nanoparticles biosynthesis and improvement of nanoparticles stability, focus is needed and more research need to be done to explore potential local and common natural resources available even though they are still at early stages, and considerable research is needed in this direction. Utilization of local resources will reduce their development cost that can lead to economically competitive with conventional methods as an alternative to the large-scale of nanoparticles production. As prerequisite to the success of this new technology, understanding the biomolecule binding mechanisms involved is a must as it brings significantly benefit to biomedical field. However, there is no clearly described on the mechanism of the biosynthesis therefore there is a need to explore the phytochemistry behind the biosynthesis. The lack of knowledge of the underlying mechanisms, the chemical components responsible for the synthesis and the nanoparticles stabilization remain as open challenge in taking benefits of natural resources such as plants for the nanoparticles synthesis. The mostly challenge faced in biosynthesis of nanoparticles is the separation of the nanoparticles from the biological material and contamination from biological cells that could lead to adverse effect in biomedical application. Therefore, it is important to understand which active groups are involved and how functional groups from the natural resources attach to the nanoparticle surface, that can produce nanoparticles with higher efficacy especially in terms of biocompatibility and bioavailability of nanoparticles. In conclusion, green technology through biosynthesis as described in this paper, provide outstanding findings that perhaps encourage researchers and newcomers to continue and extend the exploration of possibilities offered by nature and the design of innovative and safer methodologies towards the synthesis of nanomaterials, possessing desired features and exhibiting valuable properties that can be exploited in a profusion of fields.

## 7. Acknowledgements

The authors would like to express the deepest appreciation to the Ministry of Higher Education (MOHE), Malaysia for funding this project through Fundamental Research Grant Scheme (FRGS vot. K041).

Additional facilities' supports from AMMC, Universiti Teknologi Tun Hussein Onn (UTHM) are also highly appreciated.

**References**

[1] A. E. Livari, A. Aroujalian, A. Raisi, M. Fathizadeh, and Etc, "The effect of TiO2 nanoparticles on PES UF membrane fouling in water-oil separation," *Procedia Eng.*, vol. 44, pp. 1783–1785, 2012.

[2] N. Nasrollahi, V. Vatanpour, S. Aber, and N. M. Mahmoodi, "Preparation and characterization of a novel polyethersulfone (PES) ultrafiltration membrane modified with a CuO/ZnO nanocomposite to improve permeability and antifouling properties," *Sep. Purif. Technol.*, vol. 192, no. June 2017, pp. 369–382, 2018.

[3] S. M. Doke and G. D. Yadav, "Novelties of combustion synthesized titania ultrafiltration membrane in efficient removal of methylene blue dye from aqueous effluent," *Chemosphere*, vol. 117, no. 1, pp. 760–765, 2014.

[4] Y. Ying *et al.*, "Recent advances of nanomaterial-based membrane for water purification," *Appl. Mater. Today*, vol. 7, pp. 144–158, 2017.

[5] B. Molki, W. M. Aframehr, R. Bagheri, and J. Salimi, "Mixed matrix membranes of polyurethane with nickel oxide nanoparticles for CO 2 gas separation," *J. Memb. Sci.*, vol. 549, no. August 2017, pp. 588–601, 2018.

[6] C. Liao, Y. Li, and S. C. Tjong, "Visible-light active titanium dioxide nanomaterials with bactericidal properties," *Nanomaterials*, vol. 10, no. 1, 2020.

[7] T. A. Silva *et al.*, "Silver nanoparticles biosynthesis and impregnation in cellulose acetate membrane for anti-yeast therapy," *African J. Biotechnol.*, vol. 16, no. 27, pp. 1490–1500, 2017.

[8] N. Senthilkumar, E. Nandhakumar, P. Priya, D. Soni, M. Vimalan, and I. Vetha Potheter, "Synthesis, anti-bacterial, anti-artritic, anti-oxidant and in-vitro cytoxicity activities of ZnO nanoparticles using leaf extract of tectona grandis," *R. Soc. Chem.*, 2017.

[9] M. Sathishkumar, N. P. Subiraminiyam, and M. Sasikumar, "Antibacterial activities of Zinc Sulphide nanoparticles using leaf extract of Lawsonia inermis," *Nehrucolleges.Net*, 2017.

[10] Y. K. Mohanta, S. K. Panda, A. K. Bastia, and T. K. Mohanta, "Biosynthesis of silver nanoparticles from Protium serratum and investigation of their potential impacts on food safety and control," *Front. Microbiol.*, vol. 8, no. APR, pp. 1–10, 2017.

[11] S. Saif, A. Tahir, and Y. Chen, "Green Synthesis of Iron Nanoparticles and Their Environmental Applications and Implications," *Nanomaterials*, vol. 6, pp. 1–26, 2016.

[12] S. Ahmed, M. Ahmad, B. L. Swami, and S. Ikram, "A review on plants extract mediated synthesis of silver nanoparticles for antimicrobial applications: A green expertise," *J. Adv. Res.*, vol. 7, no. 1, pp. 17–28, 2016.

[13] V. Mishra, R. Sharma, N. D. Jasuja, and D. K. Gupta, "A Review on Green Synthesis of Nanoparticles and Evaluation of Antimicrobial Activity," *Int. J. Green Herb. Chem.*, vol. 3, no. 1, pp. 081–094, 2014.

[14] A. Schröfel, G. Kratošová, I. Šafařík, M. Šafaříková, I. Raška, and L. M. Shor, "Applications of biosynthesized metallic nanoparticles - A review," *Acta Biomater.*, vol. 10, pp. 4023–4042, 2014.

[15] S. Iravani, "Green synthesis of metal nanoparticles using plants," *Green Chem.*, vol. 13, pp. 2638–2650, 2011.

[16] P. Lakshmi Pravallika, G. Krishna Mohan, K. Venkateswara Rao, and K. Shanker, "Biosynthesis, characterization and acute oral toxicity studies of synthesized iron oxide nanoparticles using ethanolic extract of Centella asiatica plant," *Mater. Lett.*, vol. 236, pp. 256–259, 2019.

[17] Q. H. Tran, V. Q. Nguyen, and A. T. Le, "Silver nanoparticles: Synthesis, properties, toxicology, applications and perspectives," *Adv. Nat. Sci. Nanosci. Nanotechnol.*, vol. 4, no. 3, 2013.

[18] O. V. Kharissova, H. V. R. Dias, B. I. Kharisov, B. O. Pérez, and V. M. J. Pérez, "The greener synthesis of nanoparticles," *Trends Biotechnol.*, vol. 31, no. 4, pp. 240–248, 2013.

[19] K. Ameer, H. M. Shahbaz, and J. Kwon, "Green Extraction Methods for Polyphenols from Plant Matrices and Their Byproducts : A Review," vol. 16, pp. 295–315, 2017.
[20] M. Herlekar, S. Barve, and R. Kumar, "Plant-Mediated Green Synthesis of Iron Nanoparticles," *J. Nanoparticles*, pp. 1–9, 2014.
[21] S. Baker, D. Rakshith, K. S. Kavitha, and P. Santosh, "Plants : Emerging as Nanofactories towards Facile Route in Synthesis of Plants : Emerging as Nanofactories towards Facile Route in Synthesis of Nanoparticles," *BioImpacts*, vol. 3, no. 3, pp. 111–117, 2013.
[22] V. Parthasarathi and G. Thilagavathi, "Synthesis and characterization of titanium dioxide nanoparticles and their applications to textiles for microbe resistance," *J. Text. Appar.*, vol. 6, no. 2, pp. 1–8, 2009.
[23] P. Singh, Y. Kim, D. Zhang, and D. Yang, "Biological Synthesis of Nanoparticles from Plants and Microorganisms," *Trends Biotechnol.*, vol. 34, no. 7, pp. 588–599, 2016.
[24] A. Baharlouei and G. H. S. Bonjar, "Biological control of Sclerotinia sclerotiorum ( oilseed rape isolate ) by an effective antagonist Streptomyces," *African J. Biotechnol.*, vol. 10, no. 30, pp. 5785–5794, 2011.
[25] S. A. Dahoumane, C. Jeffryes, M. Mechouet, and S. N. Agathos, "Biosynthesis of Inorganic Nanoparticles: A Fresh Look at the Control of Shape, Size and Composition," *Bioengineering*, vol. 4, no. 14, pp. 1–16, 2017.
[26] S. C. G. K. Daniel, N. Mahalakshmi, J. Sandhiya, K. Nehru, and M. Sivakumar, "Rapid synthesis of Ag nanoparticles using Henna extract for the fabrication of Photoabsorption Enhanced Dye Sensitized Solar Cell ( PE- DSSC )," *Adv. Mater. Res.*, vol. 678, pp. 349–360, 2013.
[27] S. K. Aldosary and S. N. A. El-rahman, "Green Synthesis and Antibacterial Properties of Silver Nanoparticles of Lawsonia inermis , Rhamnus frangula , Camellia sinensis and Thymus vulgaris Extracts," *J. Pure Appl. Microbiol.*, vol. 13, no. June, pp. 1279–1284, 2019.
[28] M. Sundrarajan and S. Gowri, "Green synthesis of titanium dioxide nanoparticles by nyctanthes arbor-tristis leaves extract," *Chalcogenide Lett.*, vol. 8, no. 8, pp. 447–451, 2011.
[29] A. Altemimi, N. Lakhssassi, A. Baharlouei, D. G. Watson, and D. A. Lightfoot, "Phytochemicals: Extraction, Isolation and Identification of Bioactive Compounds from Plant Extracts," *Plants*, vol. 6, no. 42, pp. 1–23, 2017.
[30] Q. W. Zhang, L. G. Lin, and W. C. Ye, "Techniques for extraction and isolation of natural products : a comprehensive review," *Chin. Med.*, vol. 13, no. 20, pp. 1–26, 2018.
[31] J. . Abdul, R. . and Nuaman, and A. . Aham, "Biological synthesis of Titanium Dioxide nanoparticles by Curcuma longa plant extract and study its biological properties," *World Scie*, vol. 49, no. 2, pp. 204–222, 2016.
[32] V. Patidar and P. Jain, "Green Synthesis of TiO2 Nanoparticle Using Moringa Oleifera Leaf Extract," *Int. Res. J. Eng. Technol.*, pp. 470–473, 2017.
[33] S. M. Roopan *et al.*, "Efficient phyto-synthesis and structural characterization of rutile TiO2 nanoparticles using Annona squamosa peel extract," *Spectrochim. Acta - Part A Mol. Biomol. Spectrosc.*, vol. 98, pp. 86–90, 2012.
[34] G. Rajakumar, A. A. Rahuman, B. Priyamvada, V. G. Khanna, D. K. Kumar, and P. J. Sujin, "Eclipta prostrata leaf aqueous extract mediated synthesis of titanium dioxide nanoparticles," *Mater. Lett.*, vol. 68, pp. 115–117, 2012.
[35] T. Santhoshkumar *et al.*, "Green synthesis of titanium dioxide nanoparticles using Psidium guajava extract and its antibacterial and antioxidant properties," *Asian Pac. J. Trop. Med.*, vol. 7, no. 12, pp. 968–976, 2014.
[36] C. . Anokwuru, G. . Anyasor, O. Ajibaye, O. Fakoya, and P. Okebugwu, "Effect of extraction solvents on phenolic, flavonoid and antioxidant activities of three nigerian medicinal plants," *Nat. Sci.*, vol. 9, no. 7, pp. 53–61, 2011.
[37] S. M. Roopan and G. Madhumitha, "Bioorganic phase in natural food: An overview," *Bioorganic Phase Nat. Food An Overv.*, pp. 1–331, 2018.

[38] E. Koffi, T. Sea, Y. Dodehe, and S. Soro, "Effect of solvent type on extraction of polyphenols from twenty three Ivorian plants," *J. Anim. Plant Sci.*, vol. 5, no. 3, pp. 550–558, 2010.

[39] J. Rajkumari *et al.*, "Synthesis of titanium oxide nanoparticles using Aloe barbadensis mill and evaluation of its antibiofilm potential against Pseudomonas aeruginosa PAO1," *J. Photochem. Photobiol. B Biol.*, vol. 201, no. October, p. 111667, 2019.

[40] G. . Rao, A. CH, V. . Rao, S. C. Chakra, and P. Tambur, "Green Synthesis of TiO2 Nanoparticles Using Aloe Vera Extract," *Int. J. Adv. Res. Phys. Sci.*, vol. 2, no. 1A, pp. 28–34, 2015.

[41] S. Pratap, G. Saxena, V. Singh, and A. Kumar, "Green synthesis of TiO 2 nanoparticles using leaf extract of Jatropha curcas L . for photocatalytic degradation of tannery wastewater," *Chem. Eng. J.*, vol. 336, no. September 2017, pp. 386–396, 2018.

[42] K. G. Rao, C. Ashok, K. V. Rao, C. Shilpa Chakra, and V. Rajendar, "Synthesis of Tio 2 Nanoparticles From Orange Fruit Waste," *Int. J. Multidiscip. Adv. Res. Trends*, vol. ISSN, no. 3, pp. 2349–7408, 2015.

[43] M. Sundrarajan *et al.*, "Journal of Photochemistry & Photobiology , B : Biology Obtaining titanium dioxide nanoparticles with spherical shape and antimicrobial properties using M . citrifolia leaves extract by hydrothermal method," *J. Photochem. Photobiol. B Biol.*, vol. 171, no. April, pp. 117–124, 2017.

[44] G. Nabi *et al.*, "A Review on Novel Eco-Friendly Green Approach to Synthesis TiO2Nanoparticles Using Different Extracts," *J. Inorg. Organomet. Polym. Mater.*, vol. 28, pp. 1552–1564, 2018.

[45] D. Hariharan, K. Srinivasan, and N. Lc, "Synthesis and Characterization of Tio2 Nanoparticles Using Cynodon Dactylon Leaf Extract for Antibacterial and Anticancer (A549 Cell Lines) Activity," *J. Nanomedicine Res.*, vol. 5, no. 6, pp. 4–8, 2017.

[46] R. Sankar, K. Rizwana, K. S. Shivashangari, and V. Ravikumar, "Ultra-rapid photocatalytic activity of Azadirachta indica engineered colloidal titanium dioxide nanoparticles," *Appl. Nanosci.*, vol. 5, no. 6, pp. 731–736, 2015.

[47] P. S. M. Kumar, A. P. Francis, and T. Devasena, "Biosynthesized and Chemically Synthesized Titania Nanoparticles : Comparative Analysis of Antibacterial Activity Biosynthesized and Chemically Synthesized Titania Nanoparticles : Comparative Analysis of Antibacterial Activity," *J. Environ. Nanotechnol*, vol. 3, no. August 2014, pp. 73–81, 2014.

[48] K. Velayutham, A. A. Rahuman, and G. Rajakumar, "Evaluation of Catharanthus roseus leaf extract-mediated biosynthesis of titanium dioxide nanoparticles against Hippobosca maculata and Bovicola ovis," *Parasitol Res*, pp. 2329–2337, 2012.

[49] R. Dobrucka, "Synthesis of titanium dioxide nanoparticles using Echinacea purpurea herba," *Iran. J. Pharm. Res.*, vol. 16, no. 2, pp. 753–759, 2017.

[50] N. Swathi, S. D, S. Rajeshkumar, and T. Lakshmi, "Green synthesis of titanium diozide nanoparticles using Cassia fistula and its antibacterial activity," *Int. J. Res. Pharm. Sci.*, vol. 10, no. 2, pp. 856–860, 2019.

[51] M. A. Farook, H. S. M. Mohamed., J. Thomas, K. Sathiyaseelan, R. Venkatesan, and R. Kaviyarasu, "Citrus limon Leaf Extract mediated synthesis of Titanium Dioxide Nanoparticles," *Life Sci. Arch.*, vol. 3, no. 1, pp. 901–908, 2017.

[52] A. Maurya, P. Chauhan, A. Mishra, and A. K. Pandey, "Surface Functionalization of TiO 2 with Plant Extracts and their Combined Antimicrobial Activities Against E . faecalis and E . Coli," *J. Res. Updat. Polym. Sci.*, vol. 1, pp. 43–51, 2012.

[53] S. Ganesan *et al.*, "Green engineering of titanium dioxide nanoparticles using Ageratina altissima (L.) King & H.E. Robines. medicinal plant aqueous leaf extracts for enhanced photocatalytic activity," *Ann. Phytomedicine An Int. J.*, vol. 5, no. 2, pp. 69–75, 2016.

[54] S. Marimuthu *et al.*, "Acaricidal activity of synthesized titanium dioxide nanoparticles using Calotropis gigantea against Rhipicephalus microplus and Haemaphysalis bispinosa," *Asian Pac. J. Trop. Med.*, vol. 6, no. 9, pp. 682–688, 2013.

[55] G. Rajakumar, A. A. Rahuman, and C. Jayaseelan, "Solanum trilobatum extract-mediated synthesis of titanium dioxide nanoparticles to control Pediculus humanus capitis , Hyalomma anatolicum anatolicum and Anopheles subpictus," *Parasitol Res*, 2013.
[56] M. Nasrollahzadeh and S. M. Sajadi, "Synthesis and characterization of titanium dioxide nanoparticles using Euphorbia heteradena Jaub root extract and evaluation of their stability," *Ceram. Int.*, vol. 41, no. 10, pp. 14435–14439, 2015.
[57] A. A. Zahir *et al.*, "Green synthesis of silver and titanium dioxide nanoparticles using Euphorbia prostrata extract showed shift from apoptosis to G0/G1 arrest followed by necrotic cell death in Leishmania donovani," *Am. Soc. Microbiol.*, no. June, pp. 1–51, 2015.
[58] M. Hudlikar, S. Joglekar, M. Dhaygude, and K. Kodam, "Green synthesis of TiO 2 nanoparticles by using aqueous extract of Jatropha curcas L . latex," *Mater. Lett.*, vol. 75, pp. 196–199, 2012.
[59] G. Rajakumar and A. A. Rahuman, "Efficacy of larvicidal activity of green synthesized titanium dioxide nanoparticles using Mangifera indica extract against blood-feeding parasites," 2014.
[60] H. Rosi and S. Kalyanasundaram, "Synthesis , characterization , structural and optical properties of titanium-dioxide nanoparticles using Glycosmis cochinchinensis Leaf extract and its photocatalytic evaluation and antimicrobial properties," *World News Nat. Sci.*, vol. 17, no. January, pp. 1–15, 2018.
[61] A. Nithya, K. Rokesh, and K. Jothivenkatachalam, "Biosynthesis , Characterization and Application of Titanium Dioxide Nanoparticles," *Nano Vis.*, vol. 3, no. 3, pp. 169–174, 2013.
[62] A. Chatterjee, D. Nishanthini, N. Sandhiya, and J. Abraham, "Biosynthesis of titanium dioxide nanoparticles using Vigna radiata," *Asian J. Pharm. Clin. Res.*, vol. 9, no. 4, pp. 4–7, 2016.
[63] ShilpaHiremath, Vidya C, M A Lourdu Antonyraj and S. S. Chandraprabha M N, "Green Synthesis of Tio 2 nanoparticles By Using Neem Leaf Extract," *Int. Rev. Appl. Biotechnol. Biochem.*, vol. 2, no. 1, pp. 11–17, 2014.
[64] S. Sathyajothi, R. Jayavel, and A. C. Dhanemozhi, "The Fabrication of Natural Dye Sensitized Solar Cell (Dssc) based on TiO2 Using Henna and Beetroot Dye Extracts," *Mater. Today Proc.*, vol. 4, no. 2, pp. 668–676, 2017.
[65] X. Zhu and K. Pathakoti, *Green synthesis of titanium dioxide and zinc oxide nanoparticles and their usage for antimicrobial applications and environmental remediation*, no. January. 2019.
[66] G V Khade, M. B. Suwarnkar, N. L. Gavade, and K. M. Garadkar, "Green synthesis of TiO 2 and its photocatalytic activity," *J Mater Sci Mater Electron*, vol. 30, no. Iv, 2015.
[67] N. Jayarambabu, K. V. Rao, and V. Rajendar, "Biogenic synthesis, characterization, acute oral toxicity studies of synthesized Ag and ZnO nanoparticles using aqueous extract of Lawsonia inermis," *Mater. Lett.*, vol. 211, pp. 43–47, 2018.
[68] D. Sharma, S. Kanchi, and K. Bisetty, "Biogenic synthesis of nanoparticles : A review," *Arab. J. Chem.*, pp. 1–25, 2015.
[69] F. D. Bălăcianu, R. Bartoş, and A. C. Nechifor, "Organic-inorganic membrane materials," *UPB Sci. Bull. Ser. B Chem. Mater. Sci.*, vol. 71, no. 3, pp. 37–54, 2009.
[70] L. G. Devi and R. Kavitha, "Applied Catalysis B : Environmental A review on non metal ion doped titania for the photocatalytic degradation of organic pollutants under UV / solar light : Role of photogenerated charge carrier dynamics in enhancing the activity," vol. 141, pp. 559–587, 2013.
[71] W. Kemp, "Infrared Spectroscopy," in *Organic Chemistry*, 1991, pp. 19–99.
[72] M. Urbano, M. D. L. de Castro, P. M. Pe´rez, J. Garcı´a-Olmo, and M. A. Go´mez-Nieto, "Ultraviolet–visible spectroscopy and pattern recognition methods for differentiation and classification of wines," *Food Chem.*, vol. 97, pp. 166–175, 2006.
[73] G. V. Andrievsky, V. K. Klochkov, A. B. Bordyuh, and G. I. Dovbeshko, "Comparative analysis of two aqueous-colloidal solutions of C60 fullerene with help of FTIR reflectance and UV-Vis spectroscopy," *Chem. Phys. Lett.*, vol. 364, no. 1–2, pp. 8–17, 2002.

[74] E. J. Cho, H. Holback, K. C. Liu, S. A. Abouelmagd, J. Park, and Y. Yeo, "Nanoparticle Characterization : State of the Art , Challenges , and Emerging Technologies," *Mol. Pharm.*, vol. 10, pp. 2093–2110, 2013.

[75] H. Dong, G. Zeng, L. Tang, and C. Fan, "ScienceDirect An overview on limitations of TiO 2 -based particles for photocatalytic degradation of organic pollutants and the corresponding countermeasures," vol. 9, 2015.

[76] J. Brant, "Aggregation and deposition characteristics of fullerene nanoparticles in aqueous systems," *J. Nanoparticle Res.*, vol. 7, pp. 545–553, 2005.

[77] N. Ahmad and S. Sharma, "Green Synthesis of Silver Nanoparticles Using Extracts of Ananas comosus," *Green Sustain. Chem.*, vol. 2, no. November, pp. 141–147, 2012.

[78] P. Singh, Y. J. Kim, C. Wang, R. Mathiyalagan, and D. C. Yang, "Microbial synthesis of Flower-shaped gold nanoparticles Microbial synthesis of Flower-shaped gold nanoparticles," *Artif. Cells, Nanomedicine, Biotechnol.*, no. May, pp. 1–6, 2015.

[79] M. Zargar *et al.*, "Green Synthesis and Antibacterial Effect of Silver Nanoparticles Using Vitex Negundo L.," *Molecules*, vol. 16, no. May 2014, pp. 6667–6676, 2011.

[80] M. F., Z. S., O. Y., Z. Z., S. H.-S., and C. S.-R., "Fouling in membrane bioreactors: An updated review," *Water Res.*, vol. 114, pp. 151–180, 2017.

[81] M. S. Akhtar, J. Panwar, and Y. S. Yun, "Biogenic synthesis of metallic nanoparticles by plant extracts," *ACS Sustain. Chem. Eng.*, vol. 1, no. 6, pp. 591–602, 2013.

[82] T. Y. Suman, S. R. Radhika Rajasree, A. Kanchana, and S. B. Elizabeth, "Biosynthesis, characterization and cytotoxic effect of plant mediated silver nanoparticles using Morinda citrifolia root extract," *Colloids Surfaces B Biointerfaces*, vol. 106, pp. 74–78, 2013.

[83] P. Kouvaris, A. Delimitis, V. Zaspalis, D. Papadopoulos, S. A. Tsipas, and N. Michailidis, "Green synthesis and characterization of silver nanoparticles produced using Arbutus Unedo leaf extract," *Mater. Lett.*, vol. 76, pp. 18–20, 2012.

[84] R. Ameta, A. Ameta, and S. Ameta, "Photocatalytic Degradation of Organic Pollutants : A Review PHOTOCATALYTIC DEGRADATION OF ORGANIC POLLUTANTS : A REVIEW," no. December, 2012.

[85] R. D. A. Jalil and A. M. Yousef, "Comparison Study on the Effect of Nano and Bulk Titanium Dioxide Particles on Seeds Germination , Growth and Chemical Composition of Wheat Invitro and Invivo," *Al-Mustansiriyah J. Sci.*, vol. 28, no. 3, pp. 85–101, 2017.

[86] R. Azimi, H. Feizi, and M. K. Hosseini, "Can Bulk and Nanosized Titanium Dioxide Particles Improve Seed Germination Features of Wheatgrass (Agropyron desertorum)," *Not. Sci. Biol.*, vol. 5, no. 3, pp. 325–331, 2013.

[87] X. Ma, J. Geiser-lee, Y. Deng, and A. Kolmakov, "Science of the Total Environment Interactions between engineered nanoparticles ( ENPs ) and plants : Phytotoxicity , uptake and accumulation," *Sci. Total Environ.*, vol. 408, no. 16, pp. 3053–3061, 2010.

[88] Z. G. Doğaroğlu and N. Köleli, "Effects of TiO2 and ZnO nanoparticles on germination and antioxidant system of wheat (triticum aestivum L.)," *Appl. Ecol. Environ. Res.*, vol. 15, no. 3, pp. 1449–1510, 2017.

# Mixed Matrix Membrane Incorporating of MgAl-$CO_3$ Layered-Double-Hydroxide for Enhanced Carbon Dioxide Separation

**N. Fajrina[1,2], N. Yusof[1,2]*, A.F. Ismail[1,2], J. Jaafar[1,2], F. Aziz[1,2], W.N.W. Salleh[1,2], N.M. Ismail[3]**

[1] Advanced Membrane Technology Research Centre (AMTEC), Universiti Teknologi Malaysia, 81310 UTM Skudai, Johor Darul Ta'zim, Malaysia.
[2] School of Chemical and Energy Engineering (SCEE), Faculty of Engineering, Universiti Teknologi Malaysia, 81310 UTM Skudai, Johor Darul Ta'zim, Malaysia.
[3] Faculty of Engineering, Universiti Malaysia Sabah, Jalan UMS, 88400, Kota Kinabalu, Sabah, Malaysia

E-mail: norhaniza@petroleum.utm.my

**Abstract.** In this study, mixed matrix membranes were fabricated by incorporating MgAl-$CO_3$ layered double hydroxides (LDH) into polysulfone (PSF) coated with low-cost commercial PEBAX polymer for enhancing $CO_2/CH_4$ separation. LDH was synthesized via a simple co-precipitation method and flat sheet membranes were fabricated by dry/wet inversion phase. The gas separation performance on pure PSF and LDH/PSF membranes were investigated. The physical property, chemical structure and membrane morphology were characterized by XRD, FTIR, and SEM. The sample membranes were tested with $CO_2$ and $CH_4$ gas for permeance performance and selectivity of $CO_2/CH_4$ was calculated. By incorporation of LDH, the $CO_2$ permeance increased about 54.6%. LDH mixed matrix membranes displayed the 18.2 GPU of $CO_2$ permeance and $CO_2/CH_4$ selectivity of 18.0. The LDH based membrane produces innovation in membrane technology by improving its morphology and permeance performance with great potential for large-scale $CO_2$ capture and separation.

## 1. Introduction

The rapid growth of the global economy and the expanding human population urgently requires a sustainable supply of energy [1]. Considering the depletion of fossil fuels, it is imperative to find appropriate alternatives to cope with the energy crisis. Harvesting energy from biogas, a kind of renewable energy source, has attracted a great deal of attention and evolves a potential way to retard the depletion of other energy sources and fight against global warming at the same time [2-4]. However, removing $CO_2$ from the as-produced biogas is necessary to increase the calorific values. Compared with conventional biogas upgrading methods such as water and amine scrubbing [5], membrane separation is considered a more cost-effective technical solution, along with low energy consumption, simple design, environmental friendliness as well as small plant footprint [6, 7]. These advantages have led to extensive exploration in the preparation of $CO_2$ membranes with high separation performance.

Polymeric membranes are processed by a simple, environmentally sustainable process, which is cost-efficient, energy-intensive and easy to scale up [8-10]. However, the substitution of traditional technologies and their positive effect would require a long time for membrane separation, as it is constrained by the intrinsic restriction in the trade of permeability and selectivity [11]. Inorganic membranes show a better performance in separation but are expensive and hard to turn into uniform films, compared to polymeric membranes [12]. A mixture of inorganic fillers distributed over a polymer matrix called a mixed matrix membrane (MMM) has shown an attractive method to boost the separating characteristics of polymers to help solve the well-known Robeson's upper bound [13]. Latest studies concentrate on a variety of new filler products in MMM, including carbon nanotube, metal-organic framework (eg: UiO-66), clay, zeolite, silica, metal oxide and carbon molecular sieve [14-19].

For MMM to achieve the best performance, the good compatibility between filler with polymer is the key factor. Layered double hydroxide (LDH) has similar properties to clay but, it can be finely tuned and has high $CO_2$ affinity resulting in easy $CO_2$ capture. Although less research can be found on $CO_2$ separation using LDH as filler, the LDH is reported to be excellent as an adsorbent for $CO_2$ capture. Gouveia *et al.* [20] have shown that the combination of high surface area and basic sites have resulted in the maximum adsorption capacity of $CO_2$ (159.1 mg/$g_{cat}$). The large specific surface area with the availability of basic sites in LDH will certainly bring an advantage as a nanofiller in MMM for $CO_2$ capture as $CO_2$ is known as acidic gas. Liu *et al.* [21] successfully fabricated a 2D-LDH membrane and built $CO_2$-selective transport nanochannels for $CO_2$ separation. The intercalator of $CO_3^{2-}$ ions have reduced the nanochannel size from 0.7 nm to 0.3 nm and thus, provided the membrane a molecule sieving ability for $CO_2/CH_4$ separation. The results showed high performance with 150 GPU of $CO_2$ permeability and 33 of $CO_2/CH_4$ selectivity.

Polysulfone (PSF) is the most used polymer due to its properties such as good chemical stability and high mechanical strength. Kiadehi *et al.* [22] successfully fabricated nanofibers (CNF)/PSF MMM. The finding showed that the increasing of CNF concentration in MMM led to the increasing of $CO_2$ permeability and $CO_2/CH_4$ selectivity with 12.04 Barrer and 12.17, respectively. Waheed *et al.* [23] fabricated MMM by incorporating PSF with mesoporous silica particles extracted from rice husk ash and the results have shown 8.46 Barrer of $CO_2$ permeability, 0.258 Barrer of $CH_4$ permeance and 32.79 $CO_2/CH_4$ selectivity using 40% filler loading. However, there is still a membrane defect, such as pinholes, which reduces $CO_2/CH_4$ selectivity. This problem can be prevented by coating the MMM with PEBAX to seal pinholes and improve the permeance of selective gas. The previous research indicated the ideal range of PEBAX is about 3% to 5% [24-27]. Researchers claimed in this range that although there is a risk of pore penetration, it also makes excellent transport of gas. Jamil *et al.* [28] showed that membrane defects on the surface can be minimized by PEBAX coating. Their finding exposed that the PEBAX on MMM has the highest $CO_2/CH_4$ selectivity with 38.45 at 5 bar in room temperature.

Up to date, the incorporation of LDH in MMM for $CO_2$ capture is still recent and as for the author's knowledge, there is no research on coated MgAl-$CO_3$ LDH/PSF MMM fabrication has been done. Herein, in this research, the compatibility of LDH with PSF was investigated. The fabricated MMM was coated with 3 wt% PEBAX to seal pinholes and improve the gas permeances.

## 2. Methodology

### *2.1. Materials*

$Al(NO_3)_3 \cdot 9H_2O$ and $Mg(NO_3)_2 \cdot 6H_2O$ were provided by Fisher Scientific (M) Sdn Bhd. Acetone, $Na_2CO_3$ and NaOH were purchased from Sigma-Aldrich (Malaysia). Polysulfone polymer (Udel-P3500) and PEBAX MH 1657 were purchased from Solvay Advanced Polymers (USA). Deionized water (DI water) was used during the whole experiment. Without further purification, all chemical compounds were used.

### *2.2. Synthesis of LDH*

MgAl-$CO_3$ LDH was prepared via the co-precipitation method. Firstly, 9.6 g of $Mg(NO_3)_2 \cdot 6H_2O$ and 4.78 g of $Al(NO_3)_3 \cdot 9H_2O$ were dissolved in 50 ml DI water by magnetic stirring at room temperature as Solution A. For solution B, 2.65 g of $Na_2CO_3$ was dissolved in 50 ml of DI water using magnetic stirring while for solution C, 8g of NaOH pellets were dissolved in 50 ml of DI water. With continuous stirring, solution A was titrated into solution B and the pH of the mixture solution was maintained at about 9-10 by adding Solution C to the above-mentioned mixture. The mixture was stirred for a further 30 min after solution A was completely added to solution B. Then, the solution was aged at 60°C for 24 hr. The sample was consequently separated by centrifuge (6000 rpm). The LDH obtained was washed with DI water repeatedly until pH approached 7 and centrifuge each time. The LDH was further washed intensively with acetone to prevent LDHNPs agglomeration. After that, the obtained sample was dried in the oven for 24 h at 60°C and then grounded using a mortar and pestle.

### *2.3. Fabrication LDH/PSF MMM*
PSF dope solution was prepared by simple mixing. 1.5 g MgAl-$CO_3$ LDH was dispersed in 23 ml NMP solvent by sonicator for 1 h. Then, the mixture was heated to 65°C. 6 g PSF was added slowly and stirred at a stirring rate of 130 rpm. The dope solution was continually stirred for a further 24 h after the PSF was completely dissolved. The casting solution was degassed by sonicator for 4 h followed by cooling at room temperature for another 24 h. For bare PSF, the procedure is the same without included the LDH powder.

Next, the membrane was prepared by dry/wet phase inversion. The dense membrane was cast at room temperature using a casting glass rod on a smooth glass plate and was then immediately immersed in the coagulant bath. The fabricated membrane was transferred and immersed in the fresh water for 24 h and then, drying it for 24 h at room temperature.

### *2.4. Coating PEBAX layer*
The single layer of 3 wt% PEBAX layer was coated by the dip-coating method to seal the membrane defect. PEBAX coating solution was prepared with a mass ratio of 70:30 of ethanol/water. In the mixed solvent at 70°C, PEBAX pellets were first added and stirred for 2 h followed by sonicated for 4 h. The ready solution was subsequently cooled to room temperature. Next, the solution was used to coating MMM with PEBAX. The dope solution of PEBAX was pour into the glass petri dish around 5 ml to damp the dish. Small amount of PEBAX solution is used to make sure the solution just coated one side surface dip, and not overflow and cover both side of membrane. Then, the shiny surface of the flat sheet membrane was faced the solution then dip coated for 30 sec. After that, the membrane was hanged for 3 min to remove the excess solution before dried in the oven for 2 h at 60°C to allow solvent vaporization.

### *2.5. Characterization Technique*
X-ray diffraction (XRD) was used to determine the crystallinity of the sample using an X-Ray Diffractometer (Model Rigaku SmartLab). The X-ray diffractometer was conducted at a 2-theta range from $3° \geq 2\theta \geq 100°$ in room temperature. To determine the functional group that exists in LDH-based MMM, the Fourier transform infrared (FTIR) analysis was tested. FTIR spectra were obtained by Bruker TENSOR27 impact spectrometer. The Scanning Electron Microscope (SEM) was used to characterize the morphology of the membrane in the lower magnification. Before testing, the samples were also frozen fractured under cryogenic conditions in liquid nitrogen and gold/platinum coat. The surface area and cross-sectional morphology analysis of the flat sheet MMM was conducted using a scanning electron microscope (TM3000, Hitachi) fitted with different magnifications.

### *2.6. Gas Permeance Test*
To assess the membrane efficiency in terms of gas permeance and selectivity, a gas permeation test was performed. The sample membrane was put inside the tailored flat sheet membrane housing before processing. At 5 bar, single gas $CO_2$ and $CH_4$ with a purity of 99.99% were then supplied from a gas cylinder to the membrane housing. The pressure was adjusted by a two-stage pressure regulator. The flow rate was measured using a soap bubble flowmeter after the gas had permeated through the membrane.

The gas permeance ($P_i/l$) and the gas selectivity ($\alpha_{i/j}$) were calculated using equations (1) and (2).

$$\left(\frac{P_i}{l}\right)=\frac{Q_i}{A\Delta P}\frac{273.15\times10^6}{T} \quad (1)$$

$$\alpha_{i/j}=\frac{(P_i/l)}{(P_j/l)} \quad (2)$$

where $i$ and $j$ are the penetrating gas $i$ and $j$, respectively, $P_i/l$ is the gas permeance across the membrane (GPU), $Q_i$ is the volumetric flow rate of gas permeated across the membrane at STP ($cm^3/s$), A is the effective surface area of membrane ($cm^2$) and can be determined by $A=\pi r^2$, where $\pi$ is Archimedes' constant ($\pi = 3.14$) and r is the radius of the flat sheet membrane (cm), $\Delta P$ is the pressure difference (cm Hg) as the gas passes through the membrane, and T is the ambient temperature (K) at which the permeation experiment is performed.

## 3. Results and Discussion

### *3.1. Characterization Analysis*

Figure 1 shows the XRD spectra of MgAl-$CO_3$ LDH, PSF and LDH/PSF MMM. From the XRD pattern, it was confirmed that MgAl-$CO_3$ LDH shows the crystalline structure while membrane samples show an amorphous phase. For LDH, there are sharp peaks at 11.38°, 22.92°, and 34.56° which are ascribed to (003), (006) and (009), respectively. The other peaks are at 38.80°, 45.76°, 60.50° and 61.80° [29]. These peaks correspond with MgAl-$CO_3$ LDH JCPDS reference patterns (JCPDS No: 00-014-0191) [30] and deduced that the LDH has been successfully synthesized. The amorphous PSF membrane shows the broad peak diffraction at $2\theta$=17.70° and the addition of peak at 23.90° corresponding to the crystalline phase of PEBAX [31]. This shows PEBAX is successfully coated on the surface of PSF. The LDH/PSF MMM has the obvious peak of LDH, PSF and PEBAX spectra. This shows that coated MMM has been successfully prepared.

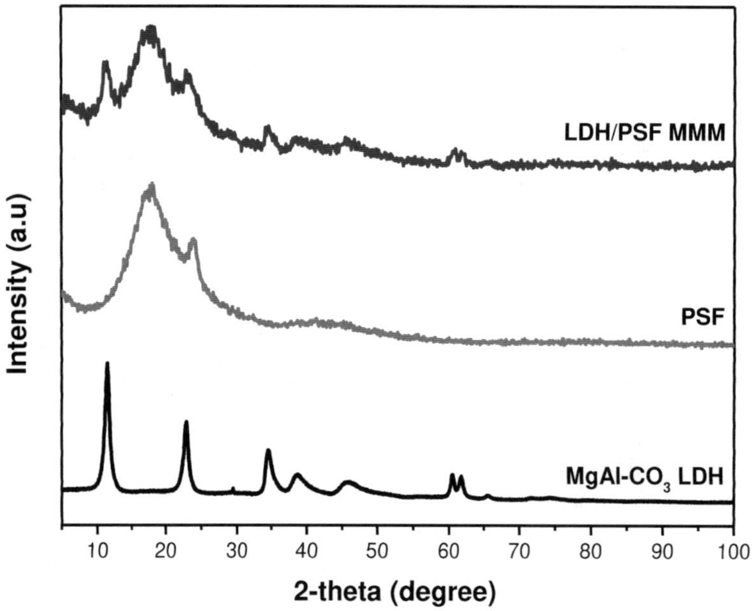


Figure 1 XRD spectra of MgAl-$CO_3$ LDH, PSF and LDH/PSF MMM

Figure 2 presents the FTIR spectra for MgAl-$CO_3$ LDH, PSF and LDH/PSF MMM. The MgAl-$CO_3$ LDH FTIR spectra show the obvious broad band at 3451.59 $cm^{-1}$ corresponds to the OH group stretching in coordination with the cations. While, at 2980.51 $cm^{-1}$, the bending assign to the H-bond between water and $CO_3^{2-}$ ions in the interlayer. For the band at 1631.33 $cm^{-1}$, it is related to the water bending in the interlayer. The sharp band at 1360.87 $cm^{-1}$ concerns the stretching of the carbonate group. The signal below the 700 $cm^{-1}$ band is related to the Mg-O and Al-O vibration in the lattice of LDH [32]. The membrane sample shows an absorption band at 1236.61 $cm^{-1}$ assign to the O=S=O group stretching vibration which is representing for PSF polymer. There is strong band at 1098.79 $cm^{-1}$ for the PEO segment for the (-C-O-C-) group. The absorption band at 3296.37 $cm^{-1}$, 1731.84 $cm^{-1}$ and 1636.37 $cm^{-1}$ corresponding to (-N–H-), (-O–C=O-) and (-H–N–C=O-) have been observed and it belongs to the functional group in the hard polyamide (PA) segment of PEBAX while 2867.27 $cm^{-1}$ in the soft and hard segment representing (-C-H-) group. Thus, it can be deduced that the PEBAX coating has been successfully prepared on the PSF surface membrane. For the composition of LDH on the MMM, the FTIR spectra show a similar pattern with PSF. Since the amount of LDH added is low around 5 wt% and MMM surface was coated with PEBAX, the FTIR spectra has high intensity absorption band for functional group and bond exist on the PEBAX.

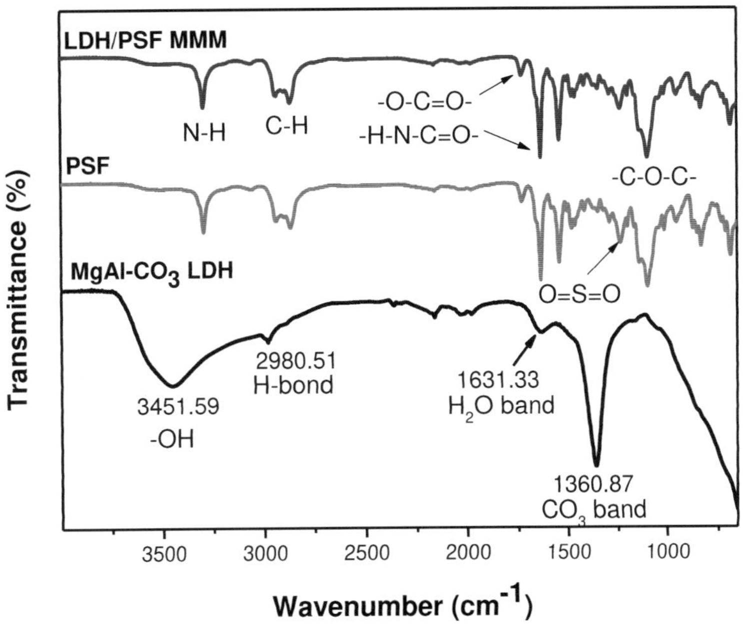


Figure 2 FTIR spectra for MgAl-$CO_3$ LDH, PSF and LDH/PSF MMM

Figure 3 displays the SEM micrographs of the PSF and LDH/PSF MMM. The membranes were shown to be invariably asymmetric, consisting of the PEBAX-coated layer on the membrane surface structure of the sponge and the finger. The size and shape of the porous membrane were adjusted when LDH was incorporated into the membrane matrix. Comparing with PSF, the MMM was less dense with longer, smoother and narrower pores. This adjustment because of the presence of LDH in the PSF membrane as the number of finger-like pores increased and grew with the addition of filler. LDH may have lowered the polymer/solvent interaction and reduced the interface strain between non-solvent to polymer dope, resulting, during process inversion, an increase in the solvent/solvent exchange rate [33]. Similar analysis carried out by Vatanpour *et al.* [34] has also shown that inserting filler into a polymer membrane increases porosity as well as macro-void structures. The SEM indicates that the one side surface of membrane is covered with a layer and it deduces the successful coating of PEBAX on the surface of LDH/PSF MMM.

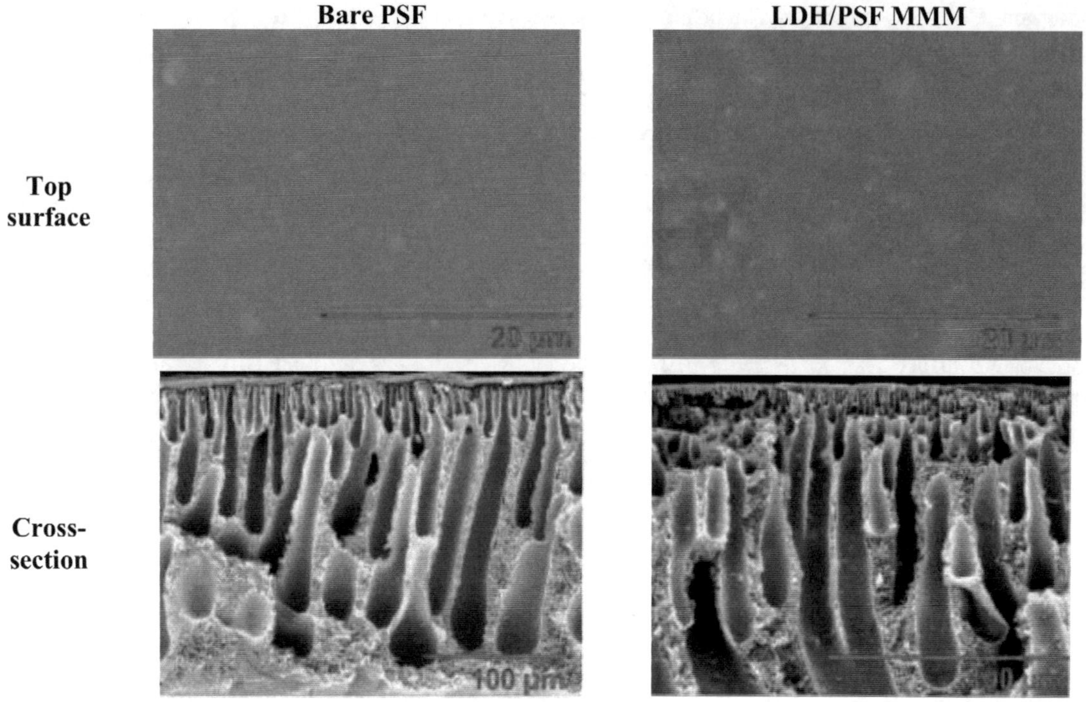


Figure 3 SEM micrographs of the PSF and LDH/PSF MMM with the scale of 20 μm (magnification x5.0k) and 100 μm (magnification x1.0k) for top surface and cross-section image, respectively.

### 3.2. *Gas Permeance Test*

Figure 4 presents the permeance and selectivity of $CO_2$ and $CH_4$ gasses for PSF and LDH/PSF MMM under 5 bar at room temperature. The results show LDH loading has a high $CO_2/CH_4$ selectivity with the better $CO_2$ permeance. The finding obtained has shown the $CO_2$ permeance increased and $CH_4$ permeance decreased with LDH loading. This meant that the PSF membrane chain packing had been interrupted by the LDH filler. This in turn enhances vacancy in the membrane, facilitates the transport of gas across the membranes. The increase in $CO_2$ permeance is because the LDH has a greater affinity with polar gas ($CO_2$) than non-polar gas ($CH_4$), which the presence of open metal sites in LDH would provide interaction between its electrostatic field and quadrupole moment of $CO_2$ while having no specific interaction with $CH_4$. Also, the structure of the LDH affects gas permeance. The interlayer channel dimensions of the 2D LDH membrane are regulated by the intercalated $CO_2^{3-}$ to 0.3 nm (3.0Å), which restricts the transport of $CH_4$ molecules of a complex diameter of 0.38 nm (3.8Å). Although the little channel can also retard the 0.33 nm (3.3Å) of $CO_2$ molecules, the LDH laminar channel will be used to relay $CO_2$ molecules, which are in $CO_2^{3-}$ shape with an ionic radius of 0.136 nm which not influenced by the sieving [21]. Furthermore, the PEBAX can also seal all the defects and pores that may be due to LDH filler at the PSF MMM surface and thus increases the membrane selectivity [28]. Thus, the gas-filler relationship, transport channel and coated membrane contributed to a large increase in selectivity between $CO_2$ and $CH_4$ up to 129.5%.

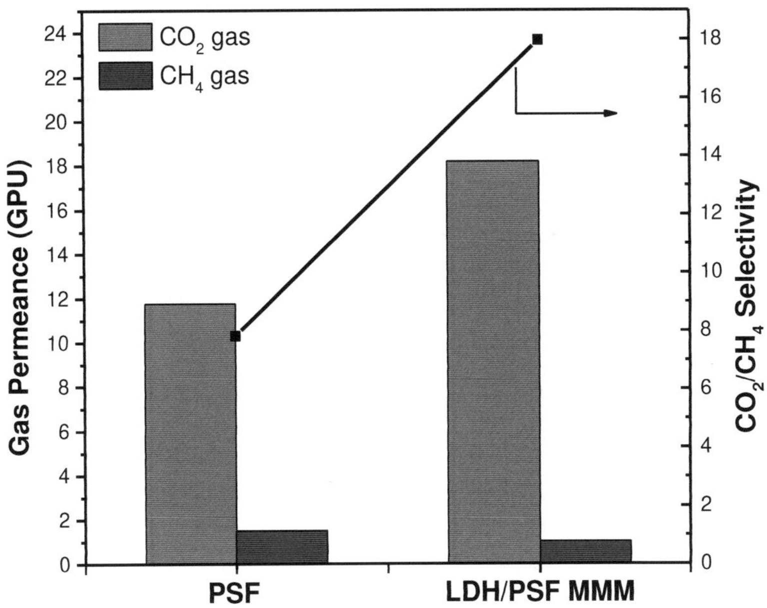


Figure 4 Permeance and selectivity of $CO_2$ and $CH_4$ gasses for PSF and LDH/PSF MMM under 5 bar at room temperature

## 4. Conclusion

The LDH/PSF MMM was successfully fabricated and the effect of the incorporation of LDH filler into the PSF membrane was investigated in this work. The addition of LDH fillers has been shown to improve $CO_2$ gas permeance by 54.5% and increase $CO_2/CH_4$ selectivity up to 18.0 compared to 7.8 for pristine PSF membrane. The morphology of the MMM membrane shows the filler altering the membrane structure to be less dense and narrow pores which provide a better channel for $CO_2$ permeance, thus explain the best performance of $CO_2/CH_4$ gas separation. For future study, the different loading of LDH and another operating parameter can be investigated as LDH filler has great potential in membrane application.

## Acknowledge

The authors would like to recognize the financial support provided by UTM-UTP-UMP-UMS Collaborative Research Grants (CRG) (Q.J130000.2451.08G26) and (Q.J130000.2451.08G72), and UTM Prototype Research Grants (UTMPR) (Q.J130000.2851.00L41) by the Malaysian Ministry of Education and Universiti Teknologi Malaysia.

## References

[1] Owusu P A, Asumadu-Sarkodie S and Dubey S 2016 A review of renewable energy sources, sustainability issues and climate change mitigation *Cogent Engineering* **3**
[2] Wasajja H, Lindeboom R E F, van Lier J B and Aravind P V 2020 Techno-economic review of biogas cleaning technologies for small scale off-grid solid oxide fuel cell applications *Fuel Processing Technology* **197** 106215
[3] Sahota S, Shah G, Ghosh P, Kapoor R, Sengupta S, Singh P, Vijay V, Sahay A, Vijay V K and Thakur I S 2018 Review of trends in biogas upgradation technologies and future perspectives *Bioresource Technology Reports* **1** 79-88

[4] Singhal S, Agarwal S, Arora S, Sharma P and Singhal N 2017 Upgrading techniques for transformation of biogas to bio-CNG: a review *International Journal of Energy Research* **41** 1657-69
[5] Baena-Moreno F M, le Saché E, Pastor-Pérez L and Reina T R 2020 Membrane-based technologies for biogas upgrading: a review *Environmental Chemistry Letters* **18** 1649-58
[6] Ma C, Wang M, Wang Z, Gao M and Wang J 2020 Recent progress on thin film composite membranes for $CO_2$ separation *Journal of $CO_2$ Utilization* **42**
[7] Liang C Z, Chung T-S and Lai J-Y 2019 A review of polymeric composite membranes for gas separation and energy production *Progress in Polymer Science* **97** 101141
[8] Li H, Feng X, Ma D, Zhang M, Zhang Y, Liu Y, Zhang J and Wang B 2018 Stable aluminum metal–organic frameworks (Al-MOFs) for balanced $CO_2$ and water selectivity *ACS Appl. Mater. Interfaces* **10** 3160-3
[9] Meis N N, Bitter J H and de Jong K P 2010 Support and size effects of activated hydrotalcites for precombustion $CO_2$ capture *Ind. Eng. Chem. Res.* **49** 1229-35
[10] Koros W J and Fleming G 1993 Membrane-based gas separation *J. Membr. Sci.* **83** 1-80
[11] Habib N, Shamair Z, Tara N, Nizami A-S, Akhtar F H, Ahmad N M, Gilani M A, Bilad M R and Khan A L 2020 Development of highly permeable and selective mixed matrix membranes based on Pebax®1657 and NOTT-300 for $CO_2$ capture *Sep. Purif. Technol.* **234** 116101
[12] Vu D Q, Koros W J and Miller S J 2003 Mixed matrix membranes using carbon molecular sieves: I. Preparation and experimental results *J. Membr. Sci.* **211** 311-34
[13] Robeson L M 2008 The upper bound revisited *J. Membr. Sci.* **320** 390-400
[14] Susanti I 2019 Polysulfone Membrane with Zeolite Filler for $CO_2/CH_4$ Gas Separation: a Review *Sci. Edu. App. J.* **1** 10-6
[15] Jiang Y, Liu C, Caro J and Huang A 2019 A new UiO-66-$NH_2$ based mixed-matrix membranes with high $CO_2/CH_4$ separation performance *Microporous Mesoporous Mater.* **274** 203-11
[16] Ismail N, Ismail A, Mustafa A, Zulhairun A, Aziz F, Bolong N and Razali A 2019 Polymer Clay Nanocomposites For Gas Separation: A Review *Environ. Contam. Rev.* **2** 01-5
[17] Lei L, Lindbråthen A, Zhang X, Favvas E P, Sandru M, Hillestad M and He X 2020 Preparation of carbon molecular sieve membranes with remarkable $CO_2/CH_4$ selectivity for high-pressure natural gas sweetening *J. Membr. Sci.* **614** 118529
[18] Zohdi S, Anbia M and Salehi S 2019 Improved $CO_2$ adsorption capacity and $CO_2/CH_4$ and $CO_2/N_2$ selectivity in novel hollow silica particles by modification with multi-walled carbon nanotubes containing amine groups *Polyhedron* **166** 175-85
[19] Mousavinejad A, Rahimpour A, Shirzad Kebria M R, Khoshhal Salestan S, Sadrzadeh M and Tavajohi Hassan Kiadeh N 2020 Nickel-Based metal–organic frameworks to improve the $CO_2/CH_4$ separation capability of thin-film pebax membranes *Ind. Eng. Chem. Res.* **59** 12834-44
[20] Gouveia L G T, Agustini C B, Perez-Lopez O W and Gutterres M 2020 $CO_2$ adsorption using solids with different surface and acid-base properties *J. Environ. Chem. Eng.* **8**
[21] Liu Y, Wu H, Min L, Song S, Yang L, Ren Y, Wu Y, Zhao R, Wang H and Jiang Z 2020 2D layered double hydroxide membranes with intrinsic breathing effect toward $CO_2$ for efficient carbon capture *J. Membr. Sci.* **598**
[22] Kiadehi A D, Rahimpour A, Jahanshahi M and Ghoreyshi A A 2015 Novel carbon nano-fibers (CNF)/polysulfone (PSf) mixed matrix membranes for gas separation *J. Ind. Eng. Chem.* **22** 199-207
[23] Waheed N, Mushtaq A, Tabassum S, Gilani M A, Ilyas A, Ashraf F, Jamal Y, Bilad M R, Khan A U and Khan A L 2016 Mixed matrix membranes based on polysulfone and rice husk extracted silica for $CO_2$ separation *Sep. Purif. Technol.* **170** 122-9
[24] Ahmadpour E, Shamsabadi A A, Behbahani R M, Aghajani M and Kargari A 2014 Study of $CO_2$ separation with PVC/Pebax composite membrane *Journal of Natural Gas Science and Engineering* **21** 518-23

[25] Sutrisna P D, Hou J, Li H, Zhang Y and Chen V 2017 Improved operational stability of Pebax-based gas separation membranes with ZIF-8: A comparative study of flat sheet and composite hollow fibre membranes *J. Membr. Sci.* **524** 266-79
[26] Wang L, Li Y, Li S, Ji P and Jiang C 2014 Preparation of composite poly(ether block amide) membrane for $CO_2$ capture *Journal of Energy Chemistry* **23** 717-25
[27] Wahab M A, Sunarti A, Ramli N H and Ahmad A 2019 24 fractional factorial design for PVDF/Pebax film composite synthesis on gas selectivity study. In: *IOP Conference Series: Materials Science and Engineering*: IOP Publishing) p 012046
[28] Jamil N, Othman N H, Shahrudin M Z, Razlan M R M, Alias N H, Marpani F, Jye L W, Goh P S and Ismail A F 2020 EEffects of Pebax Coating Concentrations on $CO_2/CH_4$ Separation of RGO/ZIF-8 PES Membranes *J. Teknol.* **82**
[29] Rezaei B, Heidarbeigy M, Ensafi A A and Dinari M 2016 Electrochemical Determination of Papaverine on Mg-Al Layered Double Hydroxide/ Graphene Oxide and CNT Modified Carbon Paste Electrode *IEEE Sen. J.* **16** 3496-503
[30] Shang S, Hanif A, Sun M, Tian Y, Ok Y S, Yu I K M, Tsang D C W, Gu Q and Shang J 2019 Novel M (Mg/Ni/Cu)-Al-$CO_3$ layered double hydroxides synthesized by aqueous miscible organic solvent treatment (AMOST) method for $CO_2$ capture *J Hazard Mater* **373** 285-93
[31] Norahim N, Faungnawakij K, Quitain A T and Klaysom C 2019 Composite membranes of graphene oxide for $CO_2/CH_4$ separation *J. Chem. Technol. Biotechnol.* **94** 2783-91
[32] Cardinale A M, Carbone C, Consani S, Fortunato M and Parodi N 2020 Layered Double Hydroxides for Remediation of Industrial Wastewater from a Galvanic Plant *Crystals* **10**
[33] Pal A, Kar S, Debnath A K, Aswal D K, Bindal R C and Tewari P K 2015 Reinforcement of nanostructured reduced graphene oxide: a facile approach to develop high-performance nanocomposite ultrafiltration membranes minimizing the trade-off between flux and selectivity *RSC Adv.* **5** 46801-16
[34] Vatanpour V, Shockravi A, Zarrabi H, Nikjavan Z and Javadi A 2015 Fabrication and characterization of anti-fouling and anti-bacterial Ag-loaded graphene oxide/polyethersulfone mixed matrix membrane *J. Ind. Eng. Chem.* **30** 342-52

# The role of g-$C_3N_4$ in round-the-clock photocatalysis for POME

**N A M Razali[1,2], W N W Salleh[1,2*], N Rosman[1,2], H Ismail[1,2], S Z N Ahmad[1,2], F Aziz[1,2], W J Lau[1,2], A F Ismail[1,2]**

[1]Advanced Membrane Technology Research Centre (AMTEC), Universiti Teknologi Malaysia, 81310, Skudai, Johor, Malaysia

[2]School of Chemical and Energy Engineering, Faculty of Engineering, Universiti Teknologi Malaysia, 81310, Skudai, Johor

*Corresponding author: hayati@petroleum.utm.my

**Abstract.** The main drawback of conventional palm oil mill effluent (POME) treatment is that the process is time and space-consuming. Besides, the treatment produces highly polluted wastewater that pollutes the environment if discharged directly. Photocatalytic process has significant potential to degrade recalcitrant organic pollutants and has recently attracted tremendous attention. However, current approaches mainly focus on visible light condition, which is still an ineffective treatment for POME. In this study, POME was successfully degraded using graphitic carbon nitride (g-$C_3N_4$) photocatalyst synthesised by calcination. The prepared photocatalyst was characterised by ultraviolet-visible diffuse reflectance (UV-Vis DRS) spectroscopy and scanning electron microscopy (SEM). The SEM results revealed the morphology of g-$C_3N_4$ photocatalyst. g-$C_3N_4$ could act as a visible-light-driven (VLD) photocatalyst with the highest photocatalytic efficiency of 71% under visible light. The present work highlights the potential of g-$C_3N_4$ towards the degradation of POME under visible light and dark condition. The highly enhanced photocatalytic performance is attributed to g-$C_3N_4$, but it does not work well in round-the-clock photocatalysis. However, g-$C_3N_4$ can work as the band alignment to drive separate photogenerated charge carriers, leading to effective photocatalytic degradation. g-$C_3N_4$ photocatalyst may be considered as an ideal candidate for treating POME.

Keywords: g-$C_3N_4$, energy-storage material, palm oil mill effluent, photocatalysis, visible-light-driven)

## 1. Introduction

In the past several decades, water pollution has become a worldwide concern, including wastewater from industries. Palm oil mill effluent (POME) is released from palm oil mills and it is the main source of water pollution [1,2]. Moreover, acidic and viscous POME can harm the aquatic ecosystem [3]. POME contains a high amount of water and suspended solids consisting of carbohydrates, amino acids, fibres, and some inorganic nutrients, such as iron, copper, potassium, magnesium, nitrogen, chromium, and cadmium [4]. POME is categorised based on its strength, which is influenced by the number and concentration of contaminants. For example, low-strength POME contains high biochemical oxygen demand/chemical oxygen demand (BOD/COD) ratio and is biodegradable [5]. POME treatment must be conducted as the effluent has a high tendency of polluting the environment [6].

Various conventional wastewater treatment methods have been used, such as biodegradation through microorganisms, and physical and chemical processes, including adsorption, filtration, and flocculation. However, the shortcomings of most of these methods are low efficiency, expensive, time-consuming, high post-treatment costs of generated secondary pollutants, and only phase transformation of contaminants occurred instead of complete mineralisation to obtain non-toxic products, such as carbon dioxide, water, and inorganic salts [7–9]. Among these methods, advanced oxidation processes (AOPs) have emerged as promising methods for removing contaminants of emerging concern from wastewater effluent [10]. Compared to other complex methods in AOPs, photocatalysis is the safest method because mild condition is used, and neither high temperatures nor hazardous materials are applied. Photocatalysis offers several advantages: environmentally friendly, high performance, low energy consumption, and no secondary pollution [11,12]. However, photocatalysis is not satisfactory owing to their low solar energy utilisation efficiency, poor selective adsorption, poor stability, and high recombination rate of electron-hole pairs [13].

Several excellent materials like titania ($TiO_2$), tungsten trioxide ($WO_3$), zinc oxide (ZnO), and ferric oxide ($Fe_2O_3$) have been explored as photocatalytic nanomaterials. Among the explored materials, $TiO_2$-based photocatalysts (3.2 eV) have excellent photoactivity and high stability but still have limitations in terms of easy recombination of photogenerated electron-hole pairs, low quantum efficiency, and the photocatalysts can be activated only by ultraviolet (UV) light [14,15]. Accordingly, numerous studies on visible-light-driven (VLD) photocatalysts have focused on air pollutants, photocatalytic hydrogen production, and wastewater treatment and disinfection [16,17]. Among the emerging VLD photocatalysts, graphitic carbon nitride (g-$C_3N_4$) is an important catalyst due to its appropriate band gap and unique light-harvesting ability that absorbs visible light less than 450 nm [19]. g-$C_3N_4$ has been demonstrated as an excellent visible light responsive photocatalyst that is inexpensive, easy to synthesise, highly stable, suitable electronic structure, and outstanding photocatalytic performance [20]. Moreover, it should be noted that g-$C_3N_4$ has the advantages of hydrophilicity, large specific surface area, inert, and environmentally friendly [21]. Unfortunately, g-$C_3N_4$ also exhibits a series of undesirable disadvantages, like low specific surface area, irregular morphology, and a hydrophobic surface that often limits photocatalytic performance [22]. Furthermore, the extensive application of g-$C_3N_4$ is limited due to poor utilisation of long-wavelength light, low quantum efficiency, and high recombination of photoexcited electrons-hole pairs [23,24].

Jiang and his team reviewed several doping techniques for g-$C_3N_4$ with metal, non-metal, co-doping, and heterojunction to enhance the light absorption, facilitate the charge separation and transportation, and prolong the charge carrier lifetime [25]. However, another limitation prevents the industrialisation innovation of photocatalysis due to light restriction. According to Lu and his co-worker, the decomposition of contaminants only occurs after light excitation, photoelectron/hole pair generation and migration, and oxidising and reductive species formation. Therefore, darkness would disrupt the catalytic reaction for most photocatalysts, as well as their photocatalytic efficiency and applications [26]. In order to achieve the catalyst with high photocatalytic activity, some efforts have been made to investigate the potential of energy-storage material (ESM) and VLD photocatalyst towards various pollutants, for example, $ZnWO_4$, $BiFeWO_6/WO_3$, $Pt/TiO_2/WO_3$, and Ag-$MoS_2$ [27–30].

In this work, the enhancement of visible light and round-the-clock photocatalysis for g-$C_3N_4$ photocatalyst prepared by calcination was investigated. The photocatalytic activity of g-$C_3N_4$ was assessed by the degradation of POME under visible light irradiation. The synthesis of material, properties, and structure of g-$C_3N_4$ are briefly discussed. The progress in improving the degradation efficiency of POME by g-$C_3N_4$-based photocatalysis is discussed. The feasibility of g-$C_3N_4$

photocatalyst for the degradation of organic pollutants is also summarised.

## 2. Experimental methodology

### *2.1. Chemicals and reagents*

Melamine (2,4,6-triamino-s-triazine) was obtained from Acros Organics. All solutions were prepared using deionised water and all chemicals chosen in this study were of analytical grade and used without further purification.

### *2.2. Preparation of g-$C_3N_4$*

For the preparation of g-$C_3N_4$, 10 g of melamine was heated in a muffle furnace at 100, 200, 300, 400, and 500 °C for 1.5 h using a heating rate of 2 °C $min^{-1}$ and then naturally cooled. The resulting powder was slightly crushed with a mortar and pestle to obtain fine powder of photocatalysts. The samples were labelled as G1 (g-$C_3N_4$ prepared at 100 °C), G2 (g-$C_3N_4$ prepared at 200 °C), G3 (g-$C_3N_4$ prepared at 300 °C), G4 (g-$C_3N_4$ prepared at 400 °C), and G5 (g-$C_3N_4$ prepared at 500 °C).

### *2.3. Characterization of photocatalysts*

For morphological examination, the samples were characterised using scanning electron microscopy (SEM). The ultraviolet-visible-near infrared (UV-Vis-NIR) spectrophotometer (UV-3600 Plus, Shimadzu) was used in the wavelength of 300-800 nm to construct the Tauc plot for band gap determination, and colour was measured using the 8025 APHA platinum-cobalt standard method at 455 nm using a UV-Vis spectrophotometer (DR5000, Hach).

### *2.4. Adsorption and photocatalytic degradation of POME*

Fresh POME was collected from an aeration pond of Felda Taib Andak palm oil mill in Kulai, Johor, Malaysia. Subsequently, the POME waste was stored in an airtight container and transported to the laboratory. The effluent was filtered to eliminate suspended solids and preserved in a freezer for storage (maximum of 3 months) to prevent microbial biodegradation of the effluent [31]. The adsorption using g-$C_3N_4$ was conducted by following dark absorption. All the experiments were conducted under the following conditions: 200 ml of POME (200 ml, 0.2 g L) was magnetically stirred for 60 min at room temperature in the dark to obtain an adsorption-desorption equilibrium. In a typical photocatalytic experiment, 0.2 g of the photocatalyst was suspended in 200 ml of POME and the degradation was observed under visible light irradiation (100 W), as shown in Figure 1.

During photodegradation, 5 ml of aqueous solution was sampled and filtered using a syringe filter every 60 min, and POME colour was measured using the platinum-cobalt (Pt-Co) standard method. The photocatalytic activity of the samples was evaluated throughout 7 h of the experiment using two sets of procedures (1 h adsorption, 7 h of light irradiation, and 1 h adsorption, 2 h of light irradiation, 1 h of dark condition) to investigate their photocatalytic activity and energy storage ability, respectively. The removal efficiency was calculated using the following Eq. (1):

$$\text{Removal efficiency (\%)} = [(C_0 - C) / C_0] \times 100\% \quad (1)$$

where $C_0$ and C are the colour of the POME in water before irradiation and the colour after a certain irradiation time (t), respectively.

## 3. Results and discussion

### *3.1. Characterisation of photocatalysts*

#### *3.1.1 SEM micrographs*

The morphology and superficial microstructure of g-$C_3N_4$ were investigated by SEM. The SEM images recorded at high magnification clearly showed different structures for the samples prepared at 200 and 400 °C. Figure 1(A) and (B) illustrate the SEM images with different magnification factors ranging from 50 to 100 μm. It is noteworthy that the SEM image of g-$C_3N_4$ prepared at 200 °C displays the formation of smaller surface, contributing to favourable photogenerated charge transfer at the interface (Figure 1(B)). As illustrated in Figure 1(A) and (B), the synthesised g-$C_3N_4$ samples showed irregular granular structure with a size of several micrometres [32,33]. The magnification of SEM images is distributed from 30 to 100 μm, as shown in Figure 1(C) and (D). There is no estimated size of both samples in Figure 1(A) and 1(C), which clearly showed irregular shapes that could be identified only by field-emission scanning electron microscopy (FESEM). As can be seen in Figure 1(C) and (D), the G4 sample is predominantly composed of mixed aggregate morphology, which is rod-like structures and short rods adhered to the irregular and small scattered particles at the surface [34,35]. The results indicate that the morphologies of the samples possess a structure composed of perforated plates due to the gases discharged from thermal treatment that partially affected the material [34,36].

**Figure 1.** SEM image of the g-C3N4, calcined at different calcination temperature (A), (B) for (g-$C_3N_4$ at 200◦C: G2), (C) and (D) for (g-$C_3N_4$ at 400◦C: G4).

#### *3.1.2 UV-Vis diffuses absorption spectra*

According to Kuznetsova and his co-worker, the electronic structure based on optical absorbance can determine the photocatalytic activity through optical diffused reflectance spectra [37]. Figure 2 presents the obtained spectra for G2. The photocatalyst was chosen as it exhibits the highest decolourisation for photocatalytic activity and the best energy storage performance. The band gap of the as-prepared sample was estimated from the UV-Vis DRS spectra via the Tauc plot in Eq. (2):

$$(\alpha h_v)^n = A (h_v - E_g) \quad (2)$$

where α is the absorption coefficient, $h_v$ is the incident photon energy, A is a constant and $E_g$ is the band gap energy (eV).

As shown in Figure 2, the sample exhibits an absorption edge in the visible light region. The band gap energy value was obtained by tracing a tendency line using the maximum inclination region of the graph. The band gap of g-$C_3N_4$ is approximately 3.05 eV.

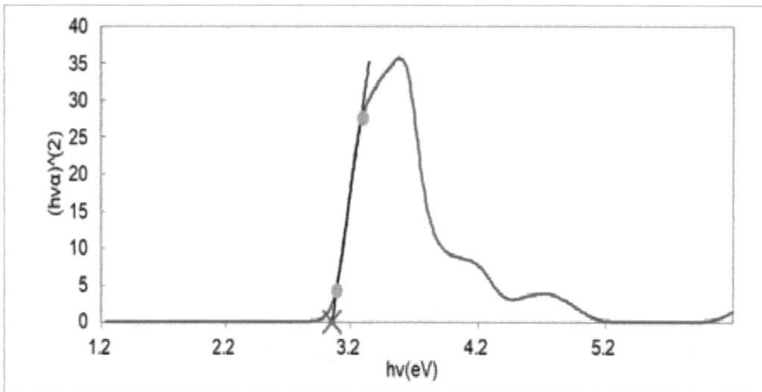


**Figure 2.** Tauc plot of g-$C_3N_4$ synthesized under different calcination treatment at 400 °C and 200 °C.

### *3.2. Photocatalysis*

The effect of the calcination temperature applied in the preparation of g-$C_3N_4$ on the photocatalytic activity for two conditions was studied to determine the photocatalyst that contributes towards visible light and energy-storage factor. The adsorption-desorption equilibrium of the photocatalysts was established by stirring the catalyst with respective pollutants for 60 min in dark condition. The experiments were carried out under light irradiation for 6 h (visible light) and also 2 h of light irradiation and 1 h of dark condition (energy storage). For this objective, these results were also compared to the measured results for g-$C_3N_4$ prepared at different temperatures of 100-500 °C labelled as G1, G2, G3, G4, and G5, as shown in Figure 3.

Upon further increment of calcination temperature, a fluctuating trend of catalytic activity was observed, whereby G1, G2, G3, G4, and G5 exhibited about 14%, 71%, 16%, 62%, and 57%, respectively. According to the results in Figure 3, the calcination temperature of g-$C_3N_4$ below 550 °C would achieve a great rate of photocatalytic reaction. This is because the effectiveness of VLD below 550 °C contributes to greater photogenerated electrons, leading to more efficient separation of charge and better photocatalytic activity [38]. From the results, G2 demonstrated the highest decolourisation that increased and decreased significantly as the calcination temperature increased. This is due to the reduced accessibility of many vacant active surface sites on g-$C_3N_4$ compared to the initial stage. There will be competition for excess vacant surface sites due to the significant repulsion between the pollutants and photocatalyst [39].

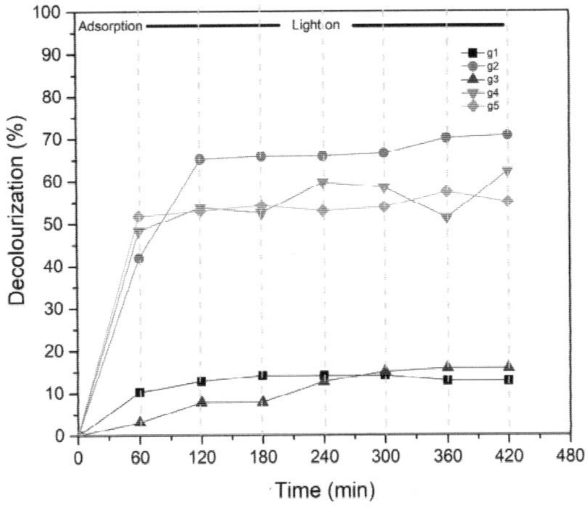


**Figure 3.** The photocatalytic degradation of POME (VLD scheme) (100 mL, 0.2g A) g-$C_3N_4$ photocatalyst.

Based on Figure 4, g-$C_3N_4$ did not demonstrate the potential of energy storage. The decolourisation of POME only reached 12%, 39%, 14%, 32%, and 14% for G1, G2, G3, G4, and G5, respectively. The findings showed that the as-prepared g-$C_3N_4$ exhibited insignificant round-the-clock photocatalysis for POME, where the photocatalytic efficiency only exceeded 30% within 420 min.

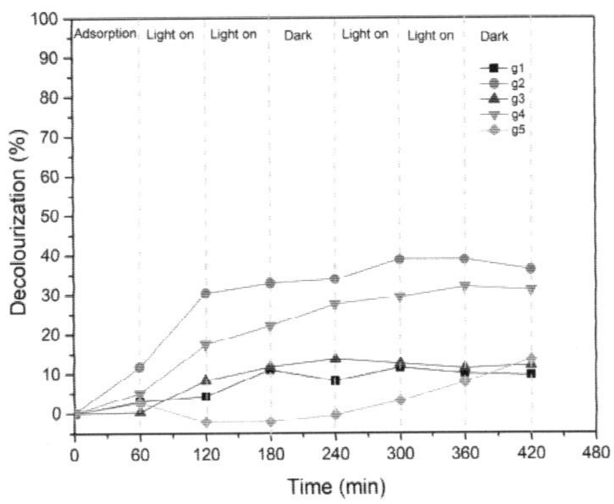


**Figure 4.** The photocatalytic degradation of POME (ESM scheme) (100 mL, 0.2g A) g-C3N4 photocatalyst.

## 4. Conclusion

This study evaluated the morphological, adsorption, and photocatalytic performance of g-$C_3N_4$. g-$C_3N_4$ photocatalysts were prepared at different calcination temperatures via calcination to study the potential of ESM and VLD photocatalysts. Improved POME degradation under visible light irradiation was observed among the g-$C_3N_4$ photocatalysts, with the best photocatalytic activity of 71%. The enhanced photocatalysis is ascribed to the strong light absorption intensity of g-$C_3N_4$, although there is not much improvement in round-the-clock photocatalysis.

## 5. References

[1] Poh P E and Chong M F 2009 Development of anaerobic digestion methods for palm oil mill effluent (POME) treatment *Bioresour. Technol.* **100** 1–9

[2] Tan Y D and Lim J S 2019 Feasibility of palm oil mill effluent elimination towards sustainable Malaysian palm oil industry *Renew. Sustain. Energy Rev.* **111** 507–22

[3] Abrams J F, Hohn S, Rixen T, Baum A and Merico A 2016 The impact of Indonesian peatland degradation on downstream marine ecosystems and the global carbon cycle *Glob. Chang. Biol.* **22** 325–37

[4] Bello M M and Abdul Raman A A 2017 Trend and current practices of palm oil mill effluent polishing: application of advanced oxidation processes and their future perspectives. *J. Environ. Manag.* **198** 170–82

[5] Abdulsalam M, Man H C, Idris A I, Yunos K F and Abidin Z Z 2018 Treatment of palm oil mill effluent using membrane bioreactor: Novel processes and their major drawbacks *Water (Switzerland)* **10**

[6] Rupani P and Singh R 2010 Review of current palm oil mill effluent (POME) treatment methods: Vermicomposting as a sustainable practice *World Appl. Sci. ...* **11** 70–81

[7] Yunus I S, Kurniawan A, Adityawarman D, Sofian I, Kurniawan A and Adityawarman D 2012 Nanotechnologies in water and air pollution treatment *Environ. Technol. Rev.* **2515**

[8] Chi G T, Churchley J, Huddersman K D, Trent S, Road S M and Cv C 2013 Pilot-Scale Removal of Trace Steroid Hormones and Pharmaceuticals and Personal Care Products from Municipal Wastewater Using a Heterogeneous Fenton ' s Catalytic Process *Int J Chem Eng* 1–10

[9] Oller I, Klamerth N, Agu A and Rodrı E M 2012 Application of solar AOPs and ozonation for elimination of micropollutants in municipal wastewater treatment plant effluents *Water Res.* **7** 4–11

[10] Rueda-Marquez J J, Levchuk I, Fernández Ibañez P and Sillanpää M 2020 A critical review on application of photocatalysis for toxicity reduction of real wastewaters *J. Clean. Prod.* **258**

[11] Abderrazak Y 2020 Photocatalysis: A step closer to the perfect synthesis *J. Organomet. Chem.* **920**

[12] Li X, Chen D, Li N, Xu Q, Li H, He J and Lu J 2020 Efficient reduction of Cr(VI) by a BMO/Bi2S3 heterojunction via synergistic adsorption and photocatalysis under visible light *J. Hazard. Mater.* **400** 123243

[13] Zhang J, Gao N, Chen F, Zhang T, Zhang G, Wang D, Xie X, Cai D, Ma X, Wu L and Wu Z 2019 Improvement of Cr (VI) photoreduction under visible-light by g-C3N4 modified by nano-network structured palygorskite *Chem. Eng. J.* **358** 398–407

[14] Fujishima, Rao and Tryk 2000 Titanium dioxide photocatalysis. *J Photochem Photobiol C* **1** 1–21

[15] Xie B, Xiong Y, Chen R, Chen J and Cai P 2005 Catalytic activities of Pd-TiO2 film towards the oxidation of formic acid *Catal. Commun.* **6** 699–704

[16] Le Y, Qi L, Wang C and Song S 2020 Hierarchical Pt/WO3 nanoflakes assembled hollow microspheres for room-temperature formaldehyde oxidation activity *Appl. Surf. Sci.* **512** 145763

[17] Zhang F, Wang X, Liu H, Liu C, Wan Y and Long Y 2019 Recent Advances and Applications of Semiconductor Photocatalytic Technology *Appl. Sci.* **9** 2489

[18] Chen D, Li T, Chen Q, Gao J, Fan B, Li J, Li X, Zhang R, Sun J and Gao L 2012 Hierarchically plasmonic photocatalysts of Ag/AgCl nanocrystals coupled with single-crystalline WO3 nanoplates *Nanoscale* **4** 5431–9

[19] Mamba G and Mishra A K 2016 Graphitic carbon nitride ( g-C 3 N 4 ) nanocomposites : A new and exciting generation of visible light driven photocatalysts for environmental pollution remediation *Appl. Catal. B Environ.* **198** 347–77

[20] Reddy K R, Reddy C V, Nadagouda M N, Shetti N P, Jaesool S and Aminabhavi T M 2019 Polymeric graphitic carbon nitride (g-C 3 N 4 )-based semiconducting nanostructured materials: Synthesis methods, properties and photocatalytic applications *J. Environ. Manage.* **238** 25–40

[21] Wu Y, Wang H, Tu W, Liu Y, Tan Y Z, Yuan X and Chew J W 2018 Quasi-polymeric construction of stable perovskite-type LaFeO3/g-C3N4 heterostructured photocatalyst for improved Z-scheme photocatalytic activity via solid p-n heterojunction interfacial effect *J. Hazard. Mater.* **347** 412–22

[22] Wang W, Fang J, Shao S, Lai M and Lu C 2017 Compact and uniform TiO2@g-C3N4 core-shell quantum heterojunction for photocatalytic degradation of tetracycline antibiotics *Appl. Catal. B Environ.* **217** 57–64

[23] Zhang M, Xu J, Zong R and Zhu Y 2014 Enhancement of visible light photocatalytic activities via porous structure of g-C 3 N 4 *Appl. Catal. B, Env.* **147** 229–35

[24] Li S, Dong G, Hailili R, Yang L, Li Y, Wang F, Zeng Y and Wang C 2016 Effective photocatalytic H 2 O 2 production under visible light irradiation at g-C 3 N 4 modulated by carbon vacancies *"Applied Catal. B, Environ.* **190** 26–35

[25] Jiang L, Yuan X, Pan Y, Liang J, Zeng G, Wu Z and Wang H 2017 Doping of graphitic carbon nitride for photocatalysis: A reveiw *Appl. Catal. B Environ.* **217** 388–406

[26] Lu Y, Zhang X, Chu Y, Yu H, Huo M, Qu J, Crittenden J C, Huo H and Yuan X 2018 Cu2O nanocrystals/TiO2 microspheres film on a rotating disk containing long-afterglow phosphor for enhanced round-the-clock photocatalysis *Appl. Catal. B Environ.* **224** 239–48

[27] Neto N F A, Nunes T B O, Li M, Longo E, Bomio M R D and Motta F V 2019 Influence of microwave-assisted hydrothermal treatment time on the crystallinity , morphology and optical properties of ZnWO 4 nanoparticles : Photocatalytic activity *Ceram. Int.*

[28] Priya A, Arunachalam P, Selvi A, Madhavan J and Al-mayouf A M 2018 Synthesis of BiFeWO 6 / WO 3 nanocomposite and its enhanced photocatalytic activity towards degradation of dye under irradiation of light *Colloids Surfaces A* **559** 83–91

[29] Khan H, Rigamonti M G and Boffito D C 2019 Enhanced photocatalytic activity of Pt-TiO2/WO3 hybrid material with energy storage ability *Appl. Catal. B Environ.* **252** 77–85

[30] Sun K, Jia F, Yang B, Lin C, Li X and Song S 2020 Synergistic effect in the reduction of Cr(VI) with Ag-MoS2 as photocatalyst *Appl. Mater. Today* **18** 100453

[31] Cheng C K, Rizauddin Derahman M and Khan M R 2015 Evaluation of the photocatalytic degradation of pre-treated palm oil mill effluent (POME) over Pt-loaded titania *J. Environ. Chem. Eng.* **3** 261–70

[32] Han X, Tian L, Jiang H, Kong L, Lv J, Shan J, Wang J and Fan X 2017 Facile transformation of low cost melamine-oxalic acid into porous graphitic carbon nitride nanosheets with high visible-light photocatalytic performance *RSC Adv.* **7** 14372–81

[33] Xu Y H, Xu G, Sun M Y and Wang K 2020 Low-temperature solvothermal–calcination preparation and enhanced photocatalytic performance of polymeric graphitic carbon nitride with disordered–ordered hybrid plane *Chem. Pap.* **74** 4067–74

[34] Zhu H L and Zheng Y Q 2018 Mesoporous Co3O4 anchored on the graphitic carbon nitride for enhanced performance supercapacitor *Electrochim. Acta* **265** 372–8

[35] Yu Z, Li F, Yang Q, Shi H, Chen Q and Xu M 2017 *Nature-Mimic Method to Fabricate Polydopamine/Graphitic Carbon Nitride for Enhancing Photocatalytic Degradation Performance* vol 5

[36] Shcherban N D, Mäki-Arvela P, Aho A, Sergiienko S A, Yaremov P S, Eränen K and Murzin D Y 2018 Melamine-derived graphitic carbon nitride as a new effective metal-free catalyst for Knoevenagel condensation of benzaldehyde with ethylcyanoacetate *Catal. Sci. Technol.* **8** 2928–37

[37] Kuznetsova M, Oliveira S A A, Rodrigues B S and Souza J S 2020 Microwave - Assisted Synthesis of Bismuth Niobate / Tungsten Oxide Photoanodes for Water Splitting *Top. Catal.*

[38] Paul D R, Sharma R, Nehra S P and Sharma A 2019 Effect of calcination temperature, pH and catalyst loading on photodegradation efficiency of urea derived graphitic carbon nitride towards methylene blue dye solution *RSC Adv.* **9** 15381–91

[39] Sia Y Y, Tan I A W and Abdullah M O 2016 Adsorption of colour, TSS and COD from palm oil mill effluent (POME) using acid-washed coconut shell activated carbon: Kinetic and mechanism studies *MATEC Web Conf.* **87**

**Acknowledgments**
The authors gratefully acknowledge the financial support from the Ministry of Education and Universiti Teknologi Malaysia under HICOE Research Grant (Project Number: R.J090301.7851.4J423) and UTM High Impact Research (Project Number: Q.J130000.2451.08G36). The authors would also like to acknowledge technical and management support from Research Management Centre (RMC), Universiti Teknologi Malaysia.

# Flow pattern, pressure drop and inclination analysis on liquid-liquid two phase flow of waxy crude oil in pipelines using PIPESIM

**A H Alpandi[1,2*], A H Mazeli[2], A Sidek[2], H Husin[1], R Junin[2] and M Z Jaafar[2]**

[1]Department of Petroleum Engineering, Universiti Teknologi Petronas, 32610 Seri Iskandar, Perak, Malaysia.
[2]School of Chemical & Energy Engineering, Faculty of Engineering, Universiti Teknologi Malaysia, 81310 Skudai, Johor Bahru, Malaysia.

*Corresponding author: amni_17009776@utp.edu.my

**Abstract**. Produced water is a water that comes out with the crude oil during the production of the well. It contains non-soluble and soluble oil or organics, dissolved and suspended solids with different chemicals used during production process. Thus, it must be properly accounted as it affects the economical productivity of crude oil and separation efficiency as a result of stubborn emulsions between crude oil and water. Thus, a simulation study was conducted using PIPESIM to predict the flow pattern and pressure drop of waxy crude oil and water flow in horizontal and inclined pipelines (i.e., -15° from horizontal). In this simulation study, water cuts were ranging from 0% to 90% while the flow rates were ranging from 2.03 to 16.21 $cm^3/s$. The study comprised fluid modelling, physical modelling and running the simulation with the most suitable multiphase flow correlation in PIPESIM. This simulation study used the waxy crude oil has 16.15% of wax content and simulation was performed at 30°C. The validity of the simulation results was accomplished by comparing the published findings. There were only two types of flow patterns that can be identified by PIPESIM; stratified wavy and dispersed flow. The investigations proved that pressure drop was greatly influenced by flow rates and flow patterns. By decreasing the inclination angle, the boundary between the stratified and dispersed flow regimes shifted to the upper left of the flow pattern map while showing a higher pressure drop than horizontal pipeline due to the combined effect of pressure difference and gravity. The simulation results can be used as a platform for better understanding on more complex cases of gas, oil and water concurrent flow in pipelines.

## 1. Introduction

Multiphase flow is defined as a combination or simultaneous flow of several phases, which might be liquid, solid or gas. The two distinct phases can be a combination of liquid-liquid, gas-liquid, or solid-liquid phases [1]. These two different phases flow and move together inside the piping system.

The investigation and research area of two-phase flow can be categorized in the form of analysis of flow patterns, pressure drop, heat transfer and void fraction [2,3,4]. The measurement and investigation for the liquids holdup and flow patterns of the dispersed phases received typically less attention for oil-water system as compared to a gas-oil system [5]. Therefore, further prediction on the flow pattern and pressure drops of waxy crude oil and water flow in both horizontal and inclined pipelines, and investigation for the liquid-liquid two-phase flow still in an ambiguity condition. Oil-water flow in a pipeline system is a common phenomenon that happens at various transportation applications all over the world.

During the transportation stage of the oil using a long-distance pipeline system, the presence of water flowing together with oil has become very serious phenomenon since the presence of water contributed to the corrosion effect [5,6,7,8]. One of the main factors that contribute to the fraction amount of oil-water in piping system is the liquid flow patterns [10]. The flow pattern of the liquid/gas consists of several categories such as bubbly flow, stratified-wavy flow, stratified flow, plug flow, intermittent flow, annular flow, slug flow and mist flow. Each type of the flow pattern shows different flow behavior. For example, a stratified flow is defined as the type of flow pattern which occurred at low liquid-liquid/gas velocities with a complete separation of the two-phases in the pipeline system. The liquid with higher density will settle down at the bottom part of the piping structure while the liquid with a slightly lower density will go to the top of the piping structure. Other than that, a bubbly flow is defined as the type of flow which the gas was dispersed in the liquid with a higher concentration in the upper half of the tube. This condition occurs in the piping system due to the buoyancy effect. According to [5], bubbly flow only occurs at a slightly higher mass flow rate in a horizontal flow.

Most of the research works mainly performed on mineral oil, paraffin oil or kerosene [11]. In Malaysia, most of the oilfields were producing waxy crude oil. This phenomenon is due to the presence of naphthenic and/or paraffin in the crude oil composition. [12] was the only researchers used Malaysia waxy crude oil in their studies related to oil-water two-phase flow in horizontal pipes.

Dealing with an oil-water mixture in a pipeline leads to unique and complex problems in the oil and gas industry due to its complicated rheological behavior [13]. This situation is worsened as the number of research work done on crude oil-water flow in pipes were very limited [12]. Thus, this simulation study has been conducted to acknowledge the findings by developing a simulation study using PIPESIM to predict the flow behaviors (i.e., flow pattern and pressure drop) of waxy crude oil-water flow in pipes at different flow rates and water cuts above wax appearance temperature (WAT).

Based on the theory, there are three forces which may affect the liquid-liquid flow either in upward and downward direction which are gravity force, inertia force and buoyancy force. Figure 1 and figure 2 shows free body diagram of forces in upward and downward flow direction.

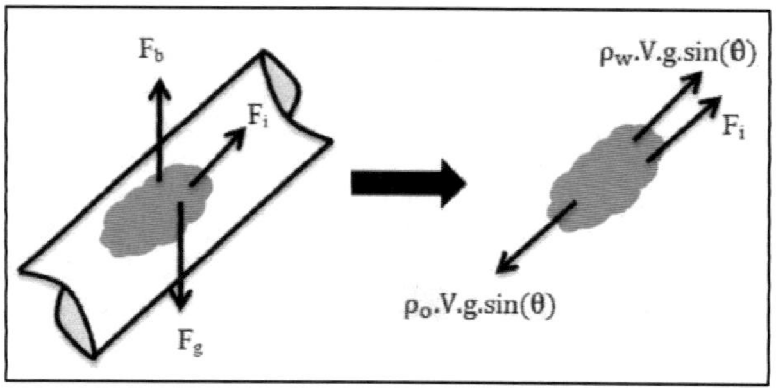


**Figure 1.** Free body diagram in upward flow direction.

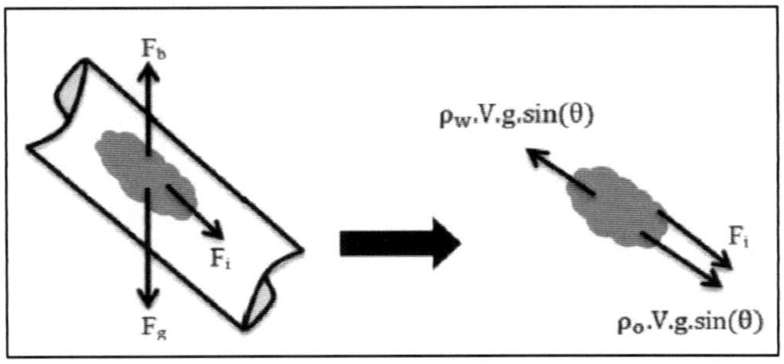


**Figure 2.** Free body diagram in downward flow direction.

Where $F_b$, $F_g$ and $F_i$ represent buoyancy force, gravity force and inertia force, respectively. V is the volume of oil bubble and θ is an inclination angle of pipeline. Meanwhile, $\rho_w$ and $\rho_o$ are water density and oil density, respectively.

In figure 1, oil bubble is flowing in upward direction and flow regimes can be predicted from the Newton's second law as in equation (1):

$$F_i + (\rho_w - \rho_o).V.g.\sin(\theta) = m_b.a \tag{1}$$

Where $a$ represents the acceleration of the bubble and $m_b$ represents the oil bubble mass. Based on the equation (1), it is understandable that the density difference between water and oil $(\rho_w - \rho_o)$ is a positive value. This will cause the bubble to accelerate and increase in the velocity of the bubble with the increment of inclination angle.

In figure 2, oil bubble is flowing in downward direction and flow regimes can be predicted from the Newton's second law as in equation (2):

$$F_i + (\rho_o - \rho_w).V.g.\sin(\theta) = m_b.a \tag{2}$$

Based on the equation (2), it is clear that the density difference between oil and water $(\rho_o - \rho_w)$ is a negative value. This will cause the bubble to decelerate and reduce in the velocity of the bubble with a decrement of inclination angle. In addition, the study clarified that all temperature during simulation remain constant (isothermal), at the WAT condition, and assume wax deposits in the pipeline.

## 2. Methodology

The simulation study was successfully conducted at Petroleum Engineering Modeling and Simulation Laboratory, Faculty of Chemical and Energy Engineering using PIPESIM software. This simulation study used pipeline and facilities analysis to determine the flow behavior characteristics such as flow pattern and pressure drop at horizontal pipeline and at inclination angle of -15°. OLGAS 7.3.1 (3 phase) was used as horizontal correlations in this study.

### *2.1. Fluid Modelling*

Fluid modelling is an elementary aspect of multiphase flow simulation. It is important to create one or more fluid models before running any simulation study. Fluid models are used to define phase behavior, provide transport and physical properties of the fluid required. Table 1 shows the fluid data used in this study from the experimental research done by [12]. Malaysian waxy-crude oil composition from [12] was described in appendix A. All the crude oil components and their percentage by weight obtained from [12] were added as input data of fluid composition in PIPESIM.

**Table 1.** General properties of crude oil [12].

| Descriptions | Crude Oil |
|---|---|
| API gravity (°API) | 41.4 |
| Density (kg/m$^3$) | 818 |
| Dynamic viscosity @ 30°C (cp) | 1.75 |
| Flash point (°C) | <19 |
| Pour point (°C) | 18 |
| WAT (°C) | 30 |
| Wax content (wt%) | 16.15 |
| Asphaltenes (wt%) | 0.06 |
| Retention time (min)* | <1 |

*Retention time is separation time between oil and water

### *2.2. Pipeline and Facilities Model*

Simple horizontal and inclined pipeline model has been developed using PIPESIM. Parameters for each component in the pipeline model was successfully be defined according to experimental work done by [12]. These were the following basic steps to build a pipeline and facilities model using PIPESIM; (1) The SI units was selected in this simulation study. (2) A source was added to the model. (3) The flow line was added to the model and the pipeline characteristics was created based on the data as shows in table 2. (4) A sink was added to the model. (5) The fluid specification and suitable horizontal flow correlation were define using OLGAS v. 7.3.1 (3-phase). (6) The model was saved as shows in figure 3 and simulation has been conducted.

**Table 2.** Pipe specifications from [12].

| Data | Pipeline |
|---|---|
| Type of pipe | Stainless steel |
| Inside diameter | 2 in. |
| Wall thickness | 0.27 in. |
| Roughness | 0.0018 in. |
| Horizontal distance | 3 meters |
| Elevation difference | 0 (horizontal)<br>-0.776 (inclined -15°) |
| Heat transfer coefficient | 0.2 |
| Ambient temperature | 30°C |

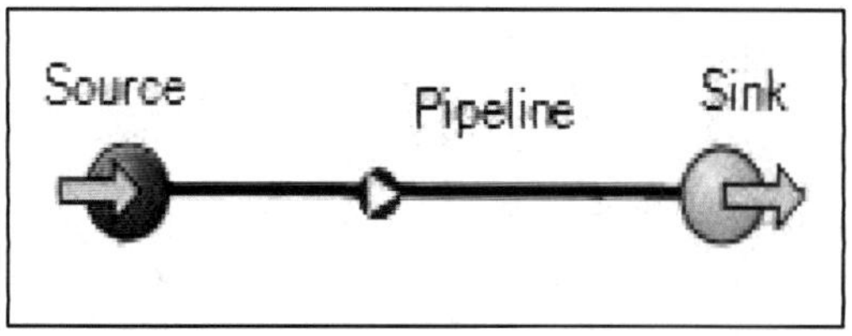


**Figure 3.** Simple model of pipeline.

For this simulation study, nine different scenarios were conducted on waxy crude oil-water two-phase flows at horizontal and inclination angle of -15° downward conditions. The water cut varies from 10 to 90%. While the flow rates of oil and water flowing in the pipelines was varied from 2.03 to 16.21 $cm^3/s$ at ambient condition. A total of 72 runs were accomplished in this simulation works in order to study the flow pattern and pressure drop of waxy crude oil-water in pipeline.

## 3. Results and Discussion

### *3.1. Flow Pattern Map*

PIPESIM is able to identify the flow pattern map of waxy crude oil-water two-phase flow in pipelines throughout respective water cuts and flow rates. The water cuts were ranging from 0.1 to 0.9 while the flow rates were ranging from 2.03 to 16.21 $cm^3/s$. Figure 4 represents the flow pattern map of waxy crude oil-water two-phase flow generated using PIPESIM.

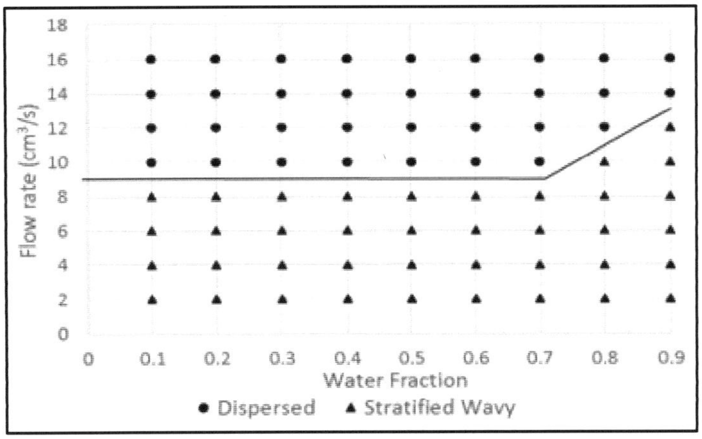


**Figure 4.** Flow pattern map generated by PIPESIM.

There were only two types of flow patterns that can be identified by using OLGAS 7.3.1 (3-phase) correlation in PIPESIM; Stratified wavy and dispersed flow. Stratified flow was clearly observed in figure 4 for flow rates ranging from 2.03 – 12.15 $cm^3/s$ for all water fractions. According to [14], stratified wavy flow can be categorized when two layers of water and oil are completely separated with water flows at the bottom while oil flows at the top of the pipeline. At low flow rates, oil and water flow tend to fully segregated due to the gravity effect occurs in the pipeline [15]. Furthermore, dispersed flow was clearly observed for flow rates ranging from 10.13 – 16.21 $cm^3/s$ for all water cuts. The respective phases were no longer continuous or segregated as the flow rate increased. [16] stated that velocity of the mixture and water fraction affect the dispersed flow either it is existed in semi or full dispersed from one phase to the other phase.

*3.1.1. Comparison of Flow Pattern Map.* A comparison study was made on flow pattern maps of waxy crude oil generated using PIPESIM with experimental work done by [12] in order to realise the objective in this research.

Based on figure 5, it was found that the experimental result of flow pattern maps were slightly differences with simulation result generated by PIPESIM for horizontal pipeline as shows in figure 4. There were five flow patterns found by [12] in their experimental works, namely stratified wavy flow (STW), stratified wavy with semi dispersed flow at interface and oil film (STSD&O), semi dispersed flow with semi emulsion at interface and thin oil film (STDE&TO), dispersion of water in oil and oil continuous with emulsion (DWE) and dispersion of oil in water with water continuous (DO). Compared to this simulation study, only two types of flow patterns could be identified using OLGAS 7.3.1 (3-phase) correlation in PIPESIM which were stratified wavy and dispersed flow.

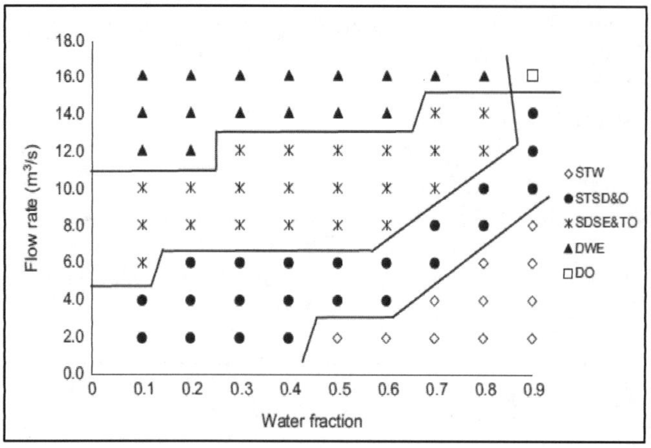


**Figure 5.** Flow pattern map of waxy crude oil-water generated by [12].

Based on the experimental work done by [12], stratified wavy with semi dispersed flow at interface and oil film (STSD&O) could be clearly observed at flow rates ranging from 2.03 to 14.18 $cm^3/s$ for all water fractions and stratified wavy flow (STW) could be clearly observed during the low and intermediate flow rates ranging from 2.03 – 8.10 $cm^3/s$ for water fraction greater than 0.5. Whereby, stratified wavy flow identified using PIPESIM was in the range from 2.03 – 12.15 $cm^3/s$ for all water fractions. For dispersed flow, semi dispersed flow with semi emulsion at interface and thin oil film (SDSE&TO) could be clearly seen in the experimental work done by [12] at water fraction ranging from 0.1 to 0.8 with flow rates from 6.08 – 14.18 $cm^3/s$.

Next, dispersion of water in oil and emulsion (DWE) could be seen at water fractions from 0.1 to 0.8 with flow rates ranging from 12.15 – 16.21 $cm^3/s$ while dispersion of oil in water with water continuous (DO) could be clearly observed at higher flow rate, 16.21 $cm^3/s$ for higher water fraction at 0.9. Whereby, the dispersed flow identified by PIPESIM was ranging from 10.13 – 16.21 $cm^3/s$ for water cut ranging from 0.1 to 0.9. According to [17], while conducted an experimental work to study oil-water flow in horizontal pipeline, they concluded that low density difference between these two phases (oil-water) and the viscosity of the oil were affecting the flow patterns notably. Thin oil film that produced inside of the wall's pipe affects the flow pattern. According to [12], the wettability inside wall of the pipeline also causes the disturbances of flow behavior.

### *3.2. Pressure Drop Analysis*

In this simulation study, there was a relationship between pressure drop, flow patterns and flow rates. This statement was supported by [15], [16] and [18]. As shown in figure 6, pressure drop obtained in this simulation work increased with flow rates and water fraction.

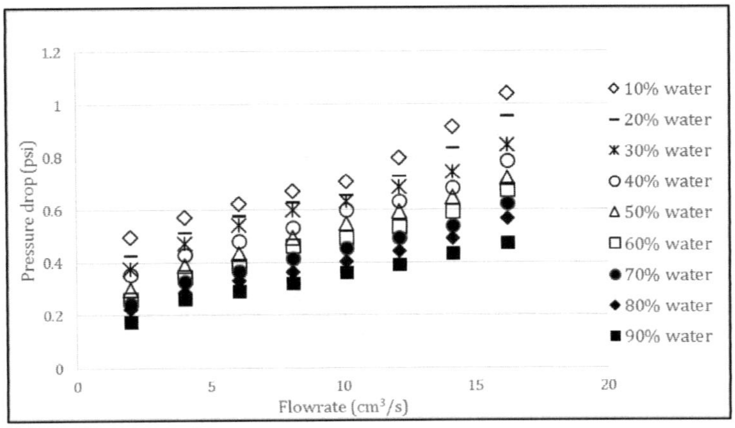


**Figure 6.** Pressure drop versus flow rate at various water fractions.

Based on figure 6 and table 3, it is shown that at the highest flow rate which is 16.21 $cm^3/s$ and lowest water fraction, $C_w = 0.1$, highest pressure drop of 1.04 psi was recorded. Whereby, at the lowest flow rate which is 2.03 $cm^3/s$ and lowest oil fraction, $C_o = 0.1$, lowest pressure drop of 0.174 psi was recorded.

**Table 3.** Statistics of pressure drop for different flow rate and oil-water fraction.

| Flow rate ($cm^3/s$) | Minimum pressure (psi) | Maximum pressure (psi) | Average | Standard deviation |
|---|---|---|---|---|
| 2.03 | 0.174 | 0.498 | 0.317444 | ±0.098224 |
| 4.05 | 0.263 | 0.573 | 0.400111 | ±0.099238 |
| 6.08 | 0.293 | 0.623 | 0.442333 | ±0.101735 |
| 8.10 | 0.321 | 0.672 | 0.498556 | ±0.112938 |
| 10.13 | 0.362 | 0.708 | 0.539333 | ±0.111575 |
| 12.15 | 0.392 | 0.798 | 0.588889 | ±0.127232 |
| 14.18 | 0.435 | 0.913 | 0.652889 | ±0.148746 |
| 16.21 | 0.472 | 1.040 | 0.741778 | ±0.172892 |

*3.2.1. Comparison of Pressure Drop.* A comparison study was made on pressure drop of waxy crude oil generated using PIPESIM with experimental work done by [12] as shows in table 4.

**Table 4.** Comparison of pressure drop between simulation work and experimental work done by [12].

| Flow rate ($cm^3/s$) | Minimum pressure (psi) | | Maximum pressure (psi) | |
|---|---|---|---|---|
| | Simulation | Experiment | Simulation | Experiment |
| 2.03 | 0.174 | 0.44 | 0.498 | 1.36 |
| 4.05 | 0.263 | 0.64 | 0.573 | 2.59 |
| 6.08 | 0.293 | 0.76 | 0.623 | 1.86 |
| 8.10 | 0.321 | 0.87 | 0.672 | 2.02 |
| 10.13 | 0.362 | 1.01 | 0.708 | 2.16 |
| 12.15 | 0.392 | 1.17 | 0.798 | 2.39 |
| 14.18 | 0.435 | 1.33 | 0.913 | 2.87 |
| 16.21 | 0.472 | 1.56 | 1.040 | 3.86 |

Based on table 4, it was noticeable that the results of pressure drop obtained from simulation work were less than the results that acquired by [12], but there are similarities in term of curve generated. There might be some other elements that caused the disturbances in waxy-crude oil-water flow behavior in pipelines. Pipe roughness, liquid rheology and pipe wettability were the elements that might affect the pressure drop. Stainless steel pipes was used by [12] while conducting their experimental work to study the flow behavior of waxy crude oil-water two-phase flow, therefore pipe roughness and pipe wettability could be the contributing factor which may affect the value of pressure drop.

Inside wall of the stainless-steel pipe can be oil wet and water wet. According to [19], pressure drop can either be higher or lower when oil-water flows together in pipeline, depending on the types of wettability. The pressure drop is lower when the inside wall of the pipeline is a water wet and vice versa [19]. Concurrently, another major contributor to the pressure loss in pipeline is pipe roughness. [20] stated that pressure drop is created inside the pipeline of liquid-liquid flow due to the interference from pipe roughness may cause liquid not moving smoothly inside the pipeline.

### *3.3. Inclination Analysis*

*3.3.1. Flow Pattern Map.* Further analysis was made by making the horizontal pipeline inclined -15° downward in order to analyse the flow pattern map of waxy crude oil-water in pipeline. Figure 7 shows the flow pattern maps of waxy crude oil-water flow in the inclination angle of -15° downward generated by PIPESIM.

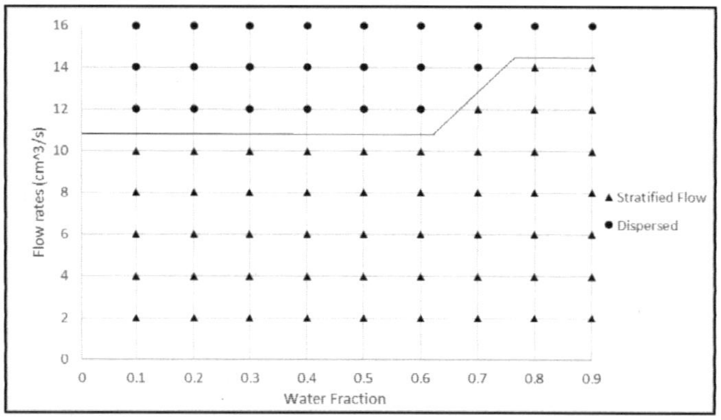


**Figure 7.** Flow pattern maps of waxy crude oil-water flow in -15⁰ downward from simulation work.

Figure 7 shows that two types of flow patterns were identified using PIPESIM; stratified wavy and dispersed flow. The boundary between stratified and dispersed flows shifts slightly to the upper left when the pipeline was inclined (-15° downward). Stratified wavy was clearly seen in this simulation study during flow rates ranging from 2.03 – 14.18 cm$^3$/s for all water fractions. As for dispersed flow, this flow pattern was clearly seen during flow rates ranging from 12.15 – 16.21 cm$^3$/s for water cut ranging from 0.1 until 0.9. According to [21], stratified wavy flow becomes dominant flow as the transition boundaries between stratified and non-stratified shifts to the upper left of the flow pattern map when decreasing the inclination angle of the pipeline.

*3.3.2. Comparison of flow pattern map (horizontal and inclined -15⁰ pipeline).* Comparison study of flow pattern maps was made between horizontal pipeline and -15° inclined pipelines. It is shown that the transition boundary between stratified and non-stratified flow shifts to the upper left of the flow patterns map. According to [21], stratified flow become supreme and the transition boundaries between stratified and non-stratified shifts to the upper left of the flow pattern map when decreasing the inclination angle of the pipeline. This is because buoyancy force will decelerate the dispersed phase and gravity force accelerates the dispersed phase for inclined downward flow. Since gravity force is greater than buoyancy force, therefore velocity of dispersed phase was decelerating. It is because wax density is greater than oil density. Thus, the differentiation between buoyancy and gravity force turns out greater as the rise of pipeline inclination angle.

*3.3.3. Pressure Drop.* Figure 8 represents the pressure drop of waxy crude oil-water two-phase flows at -15° inclination generated by PIPESIM and table 5 shows the statistics of pressure drop from simulation study.

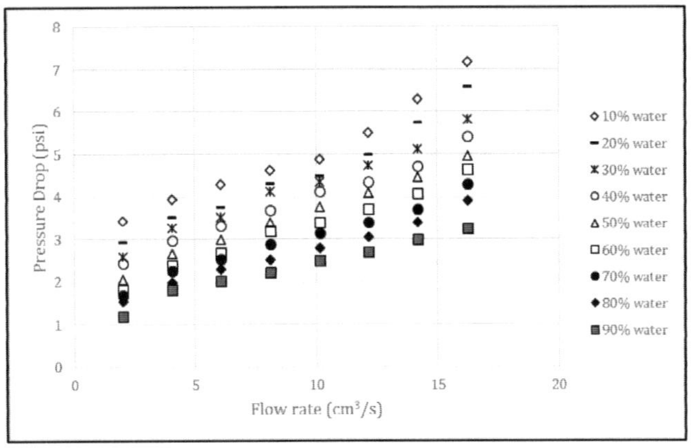


**Figure 8.** Pressure drop versus flow rate at various water fraction for -15$^{0}$ inclined pipeline generated by PIPESIM.

**Table 5.** Statistics of pressure drop for PIPESIM.

| Flow rate $(cm^3/s)$ | Minimum pressure (psi) | Maximum pressure (psi) | Average | Standard deviation |
|---|---|---|---|---|
| 2.03 | 0.217 | 0.541 | 0.360444 | ±0.098224 |
| 4.05 | 0.306 | 0.616 | 0.443111 | ±0.099238 |
| 6.08 | 0.336 | 0.666 | 0.485333 | ±0.101735 |
| 8.10 | 0.364 | 0.715 | 0.541556 | ±0.112938 |
| 10.13 | 0.405 | 0.751 | 0.582333 | ±0.111575 |
| 12.15 | 0.435 | 0.841 | 0.631889 | ±0.127232 |
| 14.18 | 0.478 | 0.966 | 0.700333 | ±0.152956 |
| 16.21 | 0.515 | 1.100 | 0.787089 | ±0.176616 |

Based on figure 8 and table 5, it is shown that at the highest flow rate which is 16.21 cm$^3$/s and lowest water fraction, $C_w = 0.1$, highest pressure drop of 1.10 psi was recorded. Whereby, at the lowest flow rate which is 2.03 cm$^3$/s and lowest oil fraction, $C_o = 0.1$, lowest pressure drop of 0.217 psi was recorded.

*3.3.4. Comparison of pressure drop (horizontal and inclined -15$^{0}$ pipeline.* Table 6 shows the comparison of pressure drop value between horizontal and -15$^{0}$ inclined pipeline obtained from simulation study using PIPESIM.

**Table 6.** Comparison of pressure drop between horizontal and -15° inclined pipeline generated by PIPESIM.

| Flow rate $(cm^3/s)$ | Minimum pressure (psi) | | Maximum pressure (psi) | |
|---|---|---|---|---|
| | Horizontal pipeline | -15° inclined pipeline | Horizontal pipeline | -15° inclined pipeline |
| 2.03 | 0.174 | 0.217 | 0.498 | 0.541 |
| 4.05 | 0.263 | 0.306 | 0.573 | 0.616 |
| 6.08 | 0.293 | 0.336 | 0.623 | 0.666 |
| 8.10 | 0.321 | 0.364 | 0.672 | 0.715 |
| 10.13 | 0.362 | 0.405 | 0.708 | 0.751 |
| 12.15 | 0.392 | 0.435 | 0.798 | 0.841 |
| 14.18 | 0.435 | 0.478 | 0.913 | 0.966 |
| 16.21 | 0.472 | 0.515 | 1.040 | 1.100 |

Based on table 6, the results of pressure drop generated by PIPESIM at -15° inclined pipelines were greater than those in horizontal pipelines. This condition occurs when the flow is driven by the combined effect of pressure difference and gravity. Gravity effect supports downward flow but has no effect on the flow rate in the horizontal condition. Downward flow can occur even in the absence of an applied pressure difference. The flow rate increases as the tilt angle of the pipeline from the horizontal is reduced to the negative direction but would reach its maximum value when the pipe is vertical.

## 4. Conclusion

A simulation study of waxy crude oil-water two-phase flow in pipeline system has been successfully conducted using PIPESIM. A total of 72 runs were accomplished in this simulation works. The water cut was varied from 10 to 90%, while the flow rates of oil and water flowing in the pipelines were varied from 2.03 to 16.21 $cm^3/s$ at ambient condition. The following conclusions were made accordingly based on the analysis of the simulation results:

(1) A simulation model was successfully developed using PIPESIM based on the experimental data obtained from [12] which all the components of crude oil listed in Appendix A were taken into account during the simulation study.

(2) There was slightly difference in terms of flow pattern generated from experimental work done by [12] and this simulation work. PIPESIM only enables to identify two types of flow patterns which are stratified wavy and dispersed flow.

(3) A flow regime was successfully generated using PIPESIM for the waxy crude oil-water two-phase flow system at 30°C which is above WAT. It was then compared with the flow regimes from the previous researchers such as [22]. In fact, Wang and Gong was the only researcher that have used actual crude oil in the oil-water two-phase flow studies. There were discrepancies in the form of flow patterns and flow regions due to the waxy crude oil compositions and characteristics. Generally, different types of fluid used in a two-phase flow system produce different flow behaviours.

(4) The highest pressure drop of 1.04 psi was recorded in horizontal pipeline at the highest flow rates which is 16.21 $cm^3/s$ and at the lowest water fraction, $C_w = 0.1$. Whereby, the lowest pressure drop of 0.174 psi was recorded in horizontal pipeline at the lowest flow rates which is 2.03 $cm^3/s$ and at the lowest oil fraction, $C_o = 0.1$.

(5) Pressure drop that was generated from this simulation work was lower than published data conducted by [12]. There might be some other elements that caused the disturbances in waxy crude oil-water flow behaviour in pipelines. Pipe roughness and pipe wettability were the other elements that might affect the value of pressure drop.

(6) By decreasing the inclination angle of the pipeline with -15° downward, the stratified flow became a dominant flow pattern. The transition boundary between stratified and non-stratified flow shifted to the upper left of the flow pattern map. Generally, buoyancy force will decelerate the dispersed phase and gravity force accelerate the dispersed phase for inclined downward flow. Since gravity force is greater than buoyancy force, the velocity of dispersed phase decelerates.
(7) The highest pressure drop of 1.10 psi was recorded in -15° inclined pipeline at the highest flow rate which is 16.21 $cm^3/s$ and lowest water fraction, $C_w = 0.1$. Whereby, the lowest pressure drop of 0.217 psi was recorded at the lowest flow rate which is 2.03 $cm^3/s$ and lowest oil fraction, $C_o = 0.1$.
(8) Pressure drops generated by PIPESIM at -15° inclined pipeline were greater than those in horizontal pipelines when the flow is driven by the combined effect of pressure difference and gravity. Gravity effect supports downward flow, but it has no effect on the flow rate in horizontal conditions. Downward flow can occur even in the absence of an applied pressure difference.

**Acknowledgement**
The authors would like to express their sincere gratitude to Universiti Teknologi Malaysia Transdisciplinary Research Grant (Q.J130000.3551.07G12) for financial support in order for us to complete this research.

**Appendixes**
Table A1 consists of Malaysian waxy crude oil composition data obtained from [12] was attached in appendix A.

**References**
[1] Yan Y, Wang L, Wang T, Wang X, Hu Y and Duan Q 2018 Application of soft computing techniques to multiphase flow measurement: A review *Flow Meas. Instrum.* **60** 30–43
[2] Bhagwat S M and Ghajar A J 2014 A flow pattern independent drift flux model based void fraction correlation for a wide range of gas-liquid two phase flow *Int. J. Multiph. Flow* **59** 186–205
[3] Ghajar A J and Bhagwat S M 2014 *Flow Patterns, Void Fraction and Pressure Drop in Gas-Liquid Two Phase Flow at Different Pipe Orientations*
[4] Bhagwat S M and Ghajar A J 2012 Similarities and differences in the flow patterns and void fraction in vertical upward and downward two phase flow *Exp. Therm. Fluid Sci.* **39** 213–27
[5] Al-Moosawy A A, Al-Hattab T . and Al-Joubouri T A 2008 Analysis of Liquid-Liquid Two Phase Flow System *Emirates J. Eng. Res.* **13** 19–26
[6] Cai J, Nesic S, Li C, Tang X, Ayello F, Ohio U, Ivan C, Cruz T and Al-Kharnis J N 2005 Experimental studies of water wetting in large-diameter horizontal oil/water pipe flows *SPE Annu. Tech. Conf. Proc.*
[7] Liu Y, Zhang B, Zhang Y, Ma L and Yang P 2016 Electrochemical polarization study on crude oil pipeline corrosion by the produced water with high salinity *Eng. Fail. Anal.* **60** 307–15
[8] Hu H and Cheng Y F 2016 Modeling by computational fluid dynamics simulation of pipeline corrosion in CO2-containing oil-water two phase flow *J. Pet. Sci. Eng.* **146** 134–41
[9] Li Q, Hu H and Cheng Y F 2016 Corrosion of pipelines in CO2-saturated oil-water emulsion flow studied by electrochemical measurements and computational fluid dynamics modeling *J. Pet. Sci. Eng.* **147** 408–15
[10] Tan J, Jing J, Hu H and You X 2018 Experimental study of the factors affecting the flow pattern transition in horizontal oil–water flow *Exp. Therm. Fluid Sci.* **98** 534–45
[11] Rodriguez O M H and Oliemans R V A 2006 Experimental study on oil-water flow in horizontal and slightly inclined pipes *Int. J. Multiph. Flow* **32** 323–43

[12] Ismail A S I, Ismail I, Zoveidavianpoor M, Mohsin R, Piroozian A, Misnan M S and Sariman M Z 2015 Experimental investigation of oil-water two-phase flow in horizontal pipes: Pressure losses, liquid holdup and flow patterns *J. Pet. Sci. Eng.* **127** 409–20

[13] Arirachakaran S, Oglesby K D, Malinowsky M S, Shoham O and Brill J P 1989 An analysis of oil/water flow phenomena in horizontal pipes *Soc. Pet. Eng. - SPE Prod. Oper. Symp. POS 1989* 155–67

[14] Sridhar S, Zhang H Q, Sarica C and Pereyra E 2011 Experiments and model assessment on high-viscosity oil/water inclined pipe flows *Proc. - SPE Annu. Tech. Conf. Exhib.* **2** 1554–63

[15] Trallero J L, Sarica C and Brill J P 1997 A study of oil/water flow patterns in horizontal pipes *SPE Prod. Facil.* **12** 165–72

[16] Vielma M, Atmaca S, Sarica C and Zhang H 2008 Characterization of oil/water flows in horizontal pipes *SPE Proj. Facil. Constr.* **3** 1–21

[17] Charles M E, Govier G W and Hodgson G W 1961 The horizontal pipeline flow of equal density oil-water mixtures *Can. J. Chem. Eng.* **39** 27–36

[18] Vuong D H, Zhang H Q, Sarica C and Li M 2009 Experimental study on high viscosity oil/water flow in horizontal and vertical pipes *Proc. - SPE Annu. Tech. Conf. Exhib.* **4** 2454–63

[19] Cai J, Li C, Tang X, Ayello F, Richter S and Nesic S 2012 Experimental study of water wetting in oil-water two phase flow-Horizontal flow of model oil *Chem. Eng. Sci.* **73** 334–44

[20] Brauner N 1991 Two-phase liquid-liquid annular flow *Int. J. Multiph. Flow* **17** 59–76

[21] Hanafizadeh P, Hojati A and Karimi A 2015 Experimental investigation of oil-water two phase flow regime in an inclined pipe *J. Pet. Sci. Eng.* **136** 12–22

[22] Wang W and Gong J 2010 Flow Regimes and Transition Characters of the High Viscosity Oil-water Two Phase Flow **5** 44–58

## Appendix A

**Table A1.** Malaysian waxy crude oil composition data obtained from [12].

| Number of components | Components | Carbon, C | Percentage (%) by weight |
|---|---|---|---|
| 1 | Nonane | $C_9H_{20}$ | 2.40 |
| 2 | Decane | $C_{10}H_{22}$ | 3.94 |
| 3 | Undecane | $CH_3(CH_2)9CH_3$ | 1.64 |
| 4 | Hexacosane | $CH_3(CH_2)24CH_3$ | 3.31 |
| 5 | Pentacosane | $CH_3(CH_2)23CH_3$ | 3.09 |
| 6 | Tetradecane | $CH_3(CH_2)12CH_3$ | 3.68 |
| 7 | Pentadecane | $CH_3(CH_2)13CH_3$ | 4.20 |
| 8 | Hexadecane | $CH_3(CH_2)14CH_3$ | 5.07 |
| 9 | Heptadecane | $CH_3(CH_2)15CH_3$ | 4.38 |
| 10 | Tetracosane | $CH_3(CH_2)22CH_3$ | 4.71 |
| 11 | Octacosane | $CH_3(CH_2)26CH_3$ | 3.89 |
| 12 | Nonadecane | $CH_3(CH_2)17CH_3$ | 4.33 |
| 13 | 9,10-Dihydrophenanthracene | $C_{14}H_{12}$ | 4.00 |
| 14 | Icosane | $C_{20}H_{42}$ | 4.56 |
| 15 | Heneicosane | $CH_3(CH_2)19CH_3$ | 4.60 |
| 16 | Benzeneacetic acid | $C_8H_8O_2$ | 3.96 |
| 17 | Triacontane | $CH_3(CH_2)28CH_3$ | 4.08 |
| 18 | Allylpentaspiro [3.0.3.0.3.0.3.0.3.1] henicosan-21 | $C_{21}H_{30}O$ | 3.38 |
| 19 | Tricosane | $CH_3(CH_2)21CH_3$ | 4.64 |
| 20 | Tetracosane | $H(CH_2)21CH_3$ | 3.94 |
| 21 | Docoane | $CH_3(CH_2)20CH_3$ | 2.72 |
| 22 | Pentaosane | $CH_3(CH_2)23CH_3$ | 4.28 |
| 23 | 2',4'-Dimethyloxanilic acid | $C_8H_{16}O_2$ | 3.32 |
| 24 | 2-Mehyl-3-phenyl-1*H*-indole | $C_{15}H_{13}N$ | 3.19 |
| 25 | Dodecahydropyrido[1,2-b] isoquinolin-6-one | $C_{13}H_{21}NO$ | 5.97 |
| 26 | 1,3-Dimethyl-4-azaphenanthrene | $C_{15}H_{13}N$ | 0.55 |
| 27 | 2-(Acetoxymethyl)-3-(methoxycarbonyl)biphenylene | $C_{17}H_{14}O_4$ | 0.95 |
| 28 | 3,3-Diisopropoxy-1,1,1,5,5,5-hexamethyltrisiloxane | $C_{21}H_{32}O_4Si_3$ | 1.23 |

# Ultrasonic Properties of Zinc-Lithium Phosphate Glass: Effect of Strontium

Nur Fatihah N.[1], **Nurhafizah H.** [1*], Ahmad N.H. [1]
[1] Advanced Optical Material Research Group (AOMRG), Department of Physics, Faculty of Science, Universiti Teknologi Malaysia, 81310 Johor Bahru, Johor, Malaysia

E-mail: nurhafizah.h@utm.my

**Abstract.** Glasses of the general formula (50–*x*)$P_2O_5$ – 30ZnO – 20$Li_2O$ – (*x*)$SrCO_3$, where (0.00 ≤ *x* ≤ 5.00 mol%) were prepared via the melt-quench technique. The mechanical properties of the glass are utilized by using ultrasonic testing. The absence of distinct and Bragg peaks in XRD spectra, confirming the amorphous nature of samples. The glass densities are found increases as well as the packing densities with respect to $SrCO_3$ concentration, which leading to the decrease in molar volume values, indicating the rigidity of the glasses. The ultrasonic results and other measured parameters attribute to the Sr ion increase in the cross-link density by the formation of P-O-Sr due to the value of Poisson's ratio obtained are between range of 0.1 to 0.2.

**Keywords.** Phosphate-based glass; Mechanical Analysis; Ultrasonic Testing.

## 1 Introduction

Phosphate glasses are known for its technological interest for electrochemical devices, which has been explored mainly for biomedical applications (i.e. bone tissue repair and fixation) as well as for wide ranging applications such as optical and nuclear waste storage. It also can easily be doped with a wide variety of ions. The phosphate glass role (network former) is used in different applications. The short-range order of phosphate glasses can be explained by the type and concentration of [$PO_4$]. The addition of any multivalent oxide to $P_2O_5$ can improves the chemical durability but mostly keep the unique physical properties of the phosphate glasses. They are generally possessing high thermal expansion coefficients, low glass transition temperatures, low softening temperatures beside low preparation temperatures and high transmission [1,2].

The addition of alkaline and alkaline earth metal ions can alter the glass structures by opening up a modified network and to reduce the bond strength. Therefore, it will disrupt the glass structure through creation of a terminal oxygen (P–O–M) bond by breaking the bridging oxygen (P–O–P) bonds resulting in depolymerisation of the phosphate glass network. The alkaline earth metals are known as shiny, silvery-white, and quite reactive metals at standard temperature and pressure. While not as reactive as the alkali metals, this family knows how to make bonds very easily. Moreover, by introducing other oxide elements such as zinc oxide (ZnO) and lithium oxide ($Li_2O$) as a glass modifier, a good quality of glass can be formed. This study is to find an optimum composition of $P_2O_5$ – ZnO – $Li_2O$ – $SrCO_3$ in order to produce strontium doped phosphate glass. Throughout this study, the nature of the glass can be determined as well as the physical and mechanical properties for industrial purpose.

## 2 Methodology

### 2.1 Melt Quench Technique

The melt quenching technique is the common way of obtaining a glass, especially for the production of bioactive glass. The required amounts of all the precursors were weighed, mixed, and heated in different types of crucibles depending on the glass composition. The materials are weighed using an electronic weighing balance, Precisa XT 220A. Then, the materials are grinded into powdered form by a ball mill to obtain more homogenous particle size in 15 minutes. The mixed quantities of these chemicals were put in alumina crucibles and placed in an electric furnace open to the atmosphere and held at 300 °C for one hour. The ground mixture is then melted in a furnace at 950 °C. Finally, the melted material is immediately transferred to an annealing furnace and poured into moulds, which held at 300 °C for three hours. This procedure was repeated for all glass samples with different mol% concentration as listed in Table 1.

**Table 1:** The composition of $(50-x)P_2O_5$–$30ZnO-20Li_2O-(x)SrCO_3$ where ($0 \leq x \leq 5$ mol%).

| Samples | Composition (mol%) | | | |
|---|---|---|---|---|
| | $P_2O_5$ | ZnO | $Li_2O$ | $SrCO_3$ |
| S1 | 50 | 30 | 20 | 0 |
| S2 | 49 | 30 | 20 | 1 |
| S3 | 48 | 30 | 20 | 2 |
| S4 | 47 | 30 | 20 | 3 |
| S5 | 46 | 30 | 20 | 4 |
| S6 | 45 | 30 | 20 | 5 |

### 2.2 X-Ray Diffraction

The amorphous state of the sample is confirmed by X-ray diffraction (XRD) measurement on Rigaku Model of Siemens Diffractometer D500 with Cu-K$\alpha$ radiation, operated at 40 kV and the glass samples are in fine powder form in 2θ range of 10-80°.

### 2.3 Physical Properties

Archimedes principle is used to measure the density of prepared glass by using Mettler Toledo electronic weighing balance. The physical parameters such as the density, molar volume density and ionic packing density of the glass sample are calculated by using different mathematical expression shown in Equation (1) to (4):

$$\rho = \left[ \frac{A}{A-B} \times (\rho_o - d) \right] + d \qquad (1)$$

where *A* represents to the weight of the glass sample in air and *B* is the weight of the sample in immersed liquid. While $\rho_0$ is the density of immersed liquid (distilled water = 1 gcm$^{-3}$) and *d* is the density of air (0.001 gcm$^{-3}$). Moreover, the molar volume can be solved by using Equation (2):

$$V_m = \frac{M}{\rho} \quad (2)$$

where $V_m$ is the molar volume, $\rho$ is the density and *M* is the molecular weight of the sample. The measurement of rigidity of the oxide system can be determined by packing density, $V_d$ which may be obtained from the Equation (3):

$$V_d = \frac{1}{V_m} \sum V_i X_i \quad (3)$$

where $V_i$ and $X_i$ represent the packing factor and mole fraction of the composition, respectively. Thus, packing factor of any oxide $D_X O_Y$ having *D* and *O* ions with ionic radii $r_m$ and $r_o$ may be solved by Equation (4):

$$V_i = \frac{4\pi N_A}{3} \left( X r_m^{\ 3} + Y r_o^{\ 3} \right) \quad (4)$$

## 2.4 Ultrasonic Measurement (Pulse echo Technique)

The glass samples with diameter between 3.62 mm to 5.19 mm were measured by using vernier callipers. Before undergoes ultrasonic scanning, the essential oil is applied to the surface of sample (scan area) as well as the ultrasonic probe to ensure that ultrasonic waves were transmitted to the sample without reflection. The pulse, which generated by the pulse oscillator were depending on the transmitting transducer. It transforms them into acoustic pulses. These acoustic pulses generated through the glass sample, then converted into electrical signals, which receive by the transducer. Hence, the output signal (amplified) from the sample is displayed on the oscilloscope. A high-power ultrasonic process control system Pulser-Receiver Panametrics NDT, a digital storage oscilloscope RIGOL DS2203E and KB-Aerotech straight beam contact transducer of diameter 0.5 mm and the central frequency of 2.25 MHz were employed for measuring the ultrasonic signals. The angle wedges KB - Aerotech made of perspex with refraction angle of 45° was used and the wedges attached to the KB - Aerotech straight beam contact transducer of diameter 0.5 mm and the central frequency of 2.25 MHz.

In pulse-echo ultrasonic test, an incident ultrasound wave emitted from a transducer on the surface of the specimen and its echo, i.e., a reflection from another surface, is detected by the same transducer at a later time, after the wave has passed through the specimen and returned [3,4]. The pulse, which generated by the pulse oscillator were depending on the transmitting transducer. It transforms them into acoustic pulses. These acoustic pulses generated through the glass sample, then converted into electrical signals, which receive by the transducer. Hence, the output signal (amplified) from the sample is displayed on the oscilloscope. The information displayed was taken to calculate the elastic constant measurements, which can be evaluated by using Equation (5) to (8).

$$\text{Longitudinal Modulus: } L = \rho {V_L}^2 \tag{5}$$

$$\text{Shear Modulus: } G = \rho {V_s}^2 \tag{6}$$

$$\text{Bulk Modulus: } K = L - \left(\frac{4}{3}\right)G \tag{7}$$

$$\text{Young's Modulus: } E = 2G(1+\sigma) \tag{8}$$

$$\text{Poisson's Ratio: } \sigma = \frac{(L-2G)}{2(L-G)} \tag{9}$$

$$\text{Microhardness: } H = \frac{(1-2\sigma)E}{[6(1+\sigma)]} \tag{10}$$

## 3 Results and Discussion

### 3.1 XRD Analysis

The existence of a broad hump in S6 revealed that, there is no significant or Bragg's peak in glass spectra. Therefore, it has been confirmed that the plane is glass. This is most likely due to the fact that density is very sensitive to both the ionic size and atomic weight of the alkaline rare earth [5]. These results revealed that addition of alkaline earth elements leads to the formation of NBOs and expands (opens up) the structure [5]. Thus, the obtained data reveals the amorphous nature of the prepared glasses and the presence of random strontium phosphate network in the samples.

### 3.2 Physical Analysis

The physical analysis, which calculated by using Equation (1) to (4) then tabulated in Table 2. Therefore, Figure 1(a) and 2(b) present the change in density, molar volume and packing density that occurred with the addition of $SrCO_3$. The increase in Sr content directly correlated with an increase in glass density [6].

Table 2: Density, molar volume and packing density of the glass.

| Mol% of $SrCO_3$ | Density, $\rho$ ($gcm^{-3}$) | Molar Volume, $V_m$ ($cm^3mol^{-1}$) | Packing Density, $V_t$ |
|---|---|---|---|
| 0 | 1.8910 | 53.6044 | 0.3861 |
| 1 | 2.3020 | 44.0588 | 0.4681 |
| 2 | 2.3796 | 42.6457 | 0.4818 |
| 3 | 2.4524 | 41.4031 | 0.4946 |
| 4 | 2.1548 | 47.1473 | 0.4325 |
| 5 | 2.1415 | 47.4672 | 0.4280 |

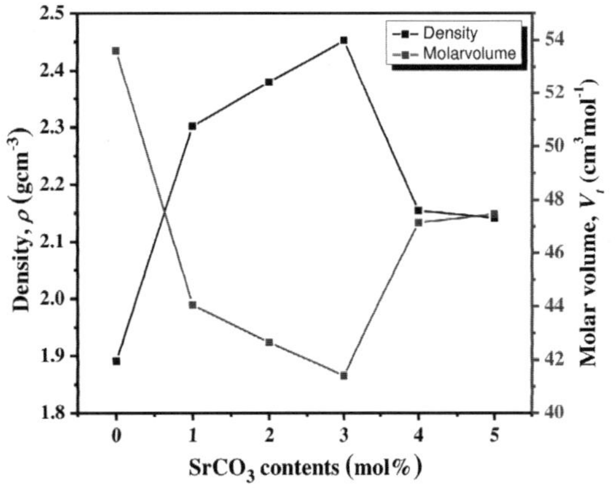


(a)

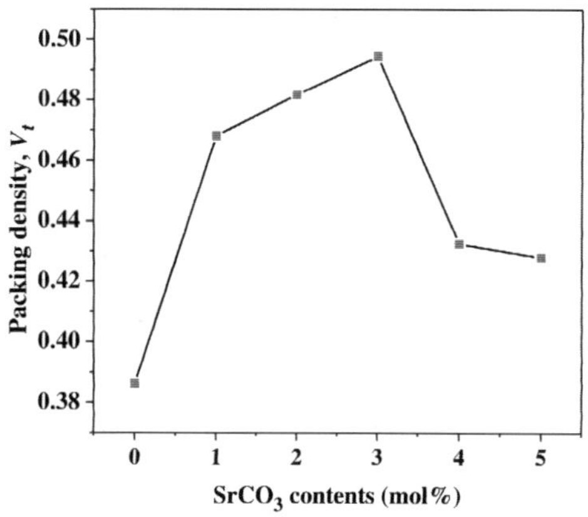


(b)

Figure 1: Density and molar volume (a) while (b) is packing density of $(50–x)P_2O_5 – 30ZnO – 20Li_2O – (x)SrCO_3$ glass system.

### 3.3 Ultrasonic Measurement (Pulse echo Technique)

The ultrasonic wave velocities obtained from existing samples are found to be sensitive to the glass system as shown in Table 3 and Table 4. It is observed as $SrCO_3$ content increases, both $v_l$ and $v_s$ also increase. The decreasing trend of $\sigma$ in Figure 4, indicates the glass network weakens. It is well reported that as Poisson's ratio decreases, the rigidity increases in glasses [18]. The de-polymerization of phosphate group increases the mechanical strength. The rigidity is generally dependent upon the bond strength and cross-link density [9]. Modi et al. [19] and Manupriya et al. [20] have also reported that a low cross-link density contained glass has higher $\sigma$ i.e. 0.2 to 0.5 whilst the high cross-link density has $\sigma$ between 0.1 and 0.2. Thus, it certainly shows that the present glass samples contain high-cross link density with addition of $SrCO_3$ to the glass system attributed to the compactness in the packing structure of the glass due to the reduction of NBO's. Elastic moduli are particularly helpful in correlating structural changes in glasses which are expected to occur as a function of composition [35]. The ultrasonic wave velocities obtained from existing samples are found to be sensitive to the glass system as shown in Table 3. It is observed from the Table 3 as $SrCO_3$ content increases, both longitudinal and shear velocity also increase. Figure 3 with aid of Table 3 display the divergence of elastic moduli with addition of $SrCO_3$ content. Longitudinal modulus, L obtained increases from 35.372 to 82.273 GPa with the increasing concentration of $SrCO_3$ up to 2 mol%. The increase in velocities has been attributed to the increase of glass network rigidness. The presence of $SrCO_3$ in zinc-lithium phosphate increased the formation of bridging oxygens (BO), resulting in high rigidity. However, addition of $SrCO_3$ up to 5 mol%, decreases the value of L from 82.273 to 21.697 GPa. It can be seen, from Table 3, all the elastic moduli values decreased with increasing of $SrCO_3$. This nonlinear variation also happens to shear, bulk as well as Young's modulus, which follow the same trend of variation as density.

Table 3: The velocities and elastic moduli of the glass system.

| Mol % of $SrCO_3$ | Longitudinal Velocity, $v_l$ ($ms^{-1}$) | Shear Velocity, $v_s$ ($ms^{-1}$) | Longitudinal Modulus, $L$ (GPa) | Shear Modulus, $G$ (GPa) | Bulk Modulus, $K$ (GPa) | Young's Modulus, $E$ (GPa) |
|---|---|---|---|---|---|---|
| 0 | 4325 | 3191 | 35.372 | 19.255 | 9. 699 | 3.7591 |
| 1 | 5190 | 3336 | 62.007 | 25.619 | 27.848 | 58.8212 |
| 2 | 5880 | 3373 | 82.273 | 27.073 | 46.176 | 67.9424 |
| 3 | 4827 | 2694 | 57.141 | 17.799 | 33.409 | 45.3447 |
| 4 | 4091 | 2651 | 36.063 | 15.143 | 15.872 | 34.4685 |
| 5 | 3183 | 2573 | 21.697 | 14.177 | 2.794 | 15.8045 |

Table 4: The Poisson's ratio and microhardness of the glass system.

| Mol % of $SrCO_3$ | Poisson's Ratio, $\sigma$ | Microhardness, $H$ (GPa) |
|---|---|---|
| 0 | -0.0974 | 0.8293 |
| 1 | 0.148 | 6.0119 |
| 2 | 0.2548 | 4.4255 |
| 3 | 0.2738 | 2.6841 |
| 4 | 0.1381 | 3.6535 |
| 5 | -0.4426 | 8.9088 |

IOP Conf. Series: Materials Science and Engineering 1142 (2021) 012009 doi:10.1088/1757-899X/1142/1/012009

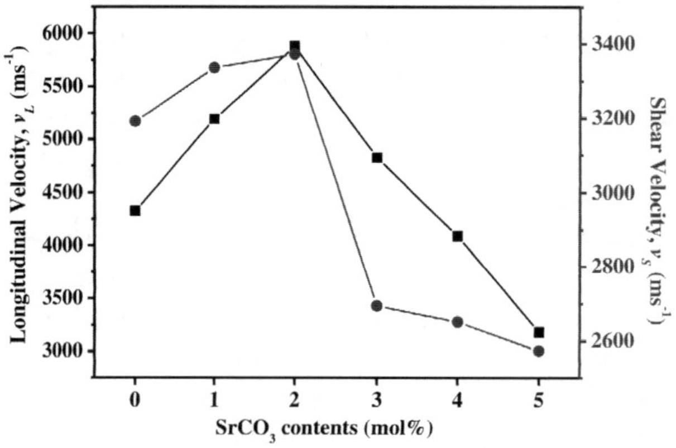


Figure 2: The variation of velocities with different mol% of $SrCO_3$

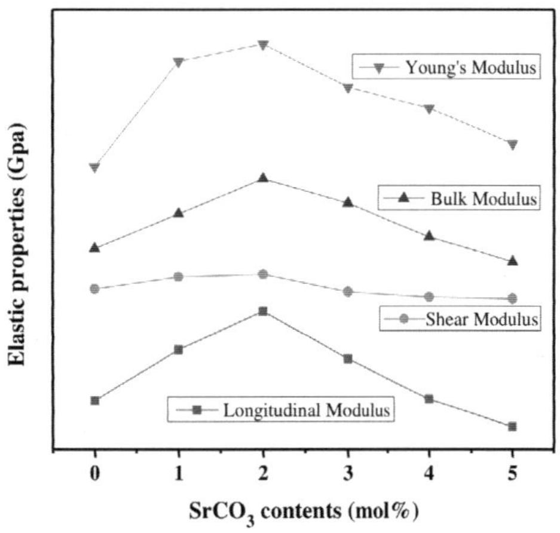


Figure 3: The variation of elastic moduli with different mol% of $SrCO_3$.

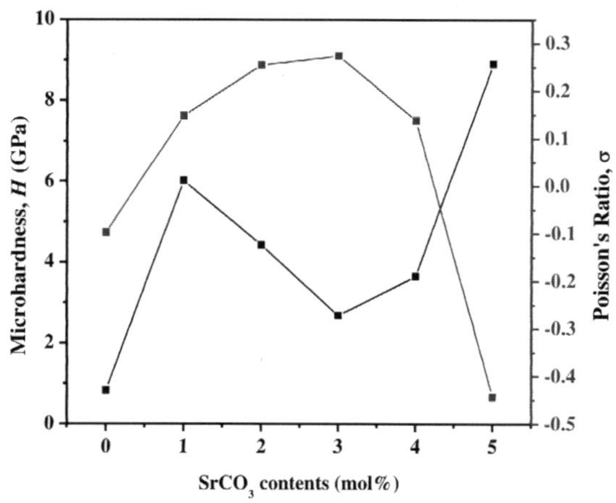


Figure 4: Microhardness ( $H$ ) and Poisson's ratio ( $\sigma$ ) varied with different mol% of $SrCO_3$.

## 4 Conclusions

The physical and mechanical properties as $P_2O_5$ is replaced by $SrCO_3$ in zinc-lithium phosphate glasses were carried out. The existence of broad hump in XRD reveals that there is no significant or Bragg's peak in glass spectra. The ultrasonic results and other measured parameters attribute to the Sr ion increase in the cross-link density by the formation of P-O-Sr. The Poisson's ratio obtained is between 0.1 and 0.2, which certainly shows that the present glass samples contain high-cross link density with addition of $SrCO_3$ to the glass system attributed to the compactness in the packing structure of the glass due to the reduction of NBO's.

## References

[1] R.K. Brow 2000 *J. Non-Cryst. Solids* 263–264 1–28

[2] J.E. Shelby 2005 *Introduction to Glass Science and Technology* 2nd edn. The Royal Society of Chemistry Cambridge
[3] Krautkrämer J. 1983 *H. Ultrasonic Testing of Materials* 3rd ed. Springer New York, NY USA.
[4] Jone N. Watson J.D. 1990 *Digital Signal Processing: Principles, Devices and Applications;* IEE Control Engineering Series 42 424p.
[5] M. Farouk A. Samir M. El Okr 2017 *Effect of alkaline earth modifier on the optical and structural properties of $Cu^{2+}$ doped phosphate glasses as a bandpass filter* Physica B: Physics of Condensed Matter.
[6] Al Qaysi M Walters NJ Foroutan F et al.2015 *Strontium- and calcium-containing, titanium-stabilised phosphate-based glasses with prolonged degradation for orthopaedic tissue engineering* Journal of Biomaterials Applications 30 (3) 300-310.
[7] Ensanya A. Abou Neel Wojciech C 2009 *Structure and properties of strontium-doped phosphate-based glasses* J. R. Soc. Interface 6 435–446.
[8] Hammad A. H. Moustafa E. B. 2020 *Study some of the structural, optical and damping properties of phosphate glasses containing borate* Journal of Non-Crystalline Solids, 544, 120209.
[9] P.K. Jha O.P. Pandey K. Singh 2014 *FTIR spectral analysis and mechanical properties of sodium phosphate glass-ceramics* Journal of Molecular Structure.
[10] L. Montagne G. Palavit G.Mairesse 1996 *Phys. Chem. Glasses* 37(5) 206–211.
[11] K. Meyer 1997 *J. Non Cryst. Solids*, 209 227–239.
[12] M. El Hezzat M. Et-tabirou L. Montagne 2003 *Phys. Chem. Glasses* 44(5) 345-348.
[13] E. Nabhan A. Nabhan N. Abd El Aal 2016 *Mechanical and Structural Properties of Zinc – Sodium - Phosphate Glasses Doped with $Cu_2O$* American Journal of Physics and Applications 4(6) 145-151.
[14] S. Vafaei M.H. Hekmatshoar 2010 *Optical and Thermal Properties of Mixed Alkali Phosphate Based Glasses* International Journal of Optics and Photonics 10(2).
[15] Kim Y.S. Choi W.G. Ryu B.K. 2014 *Effect of ZnO content change on the structure and properties of zinc borophosphate glasses* Glass Physics and Chemistry 40 (4) 408–414.
[16] Sastry S. S. Rao B. R. V. 2015 *Spectroscopic characterization of manganese-doped alkaline earth lead zinc phosphate glasses* Bulletin of Materials Science 38(2) 475–482
[17] Konidakis I. Varsamis C.P. E. Kamitsos E. I. Möncke D. Ehrt D. 2010 *Structure and Properties of Mixed Strontium–Manganese Metaphosphate Glasses* The Journal of Physical Chemistry C 114(19) 9125–9138.
[18] M. K. Halimah W. M. Daud H. A. A 2010 *Effect of AgI addition on the elastic properties of quaternary tellurite glass systems* Chalcogenide Letters 7(11) 613-620.
[19] Modi K. B. Shah S. J. Pujara N. B. Pathak T. K. Vasoya N. H. Jhala I. G. 2013 *Infrared spectral evolution, elastic, optical and thermodynamic properties study on mechanically milled (0.5)NiO – (0.5)ZnO – $5Fe_2O_4$ spinel ferrite* Journal of Molecular Structure 1049 250–262.
[20] Manupriya K.S. Thind G. Sharma V. Rajendran K. Singh A.V.G. Devi S. Aravindan *2006 Structural and acoustic investigations of calcium borate glasses* Phys. Stat. Sol 203(10) 2356-2364.

# Fabrication and characterization of hydroxyapatite based cow bone polysulfone mixed matrix membrane

**Siti Nor Suhaida Rasman[2], Mohd Riduan Jamalludin[12*], Suraya Najieha Kamarudin[2], Siti Khadijah Hubadillah[3], Mohammad Arif Budiman Pauzan[4], Mohd Hafiz Dzarfan Othman[4]**

[1]Frontier Materials Research, Centre of Excellent (FrontMate), Universiti Malaysia Perlis (UniMAP), Malaysia.
[2]Faculty of Mechanical Engineering Technology, Universiti Malaysia Perlis (UniMAP), Kampus UniCITI Alam, Sungai Chuchuh, Padang Besar, 02100, Perlis, Malaysia.
[3]School of Technology Management and Logistic, Universiti Utara Malaysia, Sintok, 06010, Kedah, Malaysia.
[4]Advanced Membrane Technology Research Centre (AMTEC), Faculty of Chemical Engineering and Energy Engineering, Universiti Teknologi Malaysia, 81310 Skudai, Johor, Malaysia.

*riduanj@unimap.edu.my

**Abstract**. In this work, polysulfone (PSf)/hydroxyapatite (HAp) mixed matrix membrane were prepared were prepared by using phase inversion technique for ultrafiltration application. The prepared PSf/HAp mixed matrix membrane were characterized in term of morphological and physicochemical properties by using SEM, contact angle measurement, membrane tensile measurement and membrane porosity. Meanwhile the performance of PSf/HAp mixed matrix membrane was conducted in term of pure water flux (PWF) by using permeability machine. The surface morphology of the MMMs showed the increasing pore number and size as the concentration of HAp increase. The increment weight ratio of the HAp in PSf membrane enhanced the membrane water flux from 113.44 $L/m^2.h.bar$ to 228.13 $L/m^2.h.bar$ due to increased porosity of the membrane. Meanwhile, the strength of the membrane was increased from 3.82 MPa to 5.2 MPa as HAp increases from 0 to 1.0 wt.% and decreased to 2.49 MPa at 1.5 wt.% HAp. This study suggested that HAp is a promising candidate as additive that can improve PSf membrane performance.

**Keywords:** polysulfone; hydroxyapatite; mixed matrix membrane; fabrication; polymer membrane; phase inversion

## 1. Introduction

Water covers 70% of the world and it is essential to the whole living things. But, nowadays the world faces a severe global crisis especially lack of fresh and clean water. In addition, in this new globalization, the increasing population, urbanization, changes of climate and inefficient management are the reasons that lead to the scarcity of freshwater globally. As being reported by The Global Risks Report (2019), the under supply of freshwater has been acknowledged as one of the largest global risks in term of potential impact over next decade [1]. Surface and ground water were assessed as adequate freshwater resources but industrialization has pushed the mankind to meet the increasing demand of water, uncover the wastewater as one of the valuable additions to the existing fresh water resources [2]. There are many conventional methods for wastewater treatment such as coagulation and flocculation, ion exchange, adsorption and etc.

Membrane process is currently used in several ways to treat wastewater from contaminant and to produce clean water as the growing need for fresh water around the world. Membrane process is a filtering method driven by a force which is pressure and it is a process that affected by the pore size of the filter [3]. Membrane is divided into several separation process which are microfiltration (MF), ultrafiltration (UF), nanofiltration (NF) and reverse osmosis (RO). Ultrafiltration commonly used to filtrate the contaminants from polluted water and the most suitable method for heavy metal removal because of the pore size range that is between 2-100nm [4]. UF can achieve more than 90% of removal efficiency with a metal concentration ranging from 10 to 112 mg/L at pH ranging from 5 to 9.5 and at 2–5 bar of pressure [5].

These days, polymeric membrane is getting attention from the researchers due to its high resistance in several aspects such as temperature and chemical. Polymer is widely used to modify the physicochemical properties of the membrane in order to increase the rejection performance [6]. Polysulfone (PSf) membranes are the most common membranes used in membrane process due to its mechanical robustness and structural and strong chemical stability [7][8]. In addition, PSf, as a resin material, is now widely used as the raw material for ultrafiltration membrane fabrication due to its outstanding mechanical properties, great chemical resistance, good thermal stability and wide pH operation range [9]. However, PSf has hydrophobic nature and susceptible towards fouling problems that can lead to the low production of flux [10]. To overcome this, embedding additive like inorganic particles to produce mixed matrix membranes (MMMs) is said to be the promising method to increase the productivity and quality of the polymer membrane [11]. Incorporation of inorganic additive into polymer matrix could provide the desired properties in the membrane such as high selectivity and high mechanical, chemical and thermal strength [12][13].

Hydroxyapatite ($Ca_{10}(PO4)_6(OH)_2$), a member of the apatite mineral, can be used for heavy metal remediation in water treatment [14]. Hydroxyapatite (HAP) consist a special chemical property that can tolerate substitution of metal cation and oxyanion in the structure at Ca2+ and PO43- position resulting in the toxic ions being removed from waste water [15]. Due to several characteristics such as low cost, non-toxic and high potential adsorption of heavy metal ions, HAp is a promising material as an adsorbent to remove contaminant in waste water [14][16][17]. In this work, fabrication of mixed matrix membranes (MMMs) PSf/HAp ultrafiltration membrane was prepared via phase inversion technique. The role of HAp for improving the hydrophilicity of the polysulfone UF membrane is discussed.

## 2. Materials and methods

### *2.1. Materials*

The cow bone waste were bought from a local butcher in Batu Pahat, Johor, Malaysia. Polysulfone (UDEL P-1700) purchased from Solvay, N-Methyl-2-Pyrrolidone (NMP) and polyvinylpyrolidone (PVP) purchased from MERCK were used as polymer, solvent and pore forming agent, respectively. Distilled water used for coagulation bath.

### *2.2. Preparation of Hap powder*

HAp were made from waste cow bones and the synthesize method were referred from [18]. The waste cow bones were crushed manually into smaller pieces. The bones then were boiled in 100°C hot water to remove excessive tissues and fats. Then, the bones were dried in oven at 100°C at 24 hr. After drying, the bones were crushed and calcined in furnace (Magna V) at 800°C for 3 hr with the heating rate of 10°C/min. The temperature of the furnace were cooled down to room temperature at 10°C/min. After calcination process, the bones were milled into powder using planetary ball mill. The final powder were sieved to obtain powder size below than 20μm.

### *2.3. Membrane preparation*

The fabrication of PSf/HAp membrane were done using phase inversion method referred from [19]. Table 1 shows the composition of the casting solution for five membranes with different mass ratio of polymer and HAp powder.

**Table 1**: Composition of the casting solution suspension

| Sample | HAp (wt.%) | PSf (wt.%) | PVP (wt.%) | NMP |
|---|---|---|---|---|
| 1 | 0 | | | 84 |
| 2 | 0.5 | | | 83.5 |
| 3 | 1.0 | 15 | 1 | 83 |
| 4 | 1.5 | | | 82.5 |
| 5 | 2.0 | | | 82 |

The flat-sheet membranes were prepared by dissolving PVP with NMP under gentle agitation. HAp powder were latter added and stirred for 6 hours to allow uniform dispersion of the powder in the solution. Lastly, the PSf pelletes were gradually added and suspension were stirred with heat 60°C until solution completely homogenous. After that, the casting solution was poured into bottle and was degassed using sonicator for 1 hour to release the bubbles. After the bubbles completely released, the dope solution was cast on a glass plate using a glass rod at room temperature and placed in coagulation bath to form a sheet of membrane due to the induction of non-solvent. The flat-sheet was then removed from coagulation bath and was dried in room temperature for 24 hour.

### *2.4. Characterization of MMMs*

The morphological properties of MMMs were characterized using JEOL JSM-6380LA scanning electron microscope (SEM) in which cross-sectional samples were initially fractured in nitrogen liquid before being spur-coated with gold, Au to enable the electron interaction with atoms in the samples. Analysis on water contact angle of membranes is necessary to determine the hydrophilicity of the membrane surface. Contact angle goniometer (IMC-159D, Imoto Machinery, Japan) with sesile drop technique was used to measure the water contact angle of the membrane. Deionized water was used as the liquid and measurement were done at 15 different location to derive an average value of water contact angle.

Tensile strength using Universal Tensile Machine with design code of D882-12 as standard tensile test referred from American Society of Testing at the strain rate of 5 mm/min. The membrane were cut into strips 5 cm long and 1 cm wide. The vernier calliper was used to measure the thickness of the membrane. At least 3 samples measurement were carried out to derive the average value.

In order to measure overall porosity, the flat-sheet membrane were cut with same length (around 2 cm). Then, the membrane were immersed in the distilled water for 24 hr. After 24 hr, the membrane were taken out from water and the excess water on the membrane surface were removed by blotting with tissue paper. The weight of wet membrane were recorded before being dried in oven at 50°C for 24 hr. The dried membrane were then being measured by analytical balance. For the measurement of the overall porosity of each sample, Equation 1 was employed.

$$porosity, \varepsilon = \frac{w_w - w_d}{\rho_w \; x \; v} \qquad (1)$$

where $\rho_w$ is density of pure water at room temperature (g/cm$^3$) and v is the volume of membrane in wet state (cm$^3$).

Pure water flux analysis was determined by an ultrafiltration cross-flow permeation testing unit feed with distilled water at pressure 2 bars. The pure water flux (PWF) will be measure by direct measurement of the permeate flow using Equation 2:

$$PWF = \frac{Q}{A \; X \; \Delta T} \qquad (2)$$

where PWF is the pure water flux (L/m$^2$.h$^{-1}$), $Q$ is the permeate volume (L), $A$ is the membrane area (m$^2$) and $\Delta T$ is the time (h).

## 3. Results and discussion

### 3.1. *Characterization of MMMs*

#### 3.1.1. *Scanning electron microscope*

The morphological structure of the membrane were examined by SEM micrography analysis as shown in Figure 1. Generally, all membrane surface showed the dispersion of HAp powder over the surface despite of agglomeration denoted by the presence of lump in certain area. The agglomeration can be defined that the HAp is naturally exists. From the figure, macrovoids also can be seen at some place among the finger-like structure of the membrane with 2.0 wt.% of HAp due to the agglomeration. The cross-section images seems to be similar for every membrane.

Therefore, the HAp additive did not have any particular influence in the MMMs morphology. However, the finger-like structure in MMMs are suprisingly more larger than the control membrane. This situation might be due to the addition HAp in MMMs. In general, asymmetrical structures were achieved for all membranes, consisting of a thin dense layer supported supported by a finger-like. This situation has common in SEM image result reported by [20]. It has been reported by [21] that the use of more additive than the suitable amount could resulted in pore blocking in the membrane and resulting in low water flux due to agglomeration of particles.

The surface of the membrane showed that the increasing number of pores from the controlled membrane to the 2.0 wt.% HAp. The increasing number of pores will increase the surface porosity. Increasing porosity will increase the water flux hence, producing higher permeate [22].

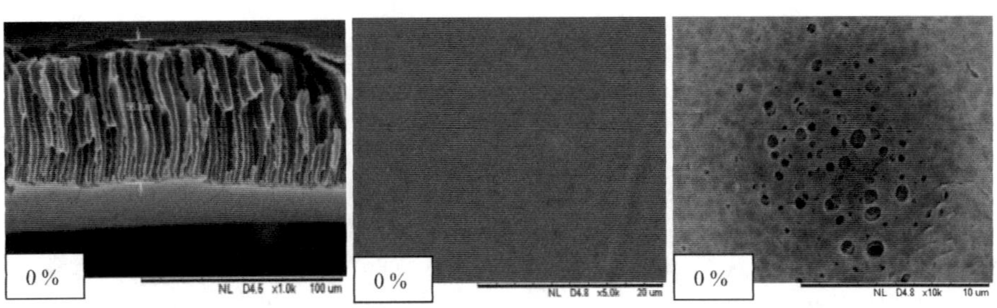

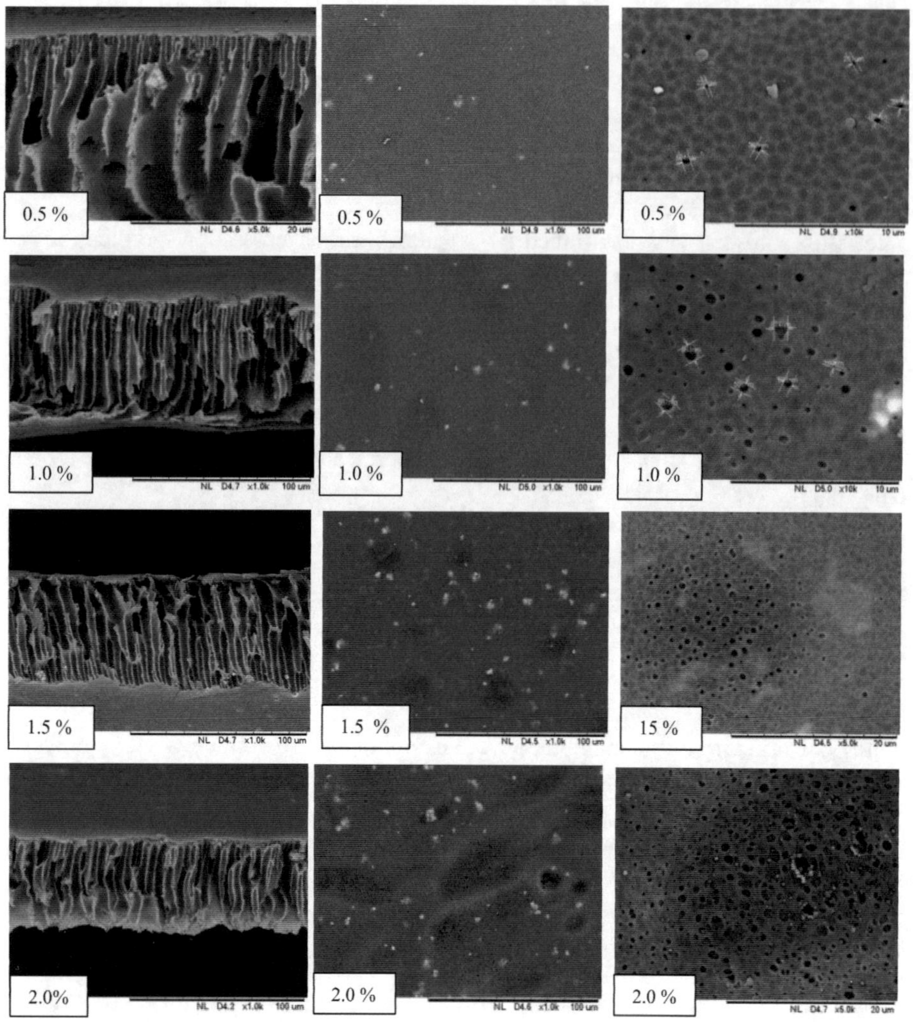


**Figure 1**: SEM images of cross section (left) and top surface (right) of membranes

*3.1.2. Contact angle analysis*

Contact angle is a method that used to measure the surface wettablity and to describe the hydrophilicity of the membrane. The higher the hydrophilicity of the membrane, the lower the value of contact angle. Figure 2 shows the value of contact angle for all samples based on HAp content. The result obviously showed that the contact angle decreased as the concentration of HAp increased from 0.5 to 1.5 wt.% which is 80.1° to 65.5° and slightly increased at HAp concentration 2.0 wt.% which is 67.2°. This increasing occurrence could be due to the presence of additive that has lowered down the surface tension of the PSf cause the water easily spread and attracted to the membrane surface [23]. The increasing hydrophilicity of the membrane might cause by the surface enrichment of hydroxyl group [19]. In this case study, HAp consist hydroxyl group that can enhance the hydrophobicity of the PSf membrane. Meanwhile the decreasing contact angle might be caused by the agglomeration of the HAp particles among the surface of the membrane. As a result, the surface (contact) area of the hydrophilic functional groups possessed by additive is reduced, hence, cause the hydrophilicity of the membrane surface decreased [24].

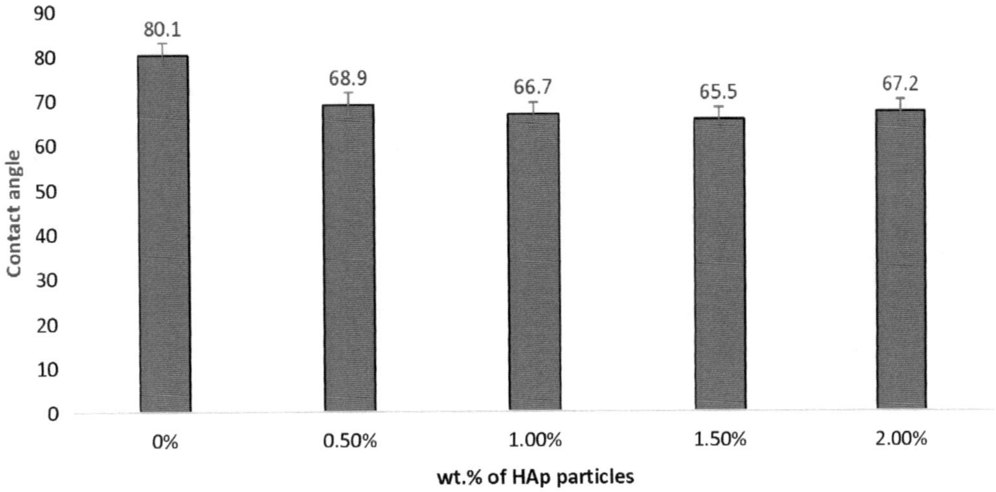


**Figure 2**: Contact angle analysis at different HAp content

*3.1.3. Tensile strength analysis*

Tensile strength analysis is used to measure the strength of the membrane with the addition of inorganic additive. The tensile analysis revealed that the membrane tensile is increased from the control membrane to membrane with 1.0 wt.% HAp and increased from 3.82 MPa to 5.2 MPa as being shown in Figure 3. However, the tensile analysis dropped at membrane with 1.5 wt.% HAp with 2.44 MPa and increased back at 2.0 wt.% HAp with 4.40 MPa. The unstability in tensile analysis might due to the structure of the membrane resulting from porosity [25]. In addition, the increasing pore also reducing the mechanical strength [26].

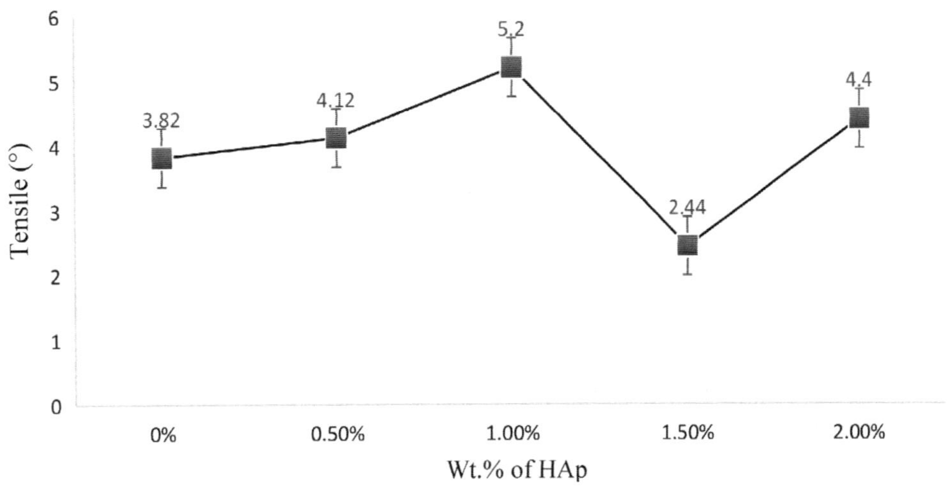


**Figure 3**:Tensile strength (MPa) with different wt.% of HAp particles.

### *3.1.4. Porosity analysis*

Membrane porosity can be defined as weight or pure water trapped in 1 $m^3$ of membrane structure [27]. Porosity is an important analysis for membrane separation because it determines the performances and properties [28]. Figure 4 demonstrated the porosity data of the prepared membrane at different concentration of HAp. The result shows that the porosity increased proportionally to the HAp concentration. These results were in line with the increase of hydrophilicity properties (via hydroxyl group in HAp) that can be seen by the large finger-like structure of MMMs than the controlled membrane [20]. However, the data for 2.0 wt.% HAp shows that the porosity value decreased. This situation can be explained due to agglomeration of HAp that occur in the finger-like structure of the membrane that cause the pore is blocked by the excessive HAp particles [21]. The increasing of solution viscosity that resulted in delaying the liquid-liquid exchanging process attributed to produce a thicker and low porosity membrane [29].

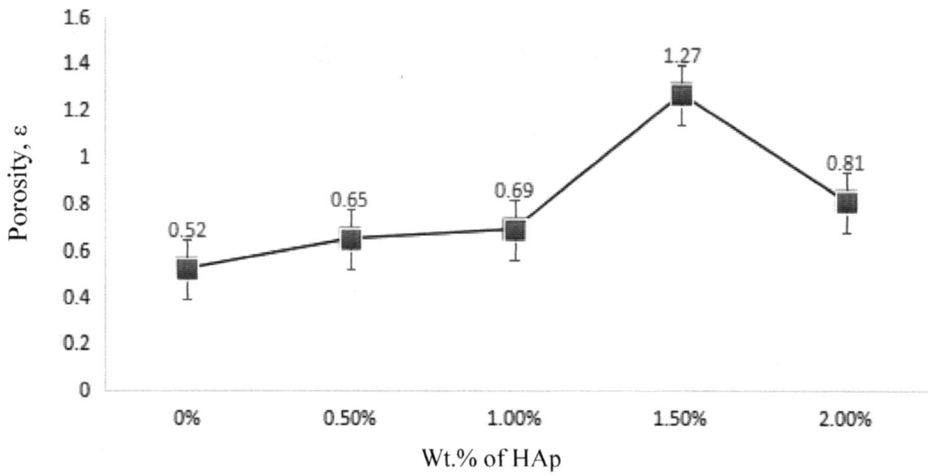


**Figure 4**: Porosity of PSf membrane with wt.% of HAp particles

### *3.2. Water permeation flux*

Figure 5 shows the result of pure water flux for each membrane. As demonstrated in the figure, the membrane flux significantly improved by the addition of HAp as additive. The PWF increases constantly from 0 wt.% to 1.5 wt.% of HAp which is from 113.44 L/$m^2$.h.bar to 228.13 L/$m^2$.h.bar. The result of PWF is in agreement with the contact angle result that conclude by adding inorganic additive HAp with increasing concentration at each membrane, the hydrophilicity can be improved which definitely affect the PWF value. The PWF value is also raised by the increasing number of pores induced by the increasing concentration of HAp in MMMs, which is in line with the increasing porosity [30]. The PWF value also corresponding to the finger-like structure in SEM observation that become larger as the concentration of HAp increased. However, the PWF decreased at concentration 2.0 wt.% and this situation might be caused by the HAp particles that creates agglomeration among the finger-like structure of the membrane and lead to pore blocking that prevent the water from flowing. The result was in line with study done by [31] and [21] about effect of inorganic additive concentration on PWF.

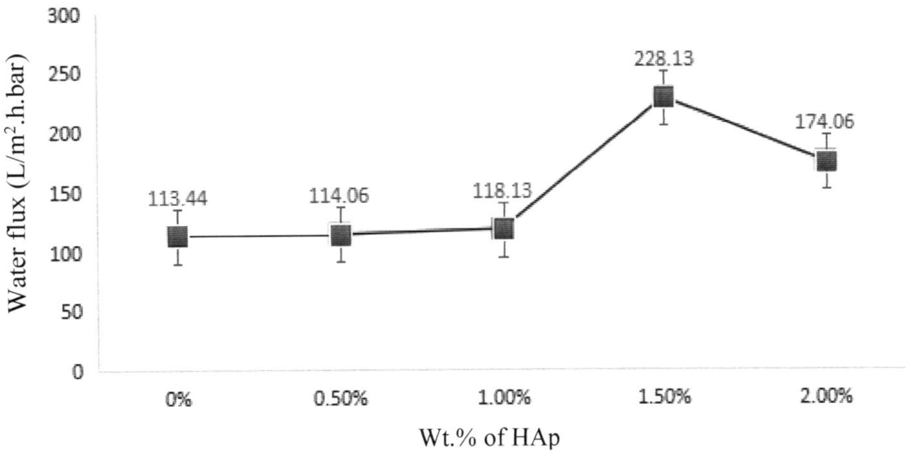


**Figure 5**: Permeation water flux result

## 4. Conclusion

In this study, flat-sheet PSf/HAp mixed matrix membranes were prepared via phase inversion technique were investigated. The effect of HAp as inorganic additive into membrane formulation for surface enhancement were observed. In addition, the morphological structures, hydrophilicity properties, porosity, tensile strength and pure water fluxes were also studied. The incorporation of HAp has been confirmed by SEM analysis that illustrates typical asymmetrical structures of the membrane. An agglomeration of HAp has been identified at certain part of membrane. Despite the agglomeration of HAp within the membrane, pure water flux analysis showed that membrane with 1.5 wt.% of HAp gives the highest permeate with 228.13 $L/m^2$.h.bar due to increasing number of pores. Corresponding with the increment of the fluxes, the porosity of each membrane sample also increases proportional to the HAp concentration. Meanwhile, the mechanical property of tensile strength was found to increase from 3.82 MPa to 5.20 MPa as the concentration of HAp increase from 0 wt.% to 1.0 wt.% and decrease at 1.5 wt.%. The addition of inorganic additive into the membrane formulation attribute to the surface enhancement of PSf membrane which is increase the hydrophilicity of the membrane. Contact angle data for 1.5 wt.% of HAp gives the lowest result which is 65.5° than the control membrane. It can be concluded that the presence of inorganic additive HAp has improved the hydrophilicity of PSf membrane.

## 5. Acknowledgment

Financial support by Ministry of Higher Education with Research Grants Scheme vot No: 9003-00759 and Collaborative Research Grant with vot No: 9023-00012 are gratefully acknowledged.

## References

[1] The Global Risk Report, (2019). The Global Risks Report. 2019. (14th Edition). World Economic Forum, Geneva.

[2] Bodzek, M., Konieczny, K., & Kwieci, A. (2012). Desalination and Water Treatment Application of membrane processes in drinking water treatment – state of art. Taylor and Francis, December, 37–41.

[3] Mai, Z. (2014). Membrane processes for water and wastewater treatment : study and modeling of interactions between membrane and organic matter Zhaohuan Mai To cite this version : HAL Id : tel-00969165 Les procédés membranaires pour le traitement de l ' eau : étude et mod
[4] Zakhar, R., Derco, J., & Čacho, F. (2018). An overview of main arsenic removal technologies. Acta Chimica Slovaca, 11(2), 107–113. https://doi.org/10.2478/acs-2018-0016.
[5] Gunatilake, S. K. (2015). Methods of Removing Heavy Metals from Industrial Wastewater. December.
[6] Lam, B., Morin-crini, N., Crini, G., Fievet, P., Morin-crini, N., Crini, G., & Fievet, P. (2017). SC. https://doi.org/10.1016/j.jclepro.2017.10.090
[7] Richards, H. L., Baker, P. G. L., & Iwuoha, E. (2012). Metal Nanoparticle Modified Polysulfone Membranes for Use in Wastewater Treatment : A Critical Review. 2012(July), 183–193. https://doi.org/10.4236/jsemat.2012.223029
[8] Harun, Z., Jamalludin, M. R., Basri, H., Shohur, M. F., Rosman, N., & Yunos, M. Z. (2013). The effect of synthetic silica on ultrafiltration PSf membrane. *Jurnal Teknologi (Sciences and Engineering)*, *65*(4), 121–125. https://doi.org/10.11113/jt.v65.2342
[9] Shen, Z., Chen, W., Xu, H., Yang, W., Kong, Q., Wang, A., & Ding, M. (2019). Fabrication of a Novel Antifouling Polysulfone Membrane with in Situ Embedment of Mxene Nanosheets.
[10] J. Wagner. Membrane Filtration Handbook: Practical Tips and Hints., Vol 127 (2011) http://doi.org/10.1007/s13398-014-0173-7.2
[11] Qadir, D., Mukhtar, H., & Keong, L. K. (2017). Mixed Matrix Membranes for Water Purification Applications. Separation and Purification Reviews, 46(1), 62–80. https://doi.org/10.1080/15422119.2016.1196460
[12] Shohur, M. F., Harun, Z., Yunos, M. Z., Hasan, S., & Jamalludin, M. R. (2013). The effect of sodium chloride (NaCl) coagulant medium of polysulfone ultrafiltration membrane. *Applied Mechanics and Materials*, *372*, 3–7. https://doi.org/10.4028/www.scientific.net/AMM.372.3
[13] Farrukh., Akhyar. M., (2018). Zeolites-Mixed-Matrix Nanofiltration Zeolites-Mixed-Matrix Nanofiltration Membranes Membranes for for the Next Generation Generation of of Water Water Purification Purification. Mmm. https://doi.org/10.5772/intechopen.75083
[14] Thanh, D. N., Novák, P., Vejpravova, J., Vu, H. N., Lederer, J., & Munshi, T. (2018). Removal of copper and nickel from water using nanocomposite of magnetic hydroxyapatite nanorods. Journal of Magnetism and Magnetic Materials, 456(November 2017), 451–460. https://doi.org/10.1016/j.jmmm.2017.11.064
[15] Dungkaew, W., Haller, K. J., Flood, A. E., & Scamehorn, J. F. (2012). Arsenic Removal by Precipitation with Calcium Phosphate Hydroxyapatite. 506, 413–416. https://doi.org/10.4028/www.scientific.net/AMR.506.413
[16] Islam, M., Chandra, P., & Patel, R. (2011). Arsenate removal from aqueous solution by cellulose-carbonated hydroxyapatite nanocomposites. 189, 755–763. https://doi.org/10.1016/j.jhazmat.2011.03.051
[17] Mirhosseini, M., Biazar, E., & Saeb, K. (2014). Removal of Arsenic from Drinking Water by Hydroxyapatite Nanoparticles. 9(2), 331–338.
[18] Hubadillah, S. K., Othman, M. H. D., Tai, Z. S., Jamalludin, M. R., Yusuf, N. K., Ahmad, A., … Harun, Z. (2020). Novel hydroxyapatite-based bio-ceramic hollow fiber membrane derived from waste cow bone for textile wastewater treatment. Chemical Engineering Journal, 379(August 2019). https://doi.org/10.1016/j.cej.2019.122396
[19] Jamalludin, M. R., Harun, Z., Hubadillah, S. K., Basri, H., Ismail, A. F., Othman, M. H. D., … Yunos, M. Z. (2016). Antifouling polysulfone membranes blended with green SiO2 from rice husk ash (RHA) for humic acid separation. Chemical Engineering Research and Design, 114, 268–279. https://doi.org/10.1016/j.cherd.2016.08.023
[20] Abdullah, N., Gohari, R. J., Yusof, N., Ismail, A. F., Juhana, J., Lau, W. J., & Matsuura, T. (2016). Polysulfone/hydrous ferric oxide ultrafiltration mixed matrix membrane: Preparation,

characterization and its adsorptive removal of lead (II) from aqueous solution. Chemical Engineering Journal, 289(Ii), 28–37. https://doi.org/10.1016/j.cej.2015.12.081
[21] Kang, Y., Obaid, M., Jang, J., Ham, M. H., & Kim, I. S. (2018). Novel sulfonated graphene oxide incorporated polysulfone nanocomposite membranes for enhanced-performance in ultrafiltration process. Chemosphere, 207, 581–589 https://doi.org/10.1016/j.chemosphere.2018.05.141
[22] Zheng, Y., Zou, S., Nanayakkara, K. G. N., Matsuura, T., & Chen, J. P. (2011). Adsorptive removal of arsenic from aqueous solution by a PVDF / zirconia blend flat sheet membrane. Journal of Membrane Science, 374(1–2), 1–11. https://doi.org/10.1016/j.memsci.2011.02.034
[23] Sabri, N. S. M., Hasbullah, H., Said, N., Ibrahim, N., Kasmani, R. M., R. Ali, R., & Rahman, S. A. (2017). Hydrophilicity Enhancement of Metal Oxide Nanoparticles Incorporated Polysulfone Ultrafiltration Membrane. Journal of Applied Membrane Science & Technology, 20(1), 29–38. https://doi.org/10.11113/amst.v20i1.26
[24] Yuliwati, E., Ismail, A. F., Matsuura, T., Kassim, M. A., & Abdullah, M. S. (2011). Effect of modi fi ed PVDF hollow fi ber submerged ultra fi ltration membrane for re fi nery wastewater treatment. 283, 214–220. https://doi.org/10.1016/j.desal.2011.03.049
[25] Ma, Y., Shi, F., Ma, J., Wu, M., Zhang, J., & Gao, C. (2012). Effect of PEG additive on the morphology and performance of polysulfone ultra fi ltration membranes. 272, 51–58. https://doi.org/10.1016/j.desal.2010.12.054
[26] Mukherjee, R., Bhunia, P., & De, S. (2016). Impact of graphene oxide on removal of heavy metals using mixed matrix membrane. Chemical Engineering Journal, 292, 284–297. https://doi.org/10.1016/j.cej.2016.02.015
[27] Yunos, M. Z., Harun, Z., Basri, H., & Ismail, A. F. (2012). Effects of water as non-solvent additive on performance of polysulfone ultrafiltration membrane. Advanced Materials Research, 488–489, 46–50. https://doi.org/10.4028/www.scientific.net/AMR.488-489.46
[28] Jamalludin, M. R., Rosli, M. U., Ishak, M. I., Khor, C. Y., Shahrin, S., Ismail, R. I., Jahidi, H. (2017). The effect of silica toward polymer membrane for water separation process. AIP Conference Proceedings, 1885. https://doi.org/10.1063/1.5002412
[29] Mahmoudi, E., Ng, L. Y., Ba-Abbad, M. M., & Mohammad, A. W. (2015). Novel nanohybrid polysulfone membrane embedded with silver nanoparticles on graphene oxide nanoplates. Chemical Engineering Journal, 277, 1–10. https://doi.org/10.1016/j.cej.2015.04.107
[30] Jamshidi Gohari, R., Lau, W. J., Matsuura, T., & Ismail, A. F. (2013). Fabrication and characterization of novel PES/Fe-Mn binary oxide UF mixed matrix membrane for adsorptive removal of As(III) from contaminated water solution. *Separation and Purification Technology*, *118*, 64–72. https://doi.org/10.1016/j.seppur.2013.06.043
[31] Rezaee, R., Nasseri, S., Mahvi, A. H., Nabizadeh, R., Mousavi, S. A., Rashidi, A., Jafari, A., & Nazmara, S. (2015). Fabrication and characterization of a polysulfone-graphene oxide nanocomposite membrane for arsenate rejection from water. Journal of Environmental Health Science and Engineering, 13(1), 1–11. https://doi.org/10.1186/s40201-015-0217-8

# Fabrication and characterization of composite hollow fibre membrane derived from hydroxyapatite cow bone and kaolin

**Suraya Najieha Kamarudin[1], Mohd Riduan Jamalludin[1,2*], Siti Nor Suhaida Rasman[1], Siti Khadijah Hubadillah[3], Mohd Arif Budiman Pauzan[4], Mohd Hafiz Dzarfan Othman[4]**

[1]Faculty of Engineering Technology, Universiti Malaysia Perlis (UniMAP), Kampus UniCITI Alam, Sungai Chuchuh, Padang Besar, 02100, Perlis, Malaysia.
[2]Frontier Materials Research, Centre of Excellence (FrontMate), Universiti Malaysia Perlis (UniMAP), 01000 Perlis, Malaysia.
[3]School of Technology Management and Logistics, Universiti Utara Malaysia, Sintok, Kedah, 06010, Malaysia
[4]Advanced Membrane Technology Research Centre (AMTEC), Faculty of Chemical and Energy Engineering, Universiti Teknologi Malaysia, 81310 Skudai, Johor, Malaysia.

riduanj@unimap.edu.my

**Abstract**. Due to its superior behaviour, membrane separation has been applied widely towards water and wastewater treatment. Compare to polymeric counterparts, ceramic membrane has become innovation technology nowadays as it can be used both separation and adsorption application. The conventional ceramic membrane made from alumina, however, in view of its high-cost output. As a result, exploring to a cheap ceramic material derived from clay and waste is gaining attention. In this study, hydroxyapatite derived from cow bone waste and kaolin were used as the main materials for the fabrication of ceramic hollow fibre membrane via phase inversion and sintering technique. The membrane composition was varied through kaolin:hydroxyapatite ration into three composition (40:0, 0:40, 20:20). Result showed the ceramic membrane fabricated from 20 wt% kaolin and 20 wt% hydroxyapatite induces excellent performance at water flux of 10168.5 L/m².h and mechanical strength at 13.33 MPa.

## 1. Introduction

Water become the world's most valuable resource because of rapidly growing populations, rising demand of water and the depletion of the quantity and quality of water supplies. Over the last century, the world's population has grown over and will increase by about 45% over the next 50 years. As long as it is properly handled, freshwater is a valuable resource of world which will be pursue to be sustainable. To secure the sustainability of local growth, the prevention of pollution from industrial, domestic and agriculture activities should be done [1]. Therefore, through treating and reusing waste water, a lot of researchers concentrated on suitable systems for receiving fresh water, the method of removing unpleasant agents, such as organic, chemicals and biological pollutants from water is water purification.

There are numerous technologies utilized, for example, adsorption, ion exchange, coagulation, and membrane technology to remove contaminants from water. Among these technologies, membrane technology is a favored technology in expelling contamination from water. Membrane technology has been identified as the primary technology for the isolation of pollutants from contaminated sources in the treatment of water and wastewater [2]. This is because membrane technology has high efficiency, low utilization of energy and high filtration performance [3]. Throughout the water industry, membrane technology is used to enhance the water quality for usage, reuse or discharge into the environment. Membranes vary from finely porous to non-porous structures and can expel pollutants like bacteria and protozoa to ions [4].

Membrane technology made tremendous strides in the 20th century, and commercial markets expanded very quickly and globally. Thus, the 21st century is known as the "water century" [5]. In certain cases, the main attractions of membrane technology are reduction of the number of processing phases, high efficiency of separation, low consumption of energy and increased quality of the finished products. In comparison, membrane has been shown to be substantially simple, providing benefits over the traditional method due to low consumption of energy, mild temperature activity, avoiding of chemical products and preservation of nutrient and other desired compounds [6]. Increased water shortages, combined with fixed developments in membrane quality, costs and demand of energy, would result in stable growth in the water industry's membranes in the near future [4]. Phase inversion is one of the most desired and widely used techniques used by most researchers through immersion precipitation as a means of making membrane for microfiltration and ultrafiltration of polymer and ceramic membranes [7]. The interest in ceramic membranes for the treatment of industrial wastewater has recently increased due to their excellent properties compared with polymeric membranes. This is due to the ceramic membranes have high thermal stability, good mechanical strength and long lifetime compared with polymeric membranes.

In this study, both hydroxyapatite based cow bone and kaolin are the main material for the fabrication of ceramic membrane. Both kaolin and cow bone waste are example of alternative ceramic membrane material which is abundantly available in Malaysia. In this study, hydroxyapatite and kaolin are combined to develop membranes for a several reasons. One of the reason is hydroxyapatite cannot stand alone as the main material of membrane since it has disadvantages such as brittleness, low tensile strength and fracture toughness. So by mixing kaolin it can increase the mechanical strength of membrane since kaolin have high strength properties. The other reason is kaolin has low ion exchange capacity that can affect the performance of the membrane. This kaolin's properties are opposite to the properties of hydroxyapatite. Due to the presence of phosphate and calcium, hydroxyapatite showed high ion exchange capacity [8].

It is possible to obtain hydroxyapatite from natural products such as egg shells, beef bones, and fish bones. Since hydroxyapatite is easily synthesized from waste materials, it is being highlighted as environment material from the aspect of the utilization of wastes [9]. Numerous studies have demonstrated hydroxyapatite's adsorption features for industrial waste water filtration and aqueous solutions. Its particles can adsorb heavy metals from the earth such as cadmium, lead, arsenic, aluminum, and nickel. Hydroxyapatite are known for its properties such as high sorption ability and low water solubility. Fu et al. (2016) [10] stated that hydroxyapatite has a good thermal and chemical stability.

Due to its excellent chemical composition, mineralogical properties and particle size, kaolin is also a preferred substrate for ceramic membranes, making it easy and possible to create good membranes. Kaolin has the characteristics of low plasticity, high refractory properties and good chemical composition [11]. Furthermore, kaolin shows hydrophilic properties, which is highly desired for the fabrication of water filtration membranes. Due to its crystalline order, mineralogical features and composition of chemical, kaolin is a favoured raw substance for porous ceramic membrane. The outstanding of kaolin elements have encouraged research to produce low-cost ceramic membranes based from kaolin [12]. Therefore, in this work, cow bone hydroxyapatite (HAp) and kaolin will be used as the materials for the membrane. The fabrication process of ceramic hollow fibre membrane will be using both phase inversion and sintering technique.

## 2. Method

### *2.1. Materials*
Kaolin were purchased from Kaolin Malaysia Sdn Bhd and cow bone hydroxyapatite (HAp) were obtained from previous work [13]. N-methyl-2-pyroolidone (NMP) (HPLC grade, Rathbone), polyethersulfone (PESf, Radel A-300, Ameco Performance, USA) and Polyethyleneglycol 30-dipolyhydroxystearate (Arlacel P135) were used respectively as solvent, polymer binder and dispersant. Tap water was also utilized as the coagulant bath and bore fluid.

### *2.2. Preparation of ceramic hollow fibre membrane derived from hydroxyapatite based cow bone and kaolin*
Figure 1 shows the preparation of ceramic hollow fibre membrane derived from hydroxyapatite based cow bone and kaolin using both phase inversion and sintering method. Ceramic suspensions at different content were prepared by mixing kaolin, cow bone hydroxyapatite powder and Arlacel P135 with NMP. Three membrane compositions were formulated as summarized in Table 1. At 190 rpm, the suspension was subsequently ball-milled for 24 hours, using four and two zirconia milling balls with diameters of 6 and 12 mm, respectively. PESf was then be used as the binding agent, and continuously milled for another 48 hours.

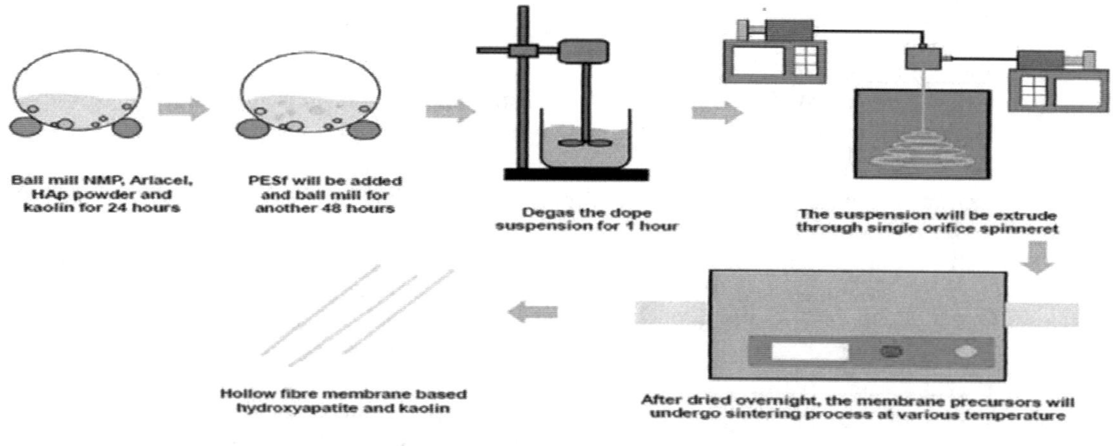


**Figure 1**. The fabrication of ceramic hollow fibre membrane derived from hydroxyapatite based cow bone and kaolin.

The prepared suspension solution was degassed under vacuum with stirring method for one hour then followed by extrusion process by using stainless steel syringe. Infusion pumps (Harvard Apparatus, PHD 2000 Programmable and Chemyx Inc., Model Nexus 6000) regulated the flow rate of the suspension, while the bore fluid rate was controlled using the bore pump (AB, Japan). Both suspensions and bore fluid flow rates were set at 10 mL/min while the air gap was 5 cm. Before cutting to the length needed, the prepared membrane precursors were dried overnight. The precursors undergo sintering process at 1200 °C temperature in a tubular furnace. Finally, the membranes were cooled down at room temperature [13].

**Table 1**. The compositions for membrane fabrication.

| Compositions | Hydroxyapatite based cow bone (wt.%) | Kaolin (wt.%) | NMP | PESf | Arlacel |
|---|---|---|---|---|---|
| A | 40 | 0 | 54 | 5 | 1 |
| B | 0 | 40 | 54 | 5 | 1 |
| C | 20 | 20 | 54 | 5 | 1 |

*2.3. Characterization and performance of ceramic hollow fibre membrane derived from hydroxyapatite based cow bone and kaolin*

The membrane's surface and cross sectional morphology were analyzed by using scanning electron microscopy (SEM, TM 3000, Hitachi). The mechanical strength of the membrane was measured via three-points boning test using an Instron Model 5544 tensile tester provided with a load cell of 1 kN. With a 30 mm distance, the membrane was placed on the sample holder. The bending strength σF was gained by using equation:

$$\sigma \mathrm{F} = \frac{8FLD_o}{\pi(D_o{}^4 - D_i{}^4)}$$

where F is the measured load at which fracture happened (N), while L, $D_o$ and $D_i$ are the membrane length, the outer and inner diameters (m).

ImageJ software was used to determine the distribution of pore size and overall porosity of the membrane. The performance of separation was carried out under a pressure of 1 bar by a cross-flow filtration system.

$$J = \frac{V}{A \times t}$$

where V is the permeate collected volume (L), A is the membrane surface area ($m^2$) and $t$ is the time of permeation (s) [14].

## 3. Results and discussion

The fabrication of composite ceramic hollow fibre membrane derived from hydroxyapatite based cow bone and kaolin were successfully developed at different composition ratio. Figure 2 illustrate the cross-sectional SEM image of composite ceramic hollow fibre membrane at different compositions with ratio cow bone:kaolin at A (40:0), B (0:40) and C (20:20) sintered at 1200 °C. Based on the SEM images, it was shown a significant change on the macrostructure of the developed ceramic membrane especially its pore size formation. For ceramic membrane prepared from cow bone waste (Figure 2) (A1 – A3), a clear finger-like voids are formed at approximately 70% of the cross section. Meanwhile, no finger-like voids was observed for ceramic membrane prepared from kaolin (Figure 2) (B1 – B3). Interestingly, a slight macrovoid formation can be observed for composite ceramic hollow fibre membrane prepared from cow bone waste and kaolin. At higher magnification of SEM images at 3000x magnification, the membrane can be linked through a process called particle packing. The unoccupied voids are filled when they reach at a point where they touch each other as more ceramic particles are introduced [15]. The particles look like bean structure and it occurred during the process of sintering. This is due to the particle of hydroxyapatite melt on the kaolin's particles. All of the composition showed the almost the same structures. This result shows that shrinkage pores occur at a certain distance in the membrane after process of sintering. This distance inhibited the development of the vermicular structure by densification of the particles in the membranes [16]. From the study of Lai et al. [17] this situation is due to dope concentration that change of the precursors resulted into pore structured disordered.

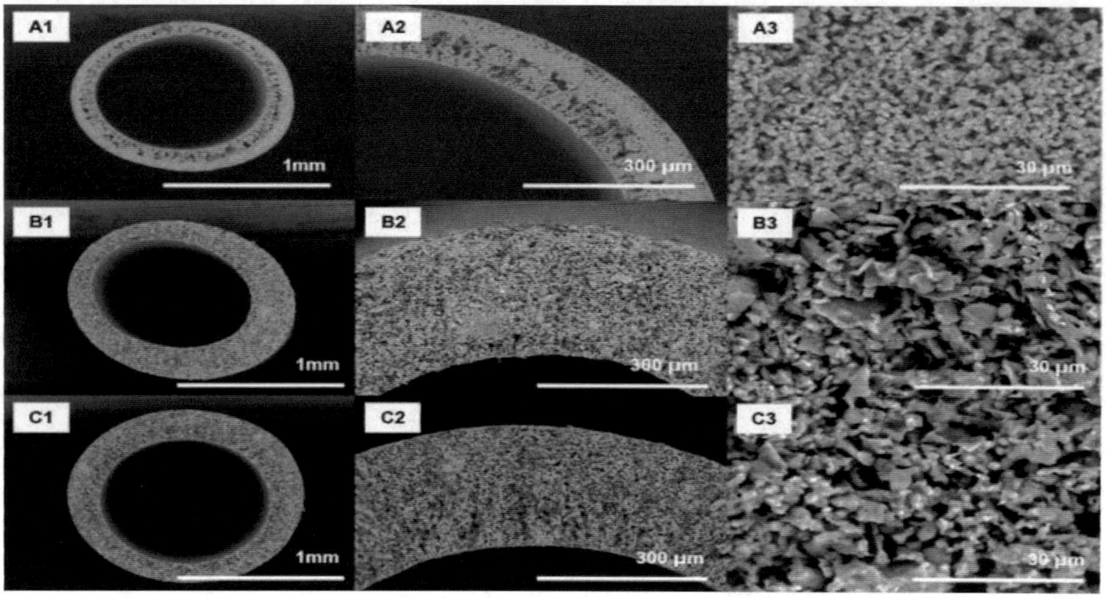


**Figure 2**. The SEM images for composite ceramic hollow fibre membrane at composition A, B and C; (1) overall cross section, (2) 300x and (3) 3000x.

Figure 3 shows the mechanical strength of composite ceramic hollow fibre membrane prepared at different composition A, B and C. The result showed that the membrane prepared from kaolin showed the highest mechanical strength of 37.18 MPa whereas membrane prepared from hydroxyapatite based cow bone showed the lowest (10.2 MPa). It should be noted here that the mechanical strength for composite ceramic hollow fibre membrane prepared from cow bone can be improved by addition of kaolin. This is according to the highest percentage of hydroxyapatite content (40 wt%) in the ceramic membrane matrix. Herein, the phenomenon occur as the brittleness properties of hydroxyapatite are the reason of this situation. In addition, this situation can be proven as composition C (20 wt% HAp and 20 wt% kaolin), showed high mechanical strength (13.33 MPa) than composition A. Since composition C contain kaolin (20 wt%) which functioning in increasing the mechanical properties of the membrane.

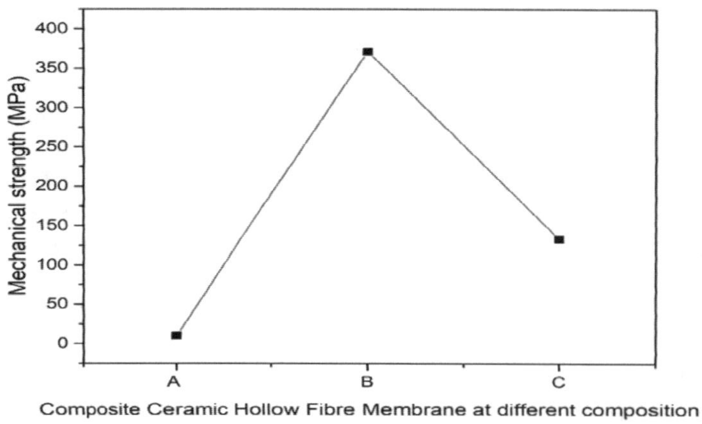


**Figure 3.** Mechanical strength of membrane prepared with different composition at the sintering temperature of 1200 °C.

In order to evaluate the performance of ceramic membrane prepared from kaolin, cow bone and composite cow bone membrane, permeate flux was conducted and presented in Figure 4. As can be see, the permeate flux for all membrane is relatively higher with value more than 8000 L/m$^2$h. In this study, the composite ceramic hollow fibre membrane prepared from kaolin and cow bone induced the highest permeate flux with value 10168.5 L/m$^2$h. This situation can be explained due to the addition of hydroxyapatite because of the rising hydrophilicity of the membrane. In addition, previous work also found that hydroxyapatite can encourage increase productivity in term of high flux [18]. It is also well known that hydroxyapatite has great potential sorption ability [19]. Interestingly, kaolin also induce hydrophilicity which can improve the permeate flux of the composite membrane. Therefore, the highest permeate flux obtained by the composite ceramic hollow fibre membrane are due to the superhydrophilicity behaviour of combining hydrophilicity properties of both materials [20].

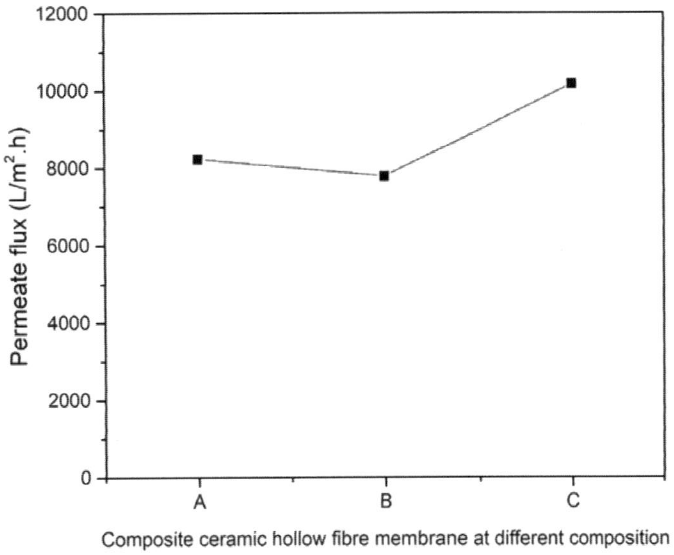


**Figure 4.** Permeate flux of membrane prepared with different composition at the sintering temperature of 1200 °C.

Nowadays, many researchers are focusing on the development of low-cost ceramic membrane from cheaper and more usable materials. Ceramic membranes have been developed from low-cost materials such as clays, animal bones waste, agricultural waste and industrial waste. There has been comprehensive research into the use of the alternative materials. Table 2 shows the comparison of ceramic membrane based from various alternative materials in this study with others previous work reported. As example for agricultural waste, Jamalludin et al. [16] successfully developed hollow fibre membrane green ceramic that made from sugarcane bagasse waste for water and oil separation due to bagasse make a good adsorbent for membrane. Hubadillah et al. [13] latest work has also successfully produced ceramic membrane in hollow fibre configuration using cow bone waste via phase inversion and sintering methods. The membrane with 60 wt% of cow bone waste hydroxyapatite resulted in highest mechanical strength (2020.5 MPa), stable flux (88.3 L/m$^2$h) and a good removal performance.

**Table 2.** Comparison of ceramic membrane based from various alternative materials.

| Membrane | Ceramic content (wt%) | Bore fluid flowrate (mL/min) | Sintering temperature (°C) | Trans membrane pressure (bar) | Flux | References |
|---|---|---|---|---|---|---|
| Kaolin | 35 | 10 | 1200 | 2 | 320 L/m$^2$h | [21] |
| Ball clay | 37.5 - 50 | 10 | 1250 | 1 | 1286 L/m$^2$h | [22] |
| Aluminum dross | 50 | 10 | 1275 | 1 | 200 mg/L | [15] |
| Sugarcane bagasse | 60 | 10 | 1000 | 2 | 466.2 L/m$^2$h | [16] |
| Cow bone waste | 69 | 10 | 1200 | 2 | 88.3 L/m$^2$h | [13] |
| Composite kaolin and cow bone | 20 | 10 | 1200 | 1 | 10168.5 L/m$^2$h | This study |

## 4. Conclusions

For fabrication of membrane, the ceramic hollow fibre membrane derived from hydroxyapatite based cow bone and kaolin were successfully prepared in this research. The membranes were characterized by using SEM, mechanical strength, imageJ software and performance by water flux. The membrane with composition C (20 wt% HAp, 20 wt% kaolin) have the mechanical strength (13.33 MPa) and it showed the highest flux with (10168.5 L/m$^2$.h) so the membrane showed its potential to be used in waste water treatment.

## 5. Acknowledgement

Financial support by Ministry of Higher Education with Research Grants Scheme vot No 9003-00759 and Collaborative Research Grant vot No 9023-00012 is gratefully acknowledged.

## References

[1] N.W. Jern, Industrial wastewater treatment, 2006. https://doi.org/No. 628.30950 N5.
[2] A. Zirehpour, A. Rahimpour, Membranes for Wastewater Treatment, Nanostructured Polym. Membr. 2 (2016) 159–207. https://doi.org/10.1002/9781118831823.ch4.
[3] V. Kumar, P.P. Sandhu, V. Ahluwalia, B.B. Mishra, S.K. Yadav, Improved upstream processing for detoxification and recovery of xylitol produced from corncob, Bioresour. Technol. 291 (2019) 121931. https://doi.org/10.1016/j.biortech.2019.121931.
[4] T. Peters, Membrane technology for water treatment, Chem. Eng. Technol. 33 (2010) 1233–1240. https://doi.org/10.1002/ceat.201000139.
[5] N.N. Li, A.G. Fane, W.S.W. Ho, T. Matsuura, Advanced Membrane Technology and Applications, 2008. https://doi.org/10.1002/9780470276280.
[6] C. de Morais Coutinho, M.C. Chiu, R.C. Basso, A.P.B. Ribeiro, L.A.G. Gonçalves, L.A. Viotto, State of art of the application of membrane technology to vegetable oils: A review, Food Res. Int. 42 (2009) 536–550. https://doi.org/10.1016/j.foodres.2009.02.010.
[7] M.F. Shohur, Z. Harun, M.Z. Yunos, S. Hasan, M.R. Jamalludin, The effect of sodium chloride (NaCl) coagulant medium of polysulfone ultrafiltration membrane, Appl. Mech. Mater. 372 (2013) 3–7. https://doi.org/10.4028/www.scientific.net/AMM.372.3.
[8] U. Iriarte-Velasco, J.L. Ayastuy, Z. Boukha, R. Bravo, M.Á. Gutierrez-Ortiz, Transition metals supported on bone-derived hydroxyapatite as potential catalysts for the Water-Gas Shift reaction, Renew. Energy. 115 (2018) 641–648. https://doi.org/10.1016/j.renene.2017.08.086.
[9] S. Choi, Y. Jeong, The removal of heavy metals in aqueous solution by

hydroxyapatite/cellulose composite, Fibers Polym. 9 (2008) 267–270. https://doi.org/10.1007/s12221-008-0042-0.
[10] L.H. Fu, Y.J. Liu, M.G. Ma, X.M. Zhang, Z.M. Xue, J.F. Zhu, Microwave-assisted hydrothermal synthesis of cellulose/hydroxyapatite nanocomposites, Polymers (Basel). 8 (2016) 1–13. https://doi.org/10.3390/polym8090316.
[11] S. Bousbih, E. Errais, F. Darragi, J. Duplay, M. Trabelsi-Ayadi, M.O. Daramola, R. Ben Amar, Treatment of textile wastewater using monolayered ultrafiltation ceramic membrane fabricated from natural kaolin clay, Environ. Technol. (United Kingdom). 0 (2020) 1–25. https://doi.org/10.1080/09593330.2020.1729242.
[12] S.K. Hubadillah, M.H.D. Othman, T. Matsuura, A.F. Ismail, M.A. Rahman, Z. Harun, J. Jaafar, M. Nomura, Fabrications and applications of low cost ceramic membrane from kaolin: A comprehensive review, Ceram. Int. 44 (2018) 4538–4560. https://doi.org/10.1016/j.ceramint.2017.12.215.
[13] S.K. Hubadillah, M.H.D. Othman, Z.S. Tai, M.R. Jamalludin, N.K. Yusuf, A. Ahmad, M.A. Rahman, J. Jaafar, S.H.S.A. Kadir, Z. Harun, Novel hydroxyapatite-based bio-ceramic hollow fiber membrane derived from waste cow bone for textile wastewater treatment, Chem. Eng. J. 379 (2020). https://doi.org/10.1016/j.cej.2019.122396.
[14] Z. Harun, M.R. Jamalludin, H. Basri, M.F. Shohur, N. Rosman, M.Z. Yunos, The effect of synthetic silica on ultrafiltration PSf membrane, J. Teknol. (Sciences Eng. 65 (2013) 121–125. https://doi.org/10.11113/jt.v65.2342.
[15] M.H.A. Aziz, M.H.D. Othman, N.A. Hashim, M.A. Rahman, J. Jaafar, S.K. Hubadillah, Z.S. Tai, Pretreated aluminium dross waste as a source of inexpensive alumina-spinel composite ceramic hollow fibre membrane for pretreatment of oily saline produced water, Ceram. Int. 45 (2019) 2069–2078. https://doi.org/10.1016/j.ceramint.2018.10.110.
[16] M.R. Jamalludin, Z. Harun, M.H.D. Othman, S.K. Hubadillah, M.Z. Yunos, A.F. Ismail, Morphology and property study of green ceramic hollow fiber membrane derived from waste sugarcane bagasse ash (WSBA), Ceram. Int. 44 (2018) 18450–18461. https://doi.org/10.1016/j.ceramint.2018.07.063.
[17] S.M. Lai, H.Y. Lai, M.Y. Chou, A facile approach for the tunable wormlike or ordered pore morphology of mesoporous silica: Effect of catalyst types and polyethylene glycol, Microporous Mesoporous Mater. 196 (2014) 31–40. https://doi.org/10.1016/j.micromeso.2014.04.043.
[18] Z.A. Rahman, N. Ali, S. Hamzah, S.M.S.T. Chik, Integration of polyethersulfone and hydroxyapatite nanofiltration membrane for vanillic acid separation, J. Teknol. 79 (2017) 97–102. https://doi.org/10.11113/jt.v79.10443.
[19] S. Zhang, C. Liu, Z. Luan, X. Peng, H. Ren, J. Wang, Arsenate removal from aqueous solutions using modified red mud, J. Hazard. Mater. 152 (2008) 486–492. https://doi.org/10.1016/j.jhazmat.2007.07.031.
[20] J. Sun, L. Wu, Polyether sulfone/hydroxyapatite mixed matrix membranes for protein purification, Appl. Surf. Sci. 308 (2014) 155–160. https://doi.org/10.1016/j.apsusc.2014.04.123.
[21] S.K. Hubadillah, M.H.D. Othman, M.A. Rahman, A.F. Ismail, J. Jaafar, Preparation and characterization of inexpensive kaolin hollow fibre membrane (KHFM) prepared using phase inversion/sintering technique for the efficient separation of real oily wastewater, Arab. J. Chem. 13 (2020) 2349–2367. https://doi.org/10.1016/j.arabjc.2018.04.018.
[22] M.H. Abd Aziz, M.H.D. Othman, N.A. Hashim, M.R. Adam, A. Mustafa, Fabrication and characterization of mullite ceramic hollow fiber membrane from natural occurring ball clay, Appl. Clay Sci. 177 (2019) 51–62. https://doi.org/10.1016/j.clay.2019.05.003.

# Activated carbon and halloysite nanotubes membrane for $CO_2$ and $CH_4$ separation

**A Gunasakaran[1], J Jafa[1], S Saalah[1], C S Sipaut[1], N Yusof[2], F Aziz[2], A F Ismail[2], M R Bilad[3], N Y Yahya[4] and N M Ismail[1,*]**

[1]Faculty of Engineering, Universiti Malaysia Sabah, Jln UMS 88400, Kota Kinabalu, Sabah, Malaysia
[2]Advanced Membrane Technology Research Center (AMTEC), Universiti Teknologi Malaysia, 81310 UTM Johor Bahru, Johor, Malaysia
[3]Chemical Engineering Department, Universiti Teknologi PETRONAS, 32610, Bandar Seri Iskandar, Perak, Malaysia
[4]Universiti Malaysia Pahang, Faculty of Engineering Technology, Lebuhraya Tun Razak, 26300 Gambang Kuantan, Pahang, Malaysia

*Corresponding author: maizura@ums.edu.my

**Abstract**. In this study, mixed matrix membranes (MMMs) were prepared where activated carbon and halloysite nanotubes with a loading of 1 wt% is incorporated into the polysulfone membrane, and the performance of each membrane was investigated. The morphological properties, mechanical strength and their correlations with the gas separation performance for CO2 and CH4 for halloysite-mixed matrix membrane (MMMs-HNT) and activated carbon-mixed matrix membrane (MMMs-AC) was studied by using Scanning Electron Microscopy (SEM-EDX), tensile test and gas permeation test. From the characterization of MMMs, SEM shows an increase of 30.77% on the thickness of the dense layer of MMMs-HNTs compared to neat membrane and MMMs-AC. The EDX results also showed that HNTs evenly distributed in the polymeric matrix without any sign of agglomeration. Elongation at the break for MMMs-HNTs also decreases to 11.38%. The gas separation performance for MMMs-HNTs increased by about 55.43% compared to MMMs-AC at 2 bar. Furthermore, MMMs-HNTs also showed in increase in the selectivity of membrane towards $CO_2$ and $CH_4$ from 0.82 to 15.83 at 2 bar. In conclusion, the addition of 1 wt% of HNTs into PSF polymeric matrix showed a better permeance of $CO_2$ and a greater selectivity compared to the neat membrane and the MMMs-AC and thus is the optimum inorganic filler for the mixed matrix membrane.

## 1. Introduction

Carbon dioxide in the atmosphere increases the greenhouse effect, which then causes global warming. Eventually, it is the source of the temperature rise, that having an effect on the world's climate. Carbon dioxide, with its most considerable involvement in climate change which causes about 60% of the heat trapped in the atmosphere [1]. Natural gas is also known as the prince of hydrocarbon, and it is kind of petroleum which often occurs in association with fossil oil. The main prevailing element in natural gas is methane, with a composition of about 70-90%, but it still contains some most substantial and lighter hydrocarbons and also some impurities like carbon dioxide.

One of the world's energy supply is natural gas and also known as the source of hydrocarbon for petrochemical feedstocks. Natural gas mostly considered clean compared to other fossil fuels, but natural gas not necessarily clean and free from contaminants found in a reservoir. Carbon dioxide should be removed to overcome problems related to process bottlenecks and to meet its standard quality specification as consumer fuel. Special priority on the removal of carbon dioxide from natural gas because it is the leading supplier for GHG emissions, a major contaminant in natural gas, and one of the challenging gas separation problems. The presence of carbon dioxide in natural gas reduces the energy content in natural gas, affects the transportation of natural gas in the pipeline to a very far distance, increases the acidity of the stream, and causes corrosion in the pipeline. Current technologies used for purification of natural gas are absorption, cryogenic separation, and adsorption. Another emerging conventional technology is membrane-based technology.

Membrane technology could become a piece of robust equipment for gas separation because it has small footprints, energy-efficient, do not require phase change, reduce the potential use of hazardous chemicals, smaller than other conventional systems, easy scale-up, flexible and lack of mechanical complexity. Polymeric membrane dominates in gas separation market due to its ability to be used in many applications and low cost [2].

Examples of polymeric membranes used in gas separation are polyamide, polytetrafluoroethylene, cellulose acetate, polysulfones, and polyethersulfones. Polysulfones (PSf) are preferable in the gas separation industry due to their toughness, stability at high temperature, high critical pressure of plasticization, higher selectivity, and good permeability in gas separation. Polysulfones are a family of high-performance thermoplastics. It is composed of a sulfonyl group, aromatic group, and ether groups. The ether group allows free rotation of the polymer chain. This improves mechanical properties and reduced the melting point. The biphenylsulfonyl group helps the polymer to be resistant towards heat, oxidation resistant and has high stiffness at elevated temperature. Despite that, due to some hindrance faced by polymeric membrane it is desirable to synthesis a more durable membrane with higher selectivity and permeability towards carbon dioxide.

A new approach so-called mixed matrix membranes (MMMs), is considered to be most practical in the future application. MMMs is a cost-effective heterogeneous membrane which involves a combination of homogenously interpenetrating inorganic particle for high selectivity and permeability, hence the polymer matrix eases the processability above the Robeson Upper-bound limit [3]. The fillers in MMMs such as carbon molecular sieve and zeolite with the molecular sieving effect can discriminate different molecules that present in feed mixture gas based on size and shape [4][5]. The polymeric membrane is mostly used in the purification of natural gas. The reason is due to their low cost, high processability, and excellent intrinsic transport properties. Due to some limitations such as suffer from a trade-off between selectivity and permeability of membrane in Robeson upper bound, that eventually constrains the upper limits of membrane separation performances [6][7]. Thus, the problem overcome by introducing MMMs that consists of a dispersion of inorganic filler within continuous polymer matrix while retaining the benefits of the polymeric membrane [8][9].

In MMMs, the porous layer only provides mechanical support, and the skin layer involves the gas separation of the membrane. The fillers improve the gas selectivity towards carbon dioxide by introducing molecular sieving effect, changing the free volume of the polymer matrix and chain rigidity of polymers [10][11][12][13]. Thus, the presence of fillers in MMMs helps in transporting the carbon dioxide gas, which then increases the performance of gas separation due to the filler's superior textural properties.

In this study, an effort was made to examine the impact of activated carbon (AC) and halloysite nanotubes (HNTs) incorporation into PSF polymer matrix on the functional group, morphological properties, tensile test, and gas separation performances in MMMs. Although research works on activated carbon and halloysite nanotubes in MMMs has been done, yet there is still a lack of studies on the comparative performances by using HNTs and AC in PSF polymeric matrix for $CO_2/CH_4$ gas separation. The main goal of this work is to fabricate an asymmetric membrane with a loading of 1 wt% of halloysite nanotubes and activated carbon (MMMs-HNTs and MMMs-AC) for $CO_2$ and $CH_4$ separation.

## 2. Methodology

### *2.1. Polymer*

Polysulfone (PSf) was used to form a continuous matrix phase in mixed matrix membrane formation. PSf was chosen for membrane fabrication because it is durable, rigid, has excellent strength and thermal properties. PSf was purchased from Sigma Aldrich that had a weight-averaged molecular weight of 35,000. The melt index of about 6.5 g/10 min. Polymer pellets were dried in the oven at 60 °C for 24 hours before the dope preparation.

### *2.2. Solvents*

N-methylpyrrolidone (NMP) [($C_5H_9NO$), 99.13 g/mol] and methanol [($C_3H_3OH$), 32.04 g/mol] were purchased from Merck, Germany. Methanol was used to remove excess solvent. Polydimethylsiloxane (PDMS) was used for the membrane coating that was supplied by Dow Corning. NMP was used due to its low toxicity. All the chemicals stated above were used without any further purification.

### *2.3. Inorganic fillers*

Inorganic fillers that were used in this study were halloysite nanotube and activated carbon. All the chemical and physical properties of the fillers were listed. Halloysite nanotubes (HNTs) was purchased from Sigma-Aldrich, Germany. The chemical composition of HNTs was $Al_2(OH)_4Si_2O_5.2H_2O$ with the molecular weight of 294.19 g/mol. The physical properties of HNTs AC are shown in Table 1. Activated carbon was a highly porous molecular sieve material, which was purchased from Sigma-Aldrich, Germany.

**Table 1.** Physical Properties of HNTs and AC

| | Values | |
|---|---|---|
| Physical properties | HNTs | AC |
| Outer diameter | 50 nm | - |
| Inner lumen diameter | 15 nm | - |
| Pore diameter | - | 2-50 nm |
| Length of tube | 1-3 µm | - |
| Surface area (S/A) | 64 $m^2/g$ | 960 $m^2/g$ |
| Pore volume | 1.26-1.34 mL/g | 1.56 mL/g |
| Refractive Index | 1.54 | - |
| Specific gravity | 2.53 $g/cm^3$ | 2.05 $g/cm^3$ |

### *2.4. Preparation of membrane solution*

In this study, three types of membrane solutions were prepared. The three membranes solution consists of a neat membrane (Neat) solution, activated carbon membrane (MMMs-AC) solution, and halloysite nanotubes membrane (MMMs-HNTs) solution.

### *2.5 Preparation of Polysulfone homogenous membrane*

The preparation of this membrane does not contain any filler (neat membrane) and acts as a control membrane. First, the 50 g of the dried polymer was added slowly into the laboratory bottle containing 150 g of NMP solvent. Then, the mixture was stirred by using the overhead stirrer at 215 rpm until a clear and viscous solution was obtained. Once a homogenous solution formed, the mixture was left in a water bath stirrer at 25 °C for 125 rpm around 40 minutes. The dope solution was left for one day in order to remove any air bubbles that formed during mixing. Due to the bubbles in the casting solution, it can cause the formation of holes during membrane formation.

*2.5.1. Preparation of mixed matrix membrane solution.* The preparation of mixed matrix membranes was quite similar as for the preparation of neat membrane; however, several additional methods were added. In this study, mixed matrix membranes were prepared with 1 wt% loading of activated carbon and halloysite nanotubes. The composition of each material for the fabrication of MMM is tabulated in Table 2. For comparison purpose, 1 wt% is the ideal loading and also to prevent agglomeration in the membrane.

**Table 2.** Composition of materials for fabrication of membrane

| Filler / Loading | (wt%) | Psf (wt%) | NMP (%) |
|---|---|---|---|
| AC | 1 | 25 | 75 |
| HNTs | 1 | 25 | 75 |

The MMMs solution was prepared by adding 0.51 g of activated carbon into a laboratory bottle containing 150 g of NMP solvent. Then the solution was mixed using a magnetic stirrer for 1 hour at 300 rpm to avoid aggregation in AC and enhance the homogeneity of the AC dispersion in the solution. After stirring, the polymer was introduced into the casting solution. Add 50 g of PSF polymer into the casting solution and was stirred using overhead stirrer at 215 rpm until a clear, viscous, and stable solution was obtained. Once a homogenous solution formed, the mixture was left in a water bath stirrer for 25 °C at 125 rpm for 40 minutes. The casting solution was kept in the bottle for one day in order to remove any air bubbles that formed during mixing. The same procedure repeated for the formation of MMMs-HNTs.

*2.5.2 Fabrication of membrane.* A smooth, dry, flat, and a dust-free rectangular glass plate was used as a casting template. Acetone was used to clean and remove any dust that present on the glass surface and glass rode. Both the sides of the glass plate were tapped by using cellophane tape. Casting or dope solution was then poured across the glass plate edge. As the glass rode was pulled towards the user, a uniform wet film of the membrane was produced across the glass plate surface. Time taken for casting each membrane was recorded. Without delay, the glass plate was transferred into the water bath. Next, the membrane was peeled off from the glass surface and immediately transferred into another basin containing a water bath hence left overnight in order to remove the solvent on the membrane. After one day, the membrane was submerged into methanol for 4 hours and then dried for three days. This method was repeated for the fabrication of MMMs-AC and MMMs-HNTs.

*2.6 Characterisation of neat and mixed matrix membrane*
Neat and mixed matrix membrane was characterized by using Fourier-Transform Infrared Spectroscopy (FTIR). FTIR was used to determine the functional group present in the three samples of the membrane. In the present work, all the infrared spectrums and functional group were recorded using Fourier Transform Infrared Radiation (FTIR), (Perkin Elmer Spectrum 100). Spectrums were obtained in the range of between 4000 $cm^{-1}$ and 400 $cm^{-1}$.

The fabricated neat and mixed matrix membranes were characterized by using Scanning electron microscopy (SEM-EDX), and tensile test. By using SEM (S-3400N, Hitachi), the morphology for all the three samples was determined. The membrane was cut into small pieces by using a knife in a dimension of 5mm X 5mm. The cross-section of the membrane surface observed under the magnification of 420x. The surface of the membrane examined under magnification of 500x, 1500x, and 10kx. For the characterization of the membrane, it was initially fractured in liquid nitrogen to obtain a clear cut of a cross-section. The purpose of immersing those membranes in liquid nitrogen was to make membrane-less brittle. The sample was then mounted on stab by using carbon tape, and then sputter-coated with gold/platinum before analyzing it by using SEM. All the elemental composition in the three samples was analyzed using elemental dispersion composition (EDX). EDX method plots the abundance of an element alongside a line. This method was conducted using SEM-EDX (S-3400N, Hitachi).

This tensile test widely used and one of the best ways to measure the mechanical properties. A universal testing machine (UTM) was used to measure mechanical strength. The membrane was clamped between grips and pulled to the opposite direction with a constant grip separation rate. From there, elongation at break and mechanical strength was measured. The sample with an area of 25 × 50 $mm^2$ was tested with a crosshead speed of 5 mm/min. The test was carried out by using a universal testing machine (Shimadzu Trapezium lite x version 1.1.2) at room temperature, and a minimum of four samples was tested.

*2.7 Gas Performance Analysis*
For gas performance analysis, the membrane samples were coated by using a PDMS solution. Once it was coated, the membranes were analysed by using $CO_2$ and $CH_4$ gas.

*2.7.1 Post-treatment before gas permeation test.* Each of the membrane samples was cut into a small circle with an outer diameter of 5.8 cm. To reduce the generation of voids and pinholes formation on the fabricated membranes, they were immersed in the coating solution consists of 3 wt% polydimethylsiloxane in n-hexane for 3-5 seconds. Then, the membranes were placed in the oven for 3 days at 60 °C in order to cure the coating. Afterwards, the coated membrane was tested for the gas separation performance of pure carbon dioxide and methane gas

*2.7.2 Gas permeation test.* The analysis test for the coated membrane was evaluated using the gas permeation unit. This test was done to study the performance of each membrane on separating the gas. Pure gas carbon dioxide ($CO_2$) and methane ($CH_4$) was used, which permeate through the membrane placed on the cell permeation test. Gas with a different feed pressure of 2, 4, 6, 8, and 10 bar was used. The test was based on the pressure on feed steam and measure the gas flux across a known area and thickness of the membrane on the permeate phase. All the test was carried out at a temperature of 25 °C. The data measurement was taken for the permeate gas [($CO_2$), ($CH_4$)] to reach a volume of 1 cm. By using a bubble flow meter, the gas for volume was measured. Thus, the pressure-normalized flux or gas permeance was calculated by using equation (1).

$$\left(\frac{P}{l}\right)_i = \left(\frac{Q_i}{(A) \times (\Delta P)}\right) \qquad (1)$$

Where $\Delta P$ is the differential pressure across the membrane (bar), A was the surface area of membrane ($cm^2$), Q was the volumetric flow rate of gas at standard temperature and pressure. The gas permeance was expressed in the unit of GPU, which stated in equation (2).

$$\text{GPU} = 1 \times 10^{-6} \left(\frac{\text{cm}^3\,(\text{STP})}{\text{cm}^2\ \text{sec cm Hg}}\right) \qquad (2)$$

The ideal selectivity of the membrane was the ratio of gas permeance, as shown in equation (3).

$$\alpha_{\mathrm{i/j}} = \frac{P_i}{P_j} \qquad (3)$$

## 3. Results and Discussion

### *3.1. Characterization of Neat membrane and mixed matrix membrane*

The addition of halloysite nanotubes and activated carbon into the polysulfone membrane are further characterized by using FTIR. The objective of this characterization is to validate the presence of the HNTs and AC. The membranes were also characterized by using SEM and EDX to study the morphology structure of membranes and the effect of the addition of activated carbon and halloysite nanotubes on the changes in the morphology structure of membranes and elemental dispersion in MMMs. On the other hand, tensile strength is studied to determine the mechanical strength of mixed matrix membranes after the effect of the addition of activated carbon and halloysite nanotubes on the changes on mechanical strength compared to the neat membrane.

*3.1.1 Functional Groups present in Different Membrane.* The FTIR spectrum for the neat membrane, which portraits a graph of the percentage of transmittance versus wavenumber is shown in Figure 1. The characteristics peak for this neat membrane were at a wavenumber of 2967.47 $cm^{-1}$ which corresponds to the symmetric and asymmetric stretching vibration of C-H that involves the entire methyl group [14].

On the other hand, characteristics peak of 1486.93 $cm^{-1}$ and 1584.58 $cm^{-1}$ contributes to aromatic C=C bond stretching in the PSF chain [14]. Phenyl ring substitution band can be seen at the peak of 833.73 $cm^{-1}$ meanwhile, amine stretching is noticed at 1105.35 $cm^{-1}$ respectively. The most important functional group detected, which is symmetric O=S=O stretching of sulfonate group at characteristics peak of 1149.76 $cm^{-1}$[15].

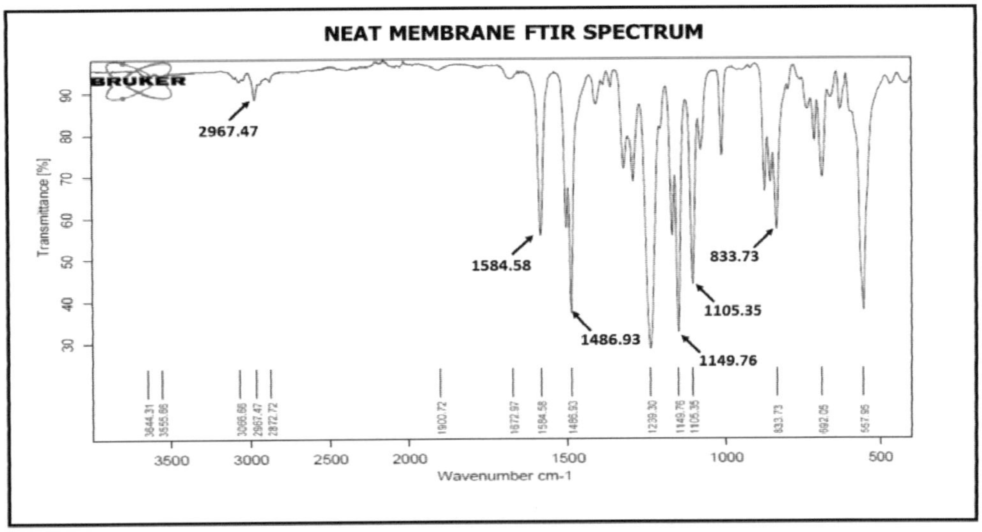


**Figure 1.** FTIR Results for Neat Membrane

Figure 2 depicts the FTIR spectrum for MMMs-AC, where activated carbon mixed with polysulfone membrane. By referring to the spectrum, it can be observed that the characteristics peak for MMMs-AC contains some peak as the neat membrane with the addition of some new functional groups from activated carbon. MMMs-AC shows a new peak at 1676.08 $cm^{-1}$ suggesting the quinones reflected in the C=O band. In characteristics peak of 1149.72 $cm^{-1}$, shows a stretching vibration of the C-O group in ester, ether, phenol and alcohol [16][17]. Meanwhile, a band at 1409.20 $cm^{-1}$ contributes to oxygen-containing functional group, which are C-O and C=O stretching vibration in the carboxylic group [18].

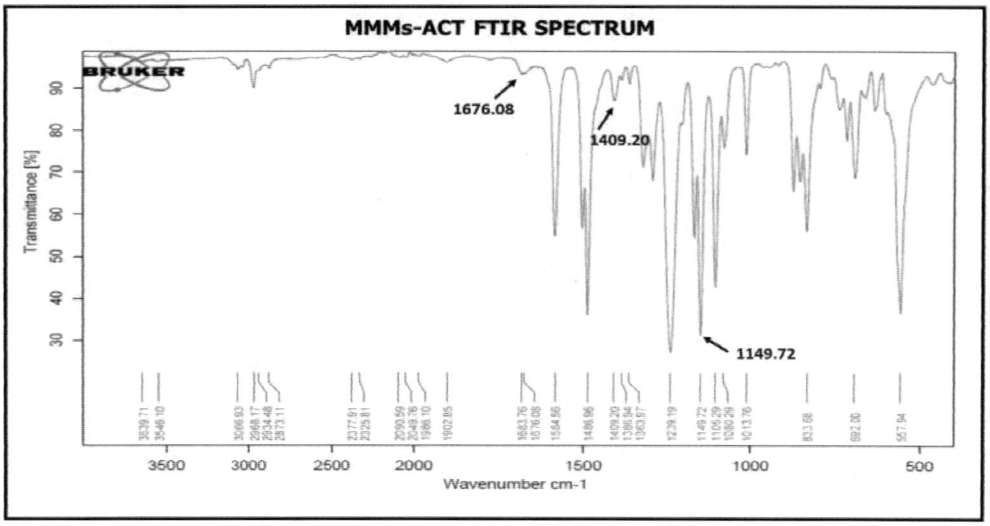


**Figure 2.** FTIR Results for MMMs-AC

Figure 3 shows the FTIR spectrum for MMMs-HNTs, where halloysite nanotubes mixed with the polysulfone membrane. From Figure 3, it can be observed that MMMs-HNTs also consist of similar functional group as the neat membrane, with a few additions of a new functional group from elements in HNTs. In MMMs-HNTs, a peak at 3616.31 $cm^{-1}$ shows a stretching vibration of Al-OH groups in the inner surface of HNTs [19]. The MMMs-HNTs spectrum is further characterized by stretching vibration of Si-O-Si group which the functional group comprises absorption peak at 1005.28 $cm^{-1}$ [20]. Therefore, this FTIR analysis results confirm the presence of elements of activated carbon and halloysite nanotubes in the mixed matrix membrane.

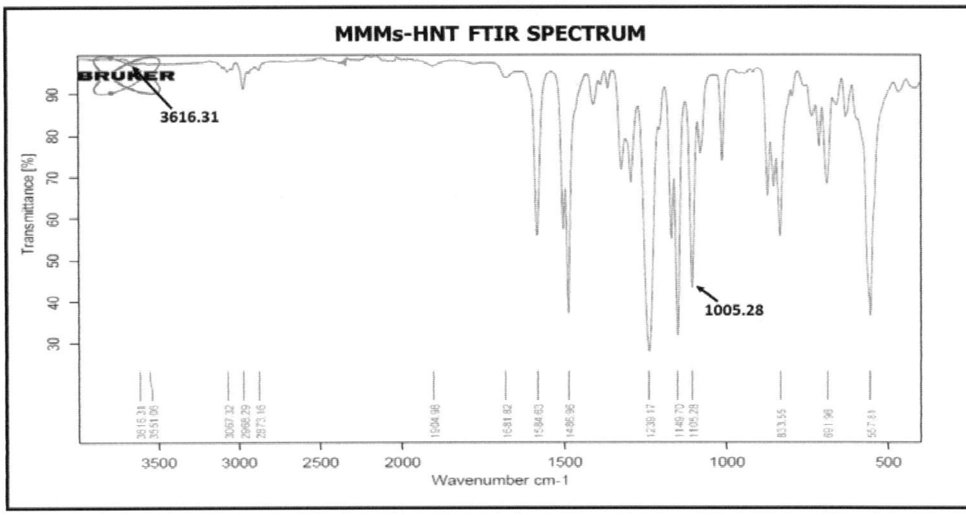


**Figure 3.** FTIR Results for MMMs-HNTs

*3.1.2 Membrane morphology by using SEM.* Membrane morphology of neat membrane, MMMs-AC, MMMs-HNTs is studied by using SEM. The effect of the addition of porous fillers (activated carbon, halloysite nanotubes) on the morphological properties is observed from the thickness of the skin layer to the development of micro-voids. Surface and cross-sectional of SEM micrographs of the different fabricated membranes are presented in Figure 4 and Figure 5.

The filler particles are homogeneously dispersed on the surface of all the membranes, as shown in Figure 4. Since the loading of fillers used was 1 wt% only, no severe agglomeration is visible (Figure 4). Figure 4(c) shows that at 1 wt% loading of halloysite nanotubes, a good dispersion of filler was observed on the membrane surface and smooth surface was formed due to the excellent compatibility of fillers with polymer matrix. The presence of halloysite nanotubes particles on membrane shown in Figure 4(c) confirmed by FTIR results for MMMs-HNTs as depicted in Figure 3. Since there are functional groups found on their exterior surface of halloysite nanotubes, this promotes better dispersion in the polymer matrix [21]. Meanwhile, in Figure 4(b), few formations of pinholes were visible, which may contribute to loss in the selectivity of gases.

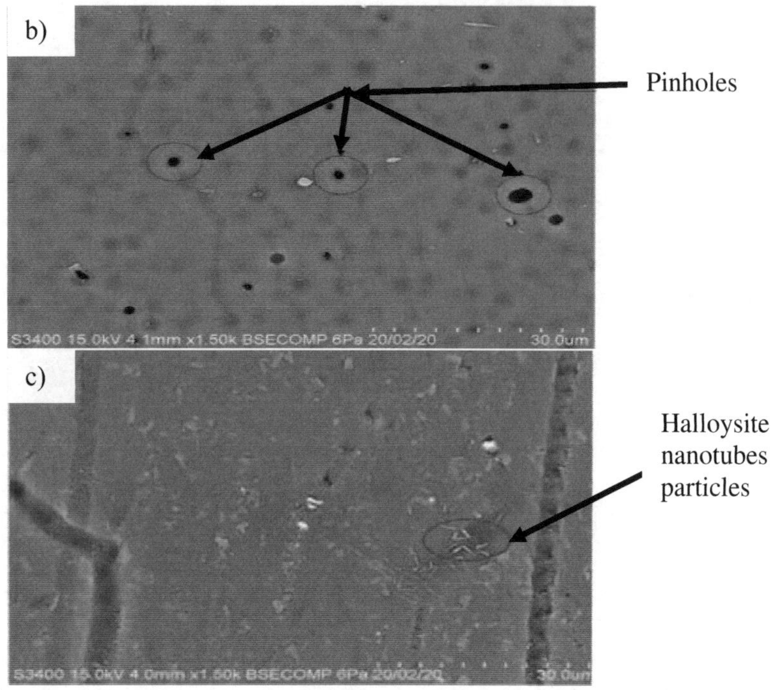


**Figure 4.** SEM images for surface area of membranes at (Magnification ×1.50k) (a) Neat membrane, (b) MMMs-AC, (c) MMMs-HNTs

In Figure 5, the typical asymmetric membrane consists of two distinct layers; which are active skin layer and porous structure. Ideally, there are two sub-structure in the porous layer of a membrane, which are spongy type and finger-like sub-structure. Hence, the sub-structure should serve as good mechanical support for the skin layer. Previous research by Rezaee et al. (2015) proves that with the addition of fillers, a morphological structure of finger-shaped pores tapering towards the skin and oblong shape image also visible throughout the cross-section can be obtained [22].

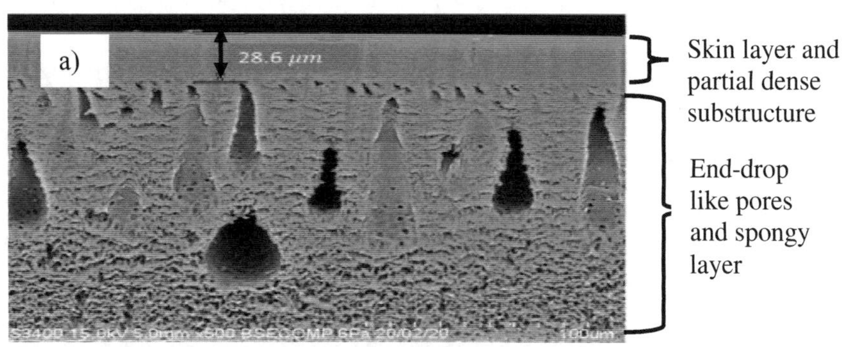

 doi:10.1088/1757-899X/1142/1/012012 

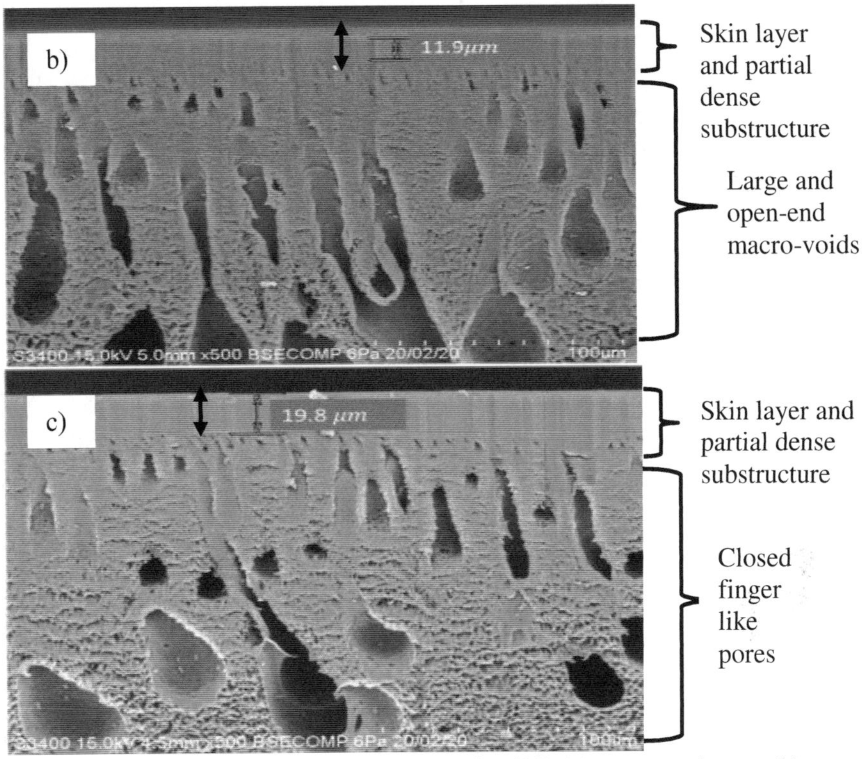


**Figure 5.** SEM cross-sectional images at (Magnification of ×500) (a) neat membrane, (b) MMMs-AC (c) MMMs-HNTs

In Figure 5(a), the prepared neat PSf membrane shows two distinctive morphologies, namely few separated closed-end drop-like pores and sponge structure exist as a significant part of the membrane at the bottom of a membrane [23]. In Figure 5(c), there is no sponge layer; hence it can be seen some of the closed-end drop like pores replaced by finger-like pores with closed ends. Hence, the formation of a finger-like structure, caused by the presence of high concentrated nascent layer, which inhibits the solvent exchange rate from the top layer [24]. Hence, the finger-like pores in HNT cross-sectional area promote the gas permeances, as shown in the results of this study for the gas permeation test.

Membrane with 1 wt% of AC loading as depicted in Figure 5(b), shows that the spongy part of the membrane is diminished and the closed-end drop like pores turn into large and open-end macro-voids which expose to the bottom surface [10]. Furthermore, as can be seen for the MMMs-AC, the number of open-end macro-voids increased. Through a simultaneous exchange of solvent, when the membrane placed in the non-solvent (water), these causes the NMP molecules to merge together and forms large macro-voids [25].

The thickness of the skin layer and partial dense layer of the membrane is decreasing from neat membrane < MMMs-HNTs < MMMs-AC with 28.6 μm, 19.8 μm, and 11.9 μm (Figure 5). Hence, the thickness of MMMs-HNTs decreased by 30.8% compared to the neat membrane, while MMMs-AC decreases by about 58.4% respectively. But, the thickness of MMMs-HNTs increases by 66.39% from MMMs-AC. On the other hand, the thickness of the neat membrane is greater, and it should be another way around, and is postulated that the reason behind this phenomenon may be due to time taken for de-mixing. Moreover, the thickness of the skin layer reduces because of the faster de-mixing process between non-solvent and solvent. The thickness of the MMMs-AC skin layer decreased compared to the neat PSF membrane and MMMs-HNTs. Since activated carbon is a greater hydrophilic filler, it causes non-solvent which is water to diffuse into PSF polymer matrix and replaced by solvent easily; hence quick de-mixing occurs between non-solvent and solvent, reduces the skin layer thickness [30].

Hence, from here, can be conclude that for thicker skin layer like MMMs-HNTs, it has have greater selectivity but lower permeability and vice-versa for thinner skin layer [26].

Under circumstances where higher pressure was used, the thin skin layer could rupture due to finger-like pores formed, which act as weak points under pressure facilitating membrane failure [27]. From the results of SEM analysis, an image in Figure 5(b) consists of a larger number of finger-like pores compared to Figure 5(c), which probably can damage the membrane if higher pressure was used. From this analysis, it can be concluded that the incorporation of halloysite nanotubes into the polymer matrix gives a better morphology structure of the membrane as compared to MMMs-AC and neat membrane. Therefore, since there are some changes in the thickness of the skin layer and overall morphological properties, these may affect the gas transport properties of different membranes.

*3.1.3 Energy Dispersive X-Ray.* EDX used to identify the elements present in sample membranes in order to prove the even distribution of activated carbon and halloysite nanotubes in MMMs. In this study, elemental analysis carried out for two membranes, which are MMMs-AC, and MMMs-HNTs. Hence, the incorporation of 1 wt% of activated carbon and halloysite nanotubes further analyzed by Energy Dispersed X-ray spectrometer (EDX) analysis.

Based on Figure 6 shows the EDX mapping for the neat membrane. An intensity of carbon, oxygen, and sulphur elements with a percentage of 60.72%, 7.97%, and 10.74% shown for a neat membrane. The polysulfone element was detected in the neat membrane and also supported by previous research [28][29].

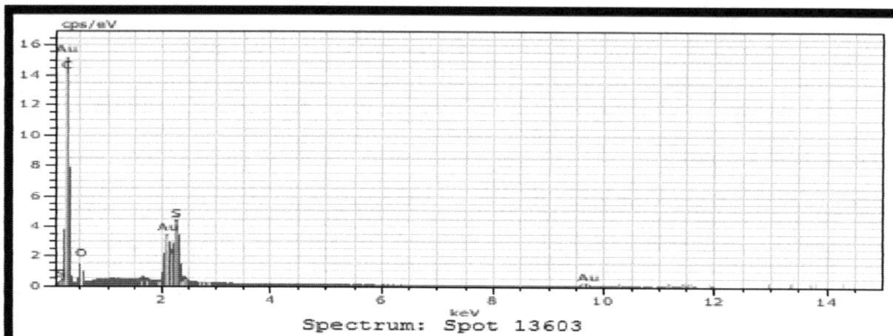


| Element | Series | unn. C [wt.%] | norm. C [wt.%] | Atom. C [at.%] | Error (1 Sigma) [wt.%] |
|---|---|---|---|---|---|
| Carbon | K-series | 37.90 | 60.72 | 84.35 | 4.49 |
| Oxygen | K-series | 4.98 | 7.97 | 8.32 | 0.79 |
| Sulfur | K-series | 6.70 | 10.74 | 5.59 | 0.27 |
| Gold | M-series | 12.84 | 20.57 | 1.74 | 0.51 |
| | Total: | 62.42 | 100.00 | 100.00 | |

**Figure 1.** Elemental spectrum and elemental composition of the neat membrane

Figure 7 displays elemental analysis for MMMs-AC. The presence of activated carbon in the polymer matrix was only verified by carbon, and oxygen with a percentage of 36.21%, and 5.17%. The results show that there is an absence of some elements of activated carbon. This phenomenon occurs may be due to agglomeration of activated carbon in the membrane.

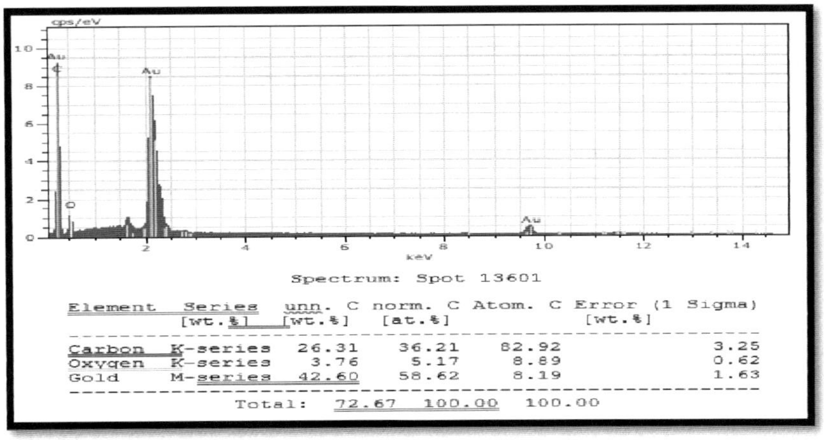


**Figure 2.** Elemental spectrum and elemental composition of MMMs-AC

Hence, in Figure 8, the EDX analysis for MMMs-HNTs shows a peak for oxygen, aluminium, and silicon with a weight percentage of 18.55%, 4.77%, and 5.66%. The present elements corresponding to the chemical structure of raw HNTs of $Al_2Si_2(OH)_4.nH_2O$ with a molar ratio of 1:1 Al to Si ratio [30][31]. Meanwhile, due to the sputtering of the gold coating during SEM application, traces of gold found in all of the three membranes.

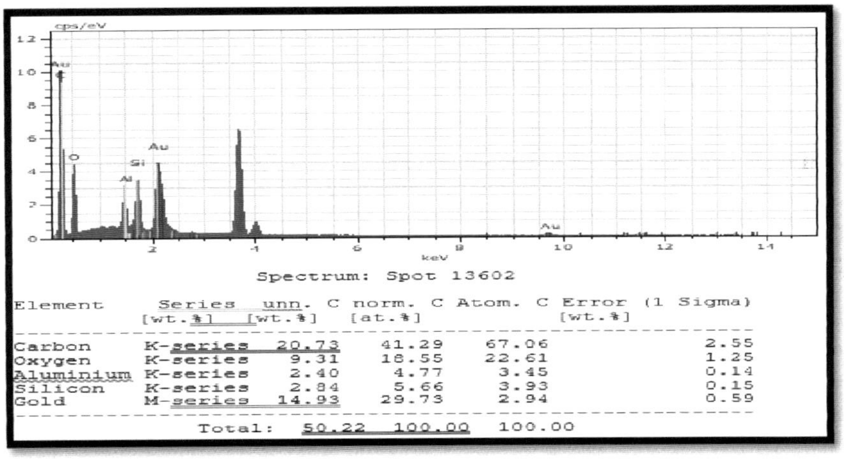


**Figure 3.** Elemental spectrum and elemental composition of MMMs-HNTs

Furthermore, in Table 3 shows the EDX elemental composition of the neat membrane, MMMs-AC, and MMMs-HNTs. It can be concluded that the elemental analysis results prove that halloysite nanotubes are scattered throughout the polymeric matrix regardless of the loadings of fillers in the membrane with uniform distribution. Thus, the EDX analysis revealed the presence of halloysite nanotubes, and no agglomeration was observed.

**Table 3.** EDX Elemental Composition of Neat membrane, MMMs-AC and MMMs-HNTs

| Elements | Composition, wt% | | |
|---|---|---|---|
| | Neat membrane | MMMs-AC | MMMs-HNTs |
| Carbon, C | 60.72 | 36.21 | 41.29 |

| | | | |
|---|---|---|---|
| Oxygen, O | 7.97 | 5.17 | 18.55 |
| Sulphur, S | 10.74 | 0.00 | 0.00 |
| Aluminium, Al | 0.00 | 0.00 | 4.77 |
| Silicon, Si | 0.00 | 0.00 | 5.66 |

*3.1.4 Mechanical strength of different membrane by using tensile test.* Mechanical properties such as tensile strength and elongation at break were analyzed for PSF neat membrane and mixed matrix membrane. Since the incorporation of fillers affects the morphology structure, consequently it will have an effect on the mechanical properties of MMMs. The fundamental reason for the addition of inorganic particles into the polymer matrix is to modify the structure of the membrane by improving the mechanical properties such as tensile strength, and elongation at break. Conversely, non-homogenous dispersion of fillers causing the formation of the defective structure of the membrane and weakens the mechanical properties of such membrane. Figure 9 shows the overall stress-strain curve using data in Table 4.

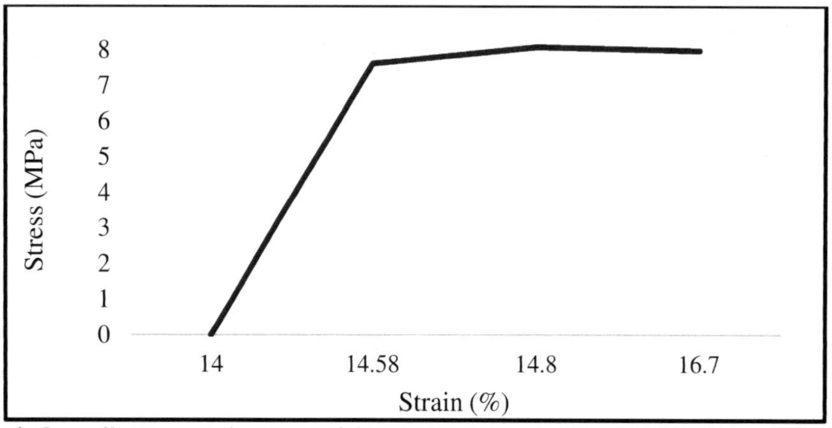

**Figure 4.** Overall stress-strain curve of the neat membrane, MMMs-HNTs and MMMs-AC

Another key factor that affects the mechanical strength is the interfacial interaction between inorganic fillers and the polymeric matrix. Figure 10 and Table 4 show the average tensile strength and the elongation at break for the neat membrane, MMMs with activated carbon and halloysite nanotubes. Among three types of membrane, MMMs-HNTs displays the highest tensile strength followed by the neat membrane and MMMs-AC of 8.09 MPa, 7.99 MPa, and 7.62 MPa respectively. Similar results by Nagla et al. (2020), where the mechanical strength for MMMs-HNTs increased at a low loading of HNTs in the polymer matrix [32]. This is most probably is due to the cylindrical tube-like structure of HNTs, which increases the surface area of interaction between the polymer matrix and HNTs. Besides that, the higher mechanical strength of MMMs-HNTs also because of the enriched hydroxyl groups found at HNTs, which improves the molecular force of attraction and hydrogen bonding between HNTs and PSF membrane [30].

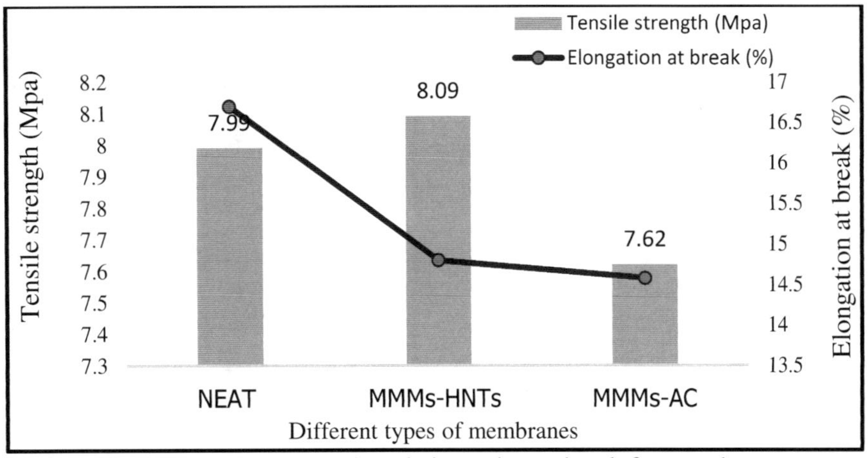


**Figure 5.** Tensile strength and elongation at break for membranes

**Table 4.** Tensile Test Results for the Membranes

| Type of membrane | Neat PSF membrane | MMMs with HNTs | MMMs with AC |
|---|---|---|---|
| Average tensile strength (MPa) | 7.99 | 8.09 | 7.62 |
| Elongation at break (%) | 16.70 | 14.8 | 14.58 |
| Elastic modulus (MPa) | 231.95 | 246.78 | 217.35 |
| Maximum load (N) | 45.06 | 46.78 | 56.88 |

Based on Setiawan et al. (2019), the elongation at break of the membrane depends on the ability of inorganic filler to restrict polymer chain movement and circumvent the phenomena of large macroscale extension. Lower elongation at break in MMMs-HNTs is due to its strong interfacial interaction between PSF polymer matrix and evenly distributed HNTs and also because of the brittleness of membrane in the presence of clay, which known as nanocomposites [4]. Moreover, Mohamed et al. (2020) also found that the elongation at break for MMMs-HNTs decreases compared to the neat membrane, which is due to the incorporation of nano-sized inorganic materials into the polymeric membrane [30].

The neat PSf membrane exhibited the largest elongation of 16.70%, which proves that the polymer chain is more flexible, where it can be stretched to a greater extent because of the weak interfacial interaction between the molecules of the polymer. Meanwhile, the mixed matrix membrane sample with 1 wt% of activated carbon content showed the second least elongation at break than MMMs-HNTs with 14.58%. MMMs-HNTs composite membrane displays significant increment in both tensile strength as well as in elongation at break with 8.09 MPa and 14.80% compared to MMMs-AC. Therefore, the mechanical strength of the membrane was much better when halloysite nanotubes are added into the PSF polymer matrix.

### *3.2 Gas Separation Performance*

The gas permeation test is carried out in order to study the gas separation performance of neat and mixed matrix membrane when $CO_2$ and $CH_4$ gas flow through it. Hence, the parameters studied are (1) effect of the addition of fillers on gas separation performance on the membrane, (2) effect of different pressure on gas separation performance on the membrane, and (3) effect of coating of the membrane using PDMS on gas separation performance on the membrane.

*3.2.1 Effect of Addition of Fillers on Gas Separation Performance on Different Types of Membranes.* Neat membrane, MMMs-AC, and MMMs-HNTs membranes are tested in order to evaluate its selectivity and permeability for the separation of $CO_2/CH_4$ gas at a fixed pressure of 2,4,6,8,10 bar which carried out at 25 °C room temperature. The permeance and selectivity values of $CO_2$ and $CH_4$ for neat and mixed matrix membrane are tabulated as a function on the effect of filler in Figure 11 and Figure 12.

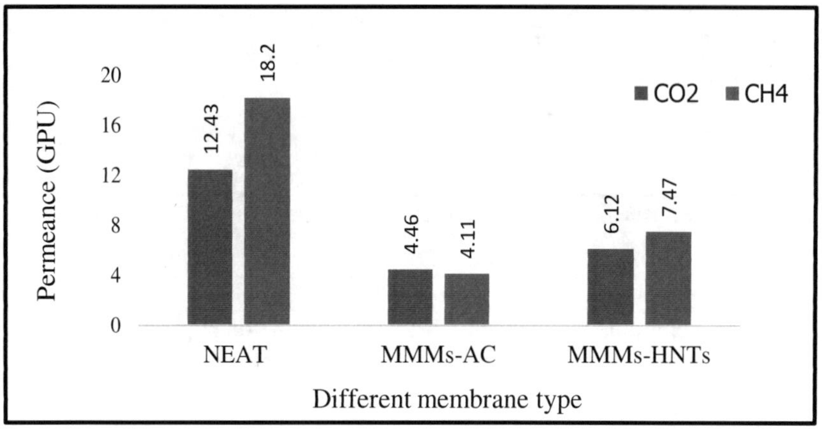


**Figure 6.** $CO_2$ and $CH_4$ permeance of different uncoated membranes at a pressure of 2 bar

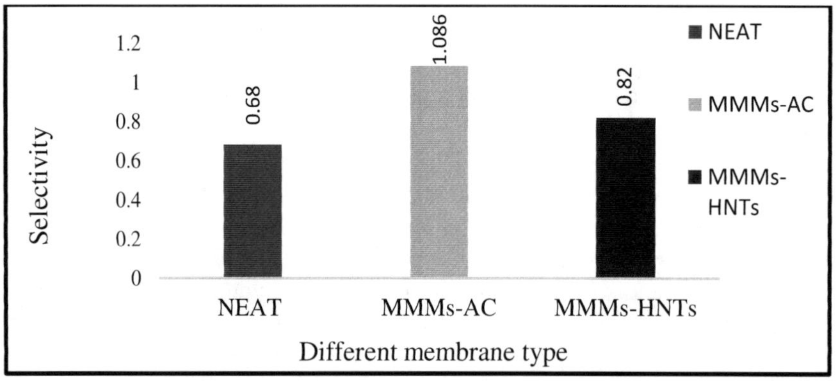


**Figure 7.** $CO_2$ and $CH_4$ selectivity of the uncoated membrane at a pressure of 2 bar

At the pressure of 2 bar, the overall results exhibit a decrease in $CO_2$ and $CH_4$ permeance from the neat membrane to MMMs with activated carbon; which is 12.43 and 18.2 to 4.46 and 4.11, showing that there is a decrease of 64.1% for $CO_2$ permeance and 77.4% for $CH_4$ permeance respectively. Nevertheless, the selectivity of the membrane increased for MMMs-AC and MMMs-HNTs compared to the neat membrane. This is because the addition of the porous fillers selectively permits the gas through the pores. Thus, the incorporation of fillers into the polymeric membrane also increases the selectivity of $CO_2$ gas relative to $CH_4$ gas.

According to Koros et al. (2003), porous inorganic fillers with specific pores usually have higher selectivity and permeability that surpasses Robeson Upper bound [33]. Consequently, when such types of fillers are added to the polymeric matrix, this improves the sieving properties, and the mixed matrix membrane tends to have higher selectivity compared to the neat membrane [34]. The fillers inside the PSf polymer matrix possess molecular sieving effect and forms a permeable channel, which allows for the better separation of gas compared to the neat membrane.

Through micro-mesoporous activated carbon in PSf polymeric matrix, more adsorbable gas ($CO_2$ transported against less adsorbable gas $CH_4$) which is by surface diffusion [35]. In the surface diffusion, the highly adsorbable gas molecule ($CO_2$) adsorbed on the membrane surface, and then move along to the permeate side on the pore surface until desorbing the gas. In this type of mechanism, the highly adsorbable gas molecules ($CO_2$) have greater permeability compared to less adsorbable gas molecules ($CH_4$) with lower permeability. This causes the selectivity to be higher compared to the neat membrane. Meanwhile, in MMMs-HNTs smaller gas ($CO_2$) transported through molecular sieving filler, whereby diffusion of larger molecules ($CH_4$) hindered by HNTs.

Although the results at 2 bar seem to be satisfactory, but the selectivity of MMMs-HNTs shows a slight decline as compared to MMMs-AC from 1.086 to 0.82. This is probably due to the formation of surface defects in MMMs-HNTs where $CH_4$ gas molecules pass through the non-selective voids, which results in poor MMMs selectivity. Thus, this section can conclude that the addition of fillers increases the selectivity of $CO_2/CH_4$ of the membrane. Hence, the results show that the MMMs with the incorporation of molecular sieving effect porous fillers, selectively separate the gases.

*3.2.2 Effect of Different Pressure on Gas Separation Performance.* The effect of different and increase of feed pressure, which is 2, 4, 6, 8 and 10 bar on gas permeability and $CO_2/CH_4$ selectivity, investigated for neat membrane and MMMs displayed in Figure 13, Figure 14, and Figure 15.

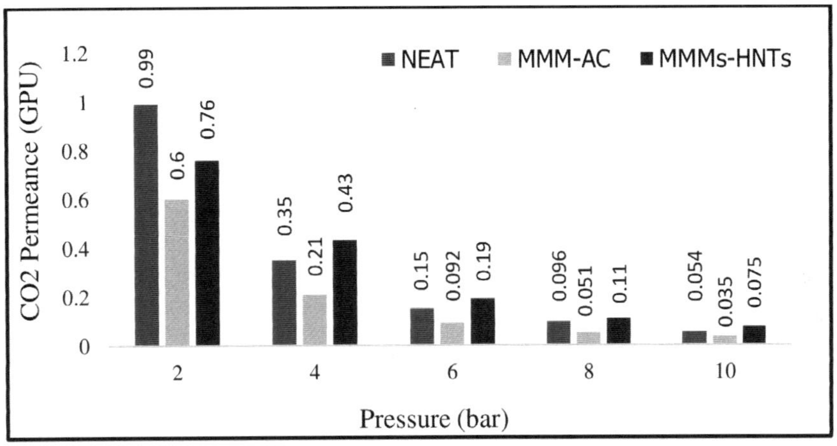


**Figure 8.** $CO_2$ permeance at different pressure

For all the MMMs in Figure 13, $CO_2$ permeance starts to decrease with increasing the feed pressure from 2 bar to 10 bar. The declining trend shows that for all fabricated MMMs, indicates there is no plasticisation occurred since no noticeable increase in $CO_2$ permeance at pressure up to 10 bar. Similar results by Shafie et al. (2019) also showed that both the $CO_2$ and $CH_4$ permeance shows a decline trend with increasing pressure that could be due to the fact of the strong interaction between polymer and fillers which induced the absorption of $CO_2$ without any formation of swelling in the membrane [36]. Hence, this proves that incorporation of HNTs and AC in the polymer phase restricts the mobility of the polymer chain and does not induced plasticization in MMMs.

Figure 14 shows the $CH_4$ permeance for pressure at 2 to 10 bar. At 2 bar, the results show a decrease of $CH_4$ permeance for MMMs-HNT of 42.16 % compared to the neat membrane at the same pressure. This proves that by adding HNTs the $CH_4$ permeance decrease, thus increasing the gas separation performance of the membrane. Moreover, this also can be explained through the structure of HNTs where the inner surface of cylindrical HNTs allows the additional transport channel for $CO_2$ gas molecules. However, HNTs does not supply any additional transport channel for $CH_4$ gas, because it is barely to be adsorbed to HNTs and also due to weak interaction between HNTs and $CH_4$ gas molecules.

Thus, this is also in agreement by the increase in separation factor, where the selectivity of MMMs-HNTs with 15.83 at 2 bar pressure compared to the neat membrane and MMMs-AC. This result also supported by SEM image, where on the MMMs-HNT, there is an absence of interfacial gaps between HNTs and polymer matrix and proper distribution of particles on the polymer matrix.

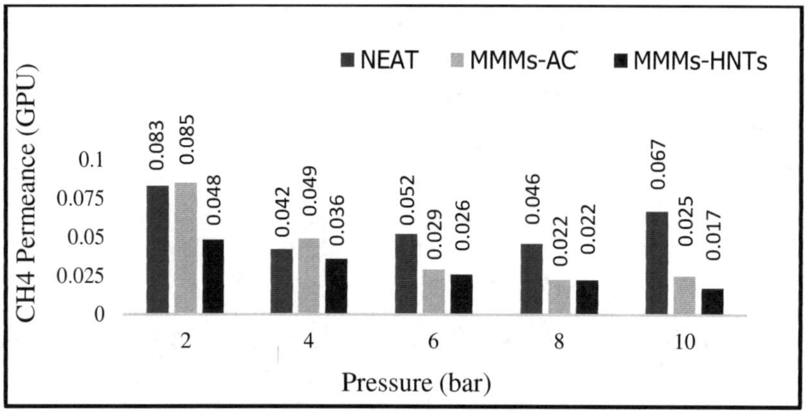


**Figure 9.** $CH_4$ permeance at different pressure

From the bar chart in Figure 15, it can be seen that the MMMs-HNTs show better selectivity towards gas as compared to MMMs-AC. At a pressure of 2 bar, the selectivity is increased to 15.83 compared to MMMs-AC with only 7.06 respectively. Hence, these results show that the incorporation of HNTs into PSF polymer matrix increases the selectivity by 10 times as compared to MMMs-AC. This is due to the fact of the unique hollow tubular structure of HNTs, which contributes to the rapid flow of gases through cylindrical of HNTs and enhance the selectivity of the membrane. In another way around, the selectivity of the neat membrane is higher, which is from 11.93 to 7.06 compared to MMMs-AC. This may be due to improper mixing of casting solution or sedimentation of activated carbon during the formation of membrane solution. Other than that, this can be supported by SEM results, which shows that MMMs-AC is not suitable to be used for gas because of a greater number of finger-like pores that probably can damage the membrane if high pressure is used.

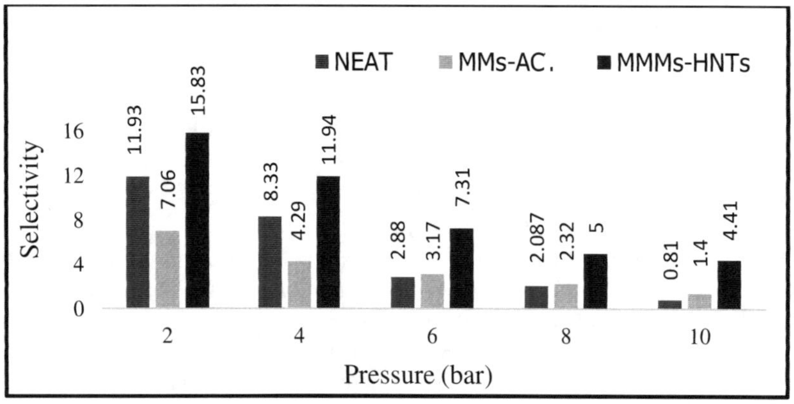


**Figure 10.** Selectivity of the membrane at different pressure

From the results shows that the selectivity is higher and decreases in permeability, it is mostly attributed due to the presence of a rigidified polymer layer around the fillers. This is probably due to the incorporation of porous fillers into the polymeric matrix. In this case, the free volume decreases; however, there is no gap formed in the interface, thus allowing stronger contact between the fillers and polymeric matrix and also restrict the movement of polymer chains [37]. Here the selectivity $CO_2/CH_4$ increases in polymer layer rigidification case because the membrane has greater resistance in transporting larger kinetic diameter of gas molecules. Thus, this analysis proves there is no plasticization occurs as pressure increases due to the stronger interaction between polymer and filler. Therefore, improvement in selectivity indicates that incorporation of halloysite nanotubes at 1 wt% is noteworthy for the $CO_2$ separation.

*3.2.3 Effect of Coating on the Performance of Different Membrane.* Usually, the permeability of $\mathbf{CO_2}$ is higher compared to $\mathbf{CH_4}$, for all the coated membrane. This is because the kinetic diameter of $\mathbf{CO_2}$ is smaller than $\mathbf{CH_4}$; however, the condensability of $\mathbf{CO_2}$ is higher than $\mathbf{CH_4}$ that exhibits higher permeability as compared to $\mathbf{CH_4}$. MMMs and the neat membrane was further modified in order to investigate the effects of PDMS coating on gas separation performance.

As depicted in Table 5, both $CO_2$ and $CH_4$ permeance for uncoated membranes is higher when compared to the modified membrane with PDMS. Thus, the difference in permeance causes both the gas penetrant to be more permeable by the formation of unselective defects in the uncoated membrane. This is the main reason why the selectivity of all uncoated membrane is lower than the coated membrane. As can be seen, the surface defects and pinholes formed on the surface layer of the membrane can be repaired through PDMS coating by forming a selective layer on the surface of the membrane and intended to reduce the blockage of gases permeation which causes a drop of both gas permeability [38].

The bar chart shows the selectivity of coated and uncoated for different membranes in Figure 16. Meanwhile, the selectivity of coated MMMs-HNTs membrane shows a tremendous improvement from 0.82 to 15.83 as compared to uncoated MMMs-HNTs. The result is also supported by previous studies done by Zulhairun et al., (2015), that shows an increment in the selectivity of gas when the MMMs are coated with PDMS [39]. The PDMS coating on membrane increases the selectivity of gas by decreasing the permeance of both $CO_2$ and $CH_4$. Moreover, a high increment in selectivity could also be related when the membrane is fully coated; there is high solubility of gases in PDMS. Although, the result for selectivity for MMMs-HNTs is acceptable but the MMMs-AC has lower selectivity compared to the neat membrane. This is maybe due to incomplete drying of the coating layer on MMMs-AC. In conclusion, it is found that the PDMS coating was able to improve the separation performances of the membrane for all the three membrane samples.

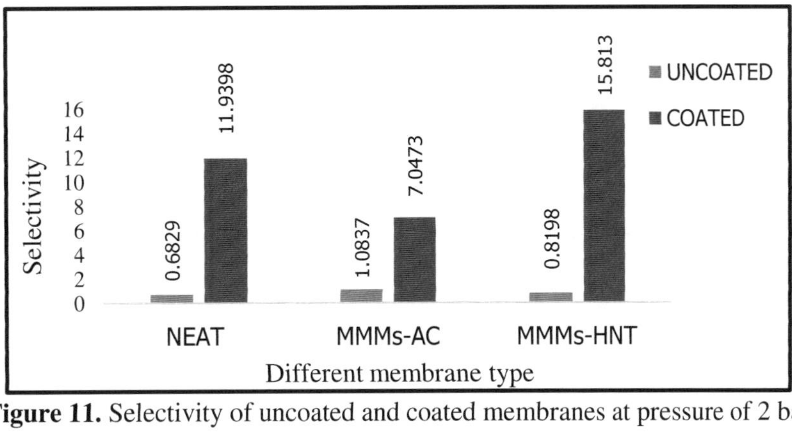


**Figure 11.** Selectivity of uncoated and coated membranes at pressure of 2 bar

**Table 5.:** Selectivity and permeance results for coated and uncoated membranes

| Membrane | Pressure | Coated membrane | | | Uncoated membrane | | |
|---|---|---|---|---|---|---|---|
| | | $P_{CO_2}$ | $P_{CH_4}$ | Selectivity, $\alpha$ ($CO_2$/ $CH_4$) | $P_{CO_2}$ | $P_{CH_4}$ | Selectivity, $\alpha$ ($CO_2$/ $CH_4$) |
| Neat | 2 | 0.99 | 0.0830 | 11.93 | 12.43 | 18.20 | 0.68 |
| MMMs-HNTs | | 0.76 | 0.048 | 15.83 | 6.12 | 7.47 | 0.82 |
| MMMs-AC | | 0.60 | 0.085 | 7.06 | 4.46 | 4.11 | 1.086 |
| Neat | 4 | 0.35 | 0.042 | 8.33 | 6.98 | 10.29 | 0.68 |
| MMMs-HNTs | | 0.43 | 0.036 | 11.94 | 3.05 | 4.71 | 0.65 |
| MMMs-AC | | 0.21 | 0.049 | 4.29 | 2.40 | 3.016 | 0.81 |
| Neat | 6 | 0.15 | 0.052 | 2.88 | 4.55 | 6.94 | 0.65 |
| MMMs-HNTs | | 0.19 | 0.026 | 7.31 | 1.81 | 3.51 | 0.52 |
| MMMs-AC | | 0.092 | 0.029 | 3.17 | 1.68 | 2.20 | 0.76 |
| Neat | 8 | 0.096 | 0.046 | 2.087 | 3.16 | 5.11 | 0.62 |
| MMMs-HNTs | | 0.11 | 0.022 | 5.00 | 1.38 | 2.78 | 0.51 |
| MMMs-AC | | 0.051 | 0.022 | 2.32 | 1.14 | 1.68 | 0.68 |
| Neat | 10 | 0.054 | 0.067 | 0.81 | 2.26 | 3.59 | 0.63 |
| MMMs-HNTs | | 0.075 | 0.017 | 4.41 | 0.90 | 1.76 | 0.51 |
| MMMs-AC | | 0.035 | 0.025 | 1.40 | 0.76 | 1.19 | 0.64 |

## *4. Conclusions*

In this work, FTIR results prove the presence of elements of activated carbon and halloysite nanotubes compared to the neat membrane and therefore these shows that we have successfully fabricated all the mixed matrix membranes, namely MMMs-HNTs and MMMs-AC. Moreover, in addition to inorganic fillers changes the morphological structure from end drop like pores to closed finger-like pores and reduced the sponge-like structure along with the decrease of the thickness of the skin layer. The mechanical strength demonstrated an enhancement in tensile strength and elongation at break for MMMs-HNTs at a low loading of HNTs in PSF polymeric matrix compared to neat membrane and MMMs-AC. The results revealed that it is due to increase in surface area of interaction between polymer matrix and HNTs and owing to its feature of enriched hydroxyl groups at HNTs improve the molecular force and hydrogen bonding between PSF polymeric matrix and HNTs.

Gas permeation results revealed that rigidification of polymer chain layer caused a decrease in permeability and an increase in selectivity. For the gas separation performance, the results showed an overall increase in the mixed matrix membrane with HNTs over the neat membrane and MMMs-AC. The addition of halloysite nanotubes into polymeric matrix increases the selectivity to 15.83 compared to the neat membrane with only 11.93 and MMMs-AC with a selectivity of 7.06, at 2 bar. This claim supported by SEM results where finger-like pores formed in the cross-sectional area of MMMs-HNTs, promotes the gas permeance. In conclusion, incorporation of HNTs into PSf polymeric matrix not only improve the tensile strength and morphological properties of the membrane but also improves $CO_2$ /$CH_4$ gas separation performance compared to activated carbon.

## *Acknowledgement*

This research is fully funded by Universiti Malaysia Sabah (UMS) with a grant code of GKP0030-2019 and UMSGreat GUG0342-1/2019. The authors would also like to acknowledge the research support from the Malaysian Ministry Education and Universiti Teknologi Malaysia under Collaborative Research Grant (CRG) grant scheme (Q.J130000.2451.08G26).

## References

[1] Kenarsari S D, Yang D, Jiang G, Zhang S, Wang J, Russell G and Fan M 2013 *RSC Adv.* **3** 22739
[2] Yampolskii Y 2012 *Macromolecules* **8** 3298
[3] Shekhawat D, Luebke D R and Pennline H W 2003 *Review of Carbon Dioxide Selective Membranes: A Topical Report* (United States: United States Department of Energy)
[4] Pal R 2007 *J. Reinf. Plast. Compos.* **26** 643
[5] Shimekit B and Mukhtar H 2012 *Natural gas purification technologies-major advances for $CO_2$ separation and future directions* (Malaysia: Universiti Teknologi Petronas) p 9
[6] Robeson M 1991 *J. Membr. Sci.* **62** 165
[7] Robeson M, Burgoyne M, Langsam C, Savoca F and Tien 1994 *Polymer* **35** 4970
[8] Rezakazemi M, Amooghin A, Rahmati M, Ismail F and Matsuura T 2014 *Prog. Polym. Sci.* **39** 817
[9] Choi W, Ingole G, Li H, Kim H, Lee K and Baek H 2016 *J. Clean. Prod.* **133** 1008
[10] Balkus J, Ferraris P and Musselman H 2010 *J. Membr. Sci.* **361** 28
[11] Zornoza B, Téllez C and Coronas J 2011 *Journal Membrane Science* **368** 100
[12] Rafiq S, Man Z, Maulud A, Muhammad N and Maitra S 2012 *Sep. Purif. Technol.* **90** 162
[13] Scholes A, Stevens W and Kentish E 2012 *Fuel* **96** 15
[14] Lock M 2018 *J. Nat. Gas Sci. Eng.* **57** 135
[15] Daramola O 2017 *Membranes* **7** 1
[16] Mopoung S, Moonsri P, Palas W and Khumpai S 2015 *Sci. World J.* 2015 1
[17] Saka C 2012 *J. Anal. Appl. Pyrolysis* **95** 21
[18] Alzaydien S 2016 *Asian J. Sci. Res.* **9** 45
[19] Ranganatha S and Venkatesha T V 2014 *RSC Advances* **4** 31230
[20] Hilliard M and Rochelle T 2010 *J. Greenh. Gas Control* **4** 707
[21] Buruga K, Kalathi J T, Kim K H, Ok Y S and Danil B 2018 J. *Ind. Eng. Chem.* **61** 169
[22] Rezaee R, Nasseri S, Mahvi A H, Nabizadeh R, Mousavi S A, Rashidi A, Jafari A and Nazmara S 2015 *J. Env. Health Sc. Eng.* **13** 61
[23] Mozia S, Grylewics A, Zgrzebnicki M, Darowna D and Czyzewski A 2019 *Polymers* **11** 671
[24] Ismail M 2015 *J. Polym. Eng.* 1
[25] Ismail M, Ismail F, Mustafa A, Zulhairun K and Nordin M 2016 *J. Polym. Eng.* **36** 65
[26] Anadao P 2013 *Appl. Surf. Sci.* 275 110
[27] Gupta M, Coyle I and Thambimuthu K 2003 *CO2 capture technologies and opportunities in Canada* (Canada: CANMET Energy Technology Centre, Natural Resources Canada)
[28] Dorosti F, Omidkhah R, Pedram Z and Moghadam F 2011 *Chem. Eng. J.* **171** 1469
[29] Mohamad B, Fong Y and Shariff A 2016 *Procedia Eng.* **148** 621
[30] Mohamed N K, Kochkodan V, Zekri A and Ahzi S 2020 *Membranes* **10(2)** 1
[31] Nyankson E Agyei-Tuffour B, Annan E, Yaya A, Mensah B, Onwona-Agyeman B, Amedalor R, Kwaku-Frimpong B and Efavi J E 2019 *Heliyon* **5** 1
[32] Kamal N, Kochkodan V, Zekri A and Ahzi S 2020 *Membranes* **10** 2
[33] Vu Q, Koros J and Miller J 2003 *J. Membr. Sci.* **211** 311
[34] Goh S 2011 *Sep. Purif. Technol.* **81** 243
[35] Aroon A, Ismail F, Matsuura T and Rahmati M 2010 *Sep. Purif. Technol.* **75** 229
[36] Shafie S N A, Liew W X, Nordin N A H M, Bilad M R, Sazali N, Putra Z A and Wirzal M D H *Adv.Polym. Technol.* 1
[37] Escorihuela, S, Valero L, Tena A., Shishatskiy S, Escolástico S, Brinkmann T and Serra M 2018 *Membranes* **8** 1
[38] Fadaly S and Aziz F 2020 *JAMST,* **24** 27
[39] Zulhairun A K, Ng B C, Ismail A F, Murali R S and Abdullah M S 2014 *Sep. Purif. Technol.* **137** 1

# Ionic Liquid Blend Thin Film Composite Membrane for Carbon Dioxide Separation

Siti Nur Alwani Shafie [1], Vinosha Viriya, [1], Nik Abdul Hadi Md Nordin*[1], Muhammad Roil Bilad [1], Mohd Dzul Hakim Wirzal [1]

[1] Department of Chemical Engineering, Universiti Teknologi PETRONAS (UTP), 32610 Seri Iskandar, Perak Darul Ridzuan, Malaysia

* Corresponding author. Tel.: +6053687581; fax: +6053656176.
E-mail address: nahadi.sapiaa@utp.edu.my (N.A.H.M. Nordin)

**Abstract**

To utilize the superior properties of ionic liquid (IL), this work developed thin film composite (TFC) membrane incorporated with [EMIM][Tf$_2$N] prepared via interfacial polymerization (IP). The effect of IL on membrane properties and $CO_2/CH_4$ separation performance were studied. The IL loading was varied from 1-5wt%. The fabricated membranes were characterized and the gas separation performance of the membranes was evaluated in term of $CO_2$ permeability and $CO_2/CH_4$ selectivity. Analysis from Fourier-transform Infrared Spectroscopy (FTIR) shows the presence of [EMIM][TF$_2$N] functional groups proving the successful attachment of the IL in the membrane. Furthermore, it is noted that with increasing IL loading up to 3vol%, the surface morphology became rougher and the membrane became more rigid based on the Differential Scanning Calorimeter (DSC) thermogram and Atomic Force Microscopy(AFM) analysis, accordingly. Based on the gas permeation test at 1 bar feed pressure, membrane with 5wt% of IL showed $CO_2$ permeability of 3424.44 GPU and $CO_2/CH_4$ selectivity of 32.15. Compared to pure TFC membrane, TFC/IL membranes showed as much as 103 times and 32 times improvement for $CO_2$ permeability and $CO_2/CH_4$ selectivity, accordingly. The improved gas separation performance shows promising results of incorporating IL in TFC membrane for $CO_2$ separation.

**Keywords:** Thin film composite membrane; interfacial polymerization; [EMIM][TF$_2$N]; $CO_2$ separation

## 1.0 Introduction

As the increase in carbon dioxide ($CO_2$) emission to the environment has become global concern, the development of $CO_2$ removal has been widely conducted including adsorption, absorption, cryogenic and membrane separation [1-3]. In the past decade, interest have been shifting on the improvement of the performance of polymeric membrane for gas separation due to the trade-off limit between permeability and selectivity [4] including the fabrication of thin film composite membrane (TFC) [5, 6]. TFC consists of ultrathin polyamide (PA) selective layer constructed on porous substrate layer. the ultrathin selective layer (less than 0.2 $\mu$m) provide minimal mass transport resistance compared to dense membrane which contribute to the effective gas permeability and selectivity.

Numerous methods for TFC preparation have been studied including dip-coating [7, 8], phase inversion [9, 10] and interfacial polymerization (IP) method [11, 12]. In dip-coating process, the porous substrate is dipped in into the polymer (PA) solution at fixed speed. For phase inversion method, the selective layer is casted on the membrane support and the solvent removal process can be conducted either via dry, wet or dry-wet phase inversion step. In IP, the ultrathin selective layer is fabricated by the reaction of monomers on top of the porous substrate [13]. While TFC preparation via IP has shown permeation improvement for water separation compared to other methods of preparation, there is not enough studies has been conducted for gas separation thus come the propose of this paper.

Furthermore, various methods have been developed to enhance the gas separation performance including by introducing $CO_2$-philic materials during the fabrication such as IL and amine solvents. Recent studies have shown the significant improvement with the incorporation of ionic liquid (IL) [14-16]. IL is a salt which contains an organic cation and inorganic and organic anion where it possesses high $CO_2$ solubility, high selectivity, negligible volatility and highly tunable according to unique application [1, 17]. However, there are still limited works on incorporating IL into TFC membrane, especially for TFC fabricated via interfacial polymerization (IP). While it is evident that IL can improved membrane separation performances [18], the IL presence during IP could alters the polymerization reaction thus affect selective layer formation where it occurs at the interface between two phases.

Therefore, this project aims to fabricate TFC membrane incorporated with [EMIM][$TF_2N$] via IP and investigate the effect of [EMIM][$TF_2N$] on membrane properties and gas separation performance. With the presence of [EMIM][$TF_2N$] which is known to have high $CO_2$ affinity, the gas separation performance of TFC-IL is expected to be improved compared to that of TFC membrane.

## 2.0 Experimental

### *2.1 Materials*

Polysulfone (PSF, Merck), poly (vinyl pyrrolidone) (PVP, Sigma-Aldrich), N-M ethyl-2-pyrrolidone (NMP, Sigma-Aldrich), Trimesoyl chloride(TMC) (98%, Sigma-Aldrich), diethylene glycol bis(3-aminopropyl) ether (DGBAmE, 98%, Sigma-Aldrich), 1-Ethyl-3-methylimidazolium bis(trifluoromethylsulfonyl)imide ([EMIM][$TF_2N$], 99.9%, Sigma-Aldrich), distilled water (prepared in lab).

### *2.2 PSF Support Fabrication*

PSF support is prepared by mixing 15wt% PSF pallet with 3wt% PVP and 82wt% of NMP. The mixture is let to stir for 24 hours before casting. The membrane support is casted using glass rod with thickness of approximately 200 μm. The membrane is immersed in water bath for removal of residual solvent for overnight. The membrane is stored in distilled water before use.

### *2.3 Interfacial Polymerization (IP)*

The TFC membranes were prepared via interfacial polymerization (IP) process. 0.28% w/v of TMC was mixed with heptane as solvent as the organic phase while 0.35% w/v of DGBAmE was mixed with water for the aqueous phase. The PSF membrane support was sandwiched between a glass plate and Viton rubber frame. Organic phase was first poured onto the membrane for at least 10 minutes. The excess solution was poured before pouring aqueous phase solution. The IP process was allowed to take place for 5 minutes before the membrane is let dried overnight in ambient condition. The TFN membrane was prepared by mixing 1vol% - 5vol% of [EMIM][$Tf_2N$] into the aqueous phase as shown in table 1. For characterization purposes, PA layer for all membrane is prepared without PSF support.

**Table 1**: Composition of TFC membrane

| Sample | [EMIM][$TF_2N$] loading in vol% (based on the total weight of monomer) |
|---|---|
| Pure PA | 0 |
| PA-IL1 | 1 |
| PA-IL2 | 2 |
| PA-IL3 | 3 |
| PA-IL4 | 4 |
| PA-IL5 | 5 |

### *2.4 Characterization*

Fourier Transform Infrared Spectrometer (FTIR) analysis was done to determine the functional group changes with the incorporation of [EMIM][TF2N]. The analysis was done using Perkin-Elmer Spectrum 1 Fourier transform Infrared spectrometer and the spectra were analyzed by the Spectra One Software.

Differential scanning calorimetry (DSC) was used to analyze the effect IL in TFC membranes from the glass transition temperature ($Tg$) data. Firstly, small cuts and cleaned membranes were placed on the DSC pans. The samples were heated from 30 °C to 800 °C at a rate of 10 °C /min. After reaching 250 °C, the samples were naturally cooled to remove the thermal history. Under the same procedure, the second scan was conducted again. To get the $Tg$ value of the TFCs, the second scan values were referred.

Lastly, Atomic Force Microscopy (AFM) analysis was conducted to study the membrane surface topology and surface roughness using Universal Scanning Probe Microscopy (USPM-AFM, Nano Navi, SII). The scanning probe was conducted on the top surface of the TFC membranes with scan surface of area 25 μm x 25 μm.

*2.5 Gas separation performance testing*

The membranes undergo $CO_2$ and $CH_4$ single gas permeation test at 1 bar feed pressure. The permeability is calculated by using following equation:

$$\frac{P}{l} = \frac{Q_{STP}}{A \times \Delta P} \quad (1)$$

where $P/l$ is the permeance [$cm^3$(STP)/(sec $cm^2$ cmHg)], A is the effective surface area [$cm^3$], $\Delta P$ is pressure gradient across feed and permeate [cmHg], $Q_{STP}$ is the permeance flow rate at standard temperature and pressure [$cm^3$(STP)/sec]. Permeability is expressed in gas permeation units, GPU as 1 GPU= $1\times10^{-6}$ $cm^3$ (STP) $cm^{-2}$ $s^{-1}$ $cmHg^{-1}$.

Besides, the selectivity of membrane, $a_{CO_2/CH_4}$ can be determined by ratio of higher permeable gas i to lesser permeable gas j. The equation for selectivity is shown in equation below

$$a_{CO_2/CH_4} = \frac{P/l_{CO_2}}{P/l_{CH_4}} = \frac{P_{CO_2}}{P_{CH_4}} \quad (2)$$

## 3.0 Results and discussion

### *3.1 Membrane characterization*

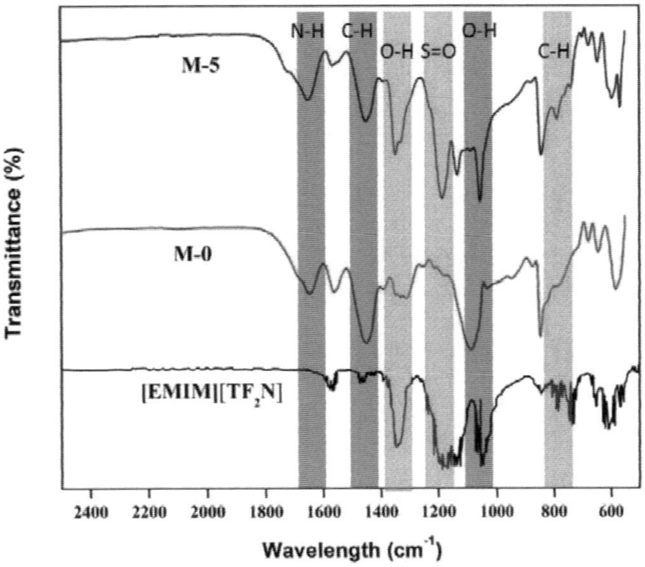


Figure 1: FTIR Spectrum for fabricated pure PA and PA-IL5. and FTIR spectra for [EMIM][TF2N] spectra is adapted from [19]

The fabricated TFC membranes were analyzed using FTIR to determine the change in functional group in PA membrane with the incorporation of [EMIM][$TF_2N$] as presented in Figure 1. For comparison, only PA and PA-IL5 were included while the FTIR spectra for [EMIM][$TF_2N$] is adopted from [19]. For PA membrane, it is noticeable that characteristic peaks of PA which include N-H group at wavelength of 1646 $cm^{-1}$, C-H group at wavelength of 1451 $cm^{-1}$ and O-H group 1087 $cm^{-1}$. Thus, this proving that the PA layer was successfully fabricated.

Upon incorporation of [EMIM][$TF_2N$], the presence of additional broad bands, can be observed at wavelength of 789 $cm^{-1}$which represents the C-H bond, S=O group at wavelength of 1184 $cm^{-1}$, C-O group presents at wavelength of 1134 $cm^{-1}$ and S=O present at wavelength of 1355 $cm^{-1}$. The presence of [EMIM][$TF_2N$] characteristic peaks indicates an additional interaction between [EMIM][$TF_2N$] and PA which confirms that the IL is successfully embedded in the PA-IL5 TFC membrane. Hence, based on these results it confirms that there is a successful attachment of the IL bonds onto the PA membrane.

**Table 2**: Glass transition temperature

| **Sample** | **Glass Transition Temperature, $T_g$ ($^oC$)** |
|---|---|
| PA | 130.16 |
| PA-IL1 | 131.01 |
| PA-IL2 | 133.06 |
| PA-IL3 | 133.42 |
| PA-IL4 | 128.83 |
| PA-IL5 | 128.13 |

For DSC, the effective thin layer of PA (without PSF support) was sent for analysis to determine the glass transition temperature ($T_g$) of PA layer of the membranes and the effect of the presence of IL on $T_g$ . The $T_g$ of the membranes are tabulated on **Table 2**. The $T_g$ of PA is recorded at 130.16 $^oC$. The $T_g$ increases with the incorporation of IL from PA till PA-IL3. As the loading increase, the interaction between the [EMIM][$TF_2N$] and polymer increased. The presence of IL increases the $T_g$ of the membrane which indicates the increase in polymer chain rigidity [20]. This is due to possible increase in interaction between PA and [EMIM][$TF_2N$] with increasing loading up to 3vol%.

However, it can be seen that the glass transition decreases at IL loading of PA-IL4 and PA-IL5. It is postulated that at higher loading (more than 3vol%) of the ionic liquid act as plasticizer which causes the polymer to relax and results in decrease in rigidity[21] and the gas separation performance (subsection 3.2). The presence of higher percentage of IL loading in the membrane decreases the $T_g$ value of the membrane which indicates the increase in polymer chain flexibility [20].

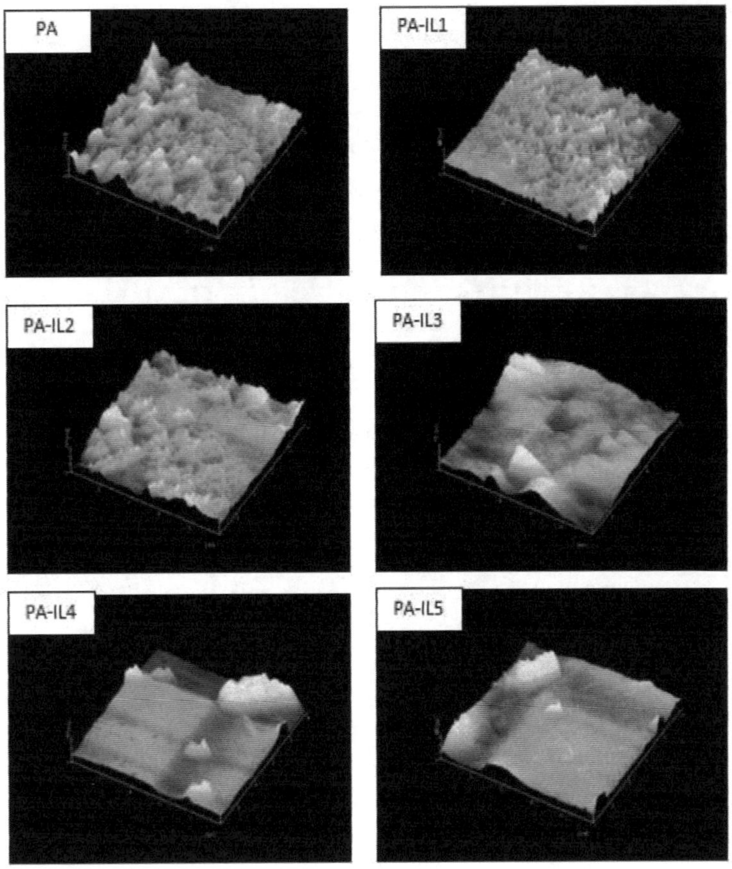


Figure 2: Cross-sectional membrane morphology at 10000x magnification at different IL loading

The surface morphologies for the prepared samples were determined by using Atomic Force Microscopy (AFM) as presented in Figure 2. The color intensity stipulates the vertical profile of the membrane surface where the pores and regions are on the dark regions and the peak points are on the bright regions. It can be observed that the membranes developed possess nodules-like-leaf form of roughness feature. The formation of the features is caused by the release of nanobubbles in the organic monomer solution during the IP process [22]. Based on the **Figure 2**, it can be observed that the roughness of the surface decreases with increasing [EMIM][$TF_2N$] loadings. Pure PA possesses the roughest surface while PA-IL5 possess the smoothest surface.

**Table 3**: Membrane surface roughness for all PA-ILTFC membranes

| Sample | Average roughness, $R_a$ (nm) |
|---|---|
| PA | 22.49 |
| PA-IL1 | 18.79 |
| PA-IL2 | 30.19 |
| PA-IL3 | 50.56 |
| PA-IL4 | 43.73 |
| PA-IL5 | 34.58 |

**Table 3** shows the roughness of membrane surface with maximum height and maximum valley at different [EMIM][$TF_2N$] loading. Based on the table, it can be observed that average roughness of the membrane decreases from PA (22.49 nm) to PA-IL1 (18.79 nm) and then increases from PA-IL1 to PA-IL3. Moreover, it can be observed that the average roughness of membrane decreases from PA-IL4 to PA-IL5, albeit still higher than PA TFC membrane. It is postulated that the decrease in surface roughness is related to the reduction in $T_g$ (**Table 2**). At higher loading (more that 3vol%), the polymer relaxes due to the lesser interaction between IL and PA resulting in decreasing surface roughness. Overall, it can be analyzed that PA-IL3 possess highest value of average roughness which is at 50.56 nm while PA-IL1 possess lowest value of the average roughness which is at 18.79 nm.

From the result, it can be postulated that the surface roughness increases with the increasing IL loading (except for PA-IL1) when compared to PA TFC membrane. The increase in roughness may be due to the increase in PA-IL interaction at the polymer surface (Table 2) and the fluctuation of the PA-IL interface interaction [23], resulting in bigger gap on the maximum and minimum height of the valley-like roughness features, thus the increase in the average surface roughness.

### 3.2 Gas permeation test

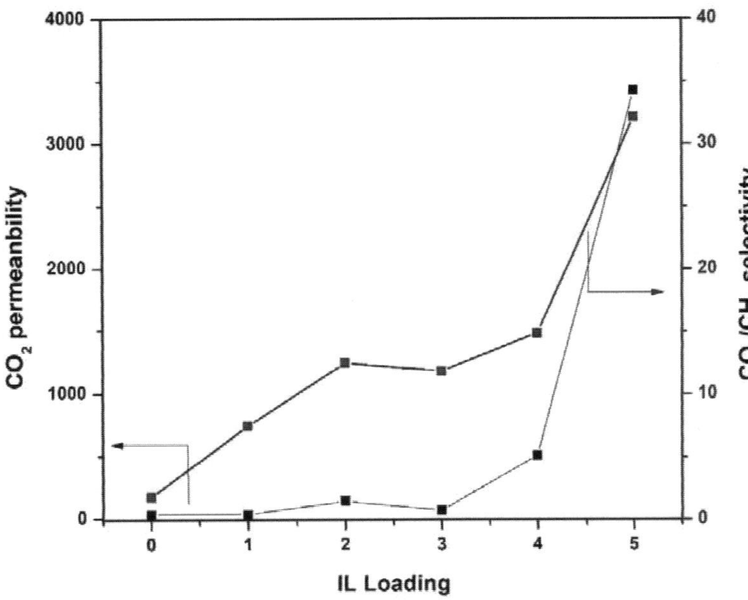


Figure 3: $CO_2$ permeability and $CO_2/CH_4$ selectivity for all PA-ILTFC membrane tested at 1 bar feed pressure

The fabricated TFC membranes were successfully tested for $CO_2$ permeability and $CO_2/CH_4$ selectivity under different loading at 1 bar feed pressure. The gas separation performance is shown in the Figure 3. Overall, with increasing [EMIM][$TF_2N$] loading, the $CO_2$ permeability increases. Pure PA membrane shows $CO_2$ permeability of 33.43 GPU. With 1vol% -3vol% of [EMIM][$TF_2N$], the $CO_2$ permeability increases slightly for about less than 2% increment. Meanwhile, the $CO_2$ permeability increases significantly with 4vol% and 5vol% of [EMIM][$TF_2N$] with increment of as much as 103 times for incorporation of 5vol% of IL. It is postulated that at higher[EMIM][$TF_2N$] loading (more than 3 vol%), the presence of IL relaxes the polymer matrix (Table 2) thus resulting in the outstanding increase in $CO_2$ permeability.

Meanwhile, for $CO_2/CH_4$ selectivity, the trend follows that of $CO_2$ permeability. The $CO_2/CH_4$ selectivity starts at 1.78 for pure PA TFC membrane. With the incorporation of 1vol% of IL, the $CO_2/CH_4$ selectivity increases 4 times while with the highest [EMIM][$TF_2N$] loading of 5vol%, the $CO_2/CH_4$ selectivity increases by 17 times compared to pure PA TFC membrane. The increase in the $CO_2/CH_4$ selectivity is due to the higher interaction between [EMIM][$TF_2N$] and $CO_2$ compared $CH_4$ thus resulting in high $CO_2/CH_4$ selectivity. As the gas solubility in IL is highly dependent on anion, the fluorine group ($CO_2$-philic group) in [EMIM][$TF_2N$] played a major role in enhancing the $CO_2$ permeability and therefore increasing the $CO_2/CH_4$ selectivity.

## 4.0 Conclusion

The PA-IL membrane were successfully fabricated and the effects on membrane properties and gas separation performance were discussed. FTIR analysis confirmed the successful fabrication of all TFC membranes with the presence of characteristic peaks of [EMIM][$TF_2N$]. The top surface morphologies show that [EMIM][$TF_2N$] reduces the membrane roughness. Thermal analysis also confirm that TFC/IL membrane become rigid until 3vol% and become more flexible with loading more than 4vol%. Based on the gas permeation test, it can be observed that PA-IL5 gives the highest gas separation performance at 1 bar feed pressure. $CO_2$ permeability increases from 33 to 3200 GPU while $CO_2/CH_4$ selectivity increases from 1.78 to 34 compared to that of pure PA TFC membrane. This shows that PA-IL TFC membrane has the opportunity to be further explored in fabricating composite membrane for gas separation application.

Acknowledgement

The authors thank the financial support provided by Yayasan Universiti Teknologi PETRONAS (YUTP) (015LC0-022) to carry out this research work. The authors also acknowledge the technical assistance provided by Chemical Engineering Department and Centralized Analytical Lab UTP during the experimental works.

## References

[1] Yokozeki, A., Shiflett, M. B., Junk, C. P., Grieco, L. M. and Foo,T. , *Physical and chemical absorptions of carbon dioxide in roomtemperature ionic liquids.* The Journal of Physical Chemistry B, 2008. **112**: p. 16654–16663.

[2] Allan, H., Gnanendran, N., *Cryogenic CO2 Capture in Natural Gas* Energy Procedia, 2009. **1**: p. 697-706.

[3] Brunetti, A., Scura, F., Barbieri, G. and Drioli, E., *Membrane technologies for $CO_2$ separation.* Journal of Membrane Science, 2010. **359**: p. 115-125.

[4] Robeson, L.M., *The upper bound revisited.* 2008. **Membrane Science**(320): p. 390-400.

[5] Wong, K.C., Goh, P.S., Ismail, A.F., *Thin Film Nanocomposite: The Next Generation Selective Membrane for $CO_2$ Removal.* Journal of Material Chemistry A, 2016. **0**: p. 1-3.

[6] Lau, W.J., Ismail, A.F., Misdan, N., Kassim, M.A., *A recent progress in thin film composite membrane: A review.* Desalination, 2010. **287**: p. 190-199.

[7] Zhou, J., Tran, M. M., Haldeman, A. T., Jin, J., Wagener, E. H., Husson, S. M. , *Perfluorocyclobutyl polymer thin-film composite membranes for $CO_2$ separations.* Journal of Membrane Science, 2014. **450**: p. 478-486.

[8] Yanagishita, H., Kitamoto, D., Haraya, K., Nakane, T., Okada, T., Matsuda, H., Idemoto,Y., Koura, N. , *Separation performance of polyimide composite membraneprepared by dip coating process.* Journal of Membrane Science, 2000. **188**(2): p. 165–172.

[9] Pinnau, I., Wind, J., Peinemann, K. V., *Ultrathin multicomponent poly(ether sulfone) membranes for gas separation made by dry/wet phase inversion.* Industrial & Engineering Chemistry Research, 1990. **29**: p. 2028-2032.

[10] Mikawa, M., Nagaoka, S., Kawakami, H. , *Gas permeation stability of asymmetric polyimide membrane with, thin skin layer:, effect of molecular weight of polyimide.* Journal of Membrane Science, 2002. **208**: p. 405-414.

[11] Lu, X., Bian, X., Shi, I.,, *Preparation and characterization of NF composite membrane.* Journal of Membrane Science, 2002. **210**: p. 3-11.

[12] Li, S., Wang, Z. , Zhang, C. , Wang, M. , Yuan, F. , Wang J. ,Wang, S. , *Interfacially polymerized thin film composite membranes containing ethylene oxide groups for $CO_2$ separation.* Journal of Membrane Science, 2013. **436**: p. 121-131.
[13] Fauzan, N.A., Mannan, H.A., Nasir, R., Mohshim, D. F., Mukhtar, H. , *Various Techniques for Preparation of Thin-Film Composite Mixed-Matrix Membranes for $CO_2$ Separation.* Chemical engineering Technology, 2019. **42**: p. 1-14.
[14] Abdelrahim, M.Y., Martins, C. F., Neves,L. A., Capasso,C., Supuran, C. T., Crespo, J. G., Barboiu, M., *Supported ionic liquid membranes immobilized with carbonic anhydrases for CO2 transport at high temperatures.* Membrane Science, 2017. **258**: p. 225-230.
[15] Bara, J.E., Lessmann, S., Gabriel, C. J., Hatakeyama, E. S., Noble, R. D., & Gin, D. L., *Synthesis and Performance of Polymerizable Room-Temperature Ionic Liquids as Gas Separation Membranes.* Industrial & Engineering Chemistry Research, 2007. **46**: p. 5397–5404.
[16] Bernado, P., Jansen J. C., Bazzareli, F., Tasseli, F., Fuoco, A., Friess, K., Izak, P., Jarmarova, V., Kacirkova, M., Clarizia, G., *Gas transport properties of Pebax®/room temperature ionic liquid gel membranes.* Separation and Purification Technology, 2012. **97**: p. 73-82.
[17] Wilkes, J.S., *Properties of ionic liquid solvents for catalysis.* Journal of Molecular Catalysis A: Chemical, 2004. **214**(1): p. 11-17.
[18] Hert, D.G., et al., *Enhancement of oxygen and methane solubility in 1-hexyl-3-methylimidazolium bis(trifluoromethylsulfonyl)imide using carbon dioxide.* Chemical Communications, 2005: p. 2603-2605.
[19] Booth, R.S., Annesley, C. J., Wong, J. W., Vogelhuber, K. M., Boatz, J. A., Stearns, J. A., *Identification of multiple conformers of the ionic liquid [emim][tf2n] in the gas phase using IR/UV action spectroscopy.* Royal Society of Chemistry, 2016. **18**: p. 17037-17043.
[20] Shafie, S.N.A., W. X. Liew, N. A. Hadi , M. R. Bilad, N Sazali, Z. A. Putra, , *CO2-Philic [EMIM][Tf2N] Modified Silica in Mixed Matrix Membrane for High Performance CO2/CH4 Separation.* Advances in Polymer Technology, 2019.
[21] Kanehashi, S., et al., *CO2 separation properties of a glassy aromatic polyimide composite membranes containing high-content 1-butyl-3-methylimidazolium bis(trifluoromethylsulfonyl)imide ionic liquid.* Journal of Membrane Science, 2013. **430**: p. 211-222.
[22] Ma, X., Yang, Z., Yao, Z., Guo, H., Xu, Z., Tang, C. Y., *Tuning roughness features of thin film composite polyamide membranes for simultaneously enhanced permeability, selectivity and anti-fouling performance.* Journal of Colloid and Interface Science, 2019. **540**: p. 382-388.
23. Yu, X., Wang, Z., Wei, Z., Yuan, S., Zhao, J., Wang, J., Wang, S., , *Novel tertiary amino containing thin film composite membranes prepared by interfacial polymerization for CO2 capture.* Journal of Membrane Science, 2010. **362**(1-2): p. 265-278.

# Study on the effect of air gap on physico-chemical and performance of PVDF hollow fibre membrane

**H S Zakria[1], M H D Othman[1], R Kamaludin[1], A Jilani[2]**

[1]School of Chemical and Energy Engineering, Faculty of Engineering, Universiti Teknologi Malaysia (UTM), 81310 Skudai, Johor, Malaysia
[2]Center of Nanotechnology, King Abdul-Aziz University, 21589 Jeddah, Saudi Arabia

hafiz@petroleum.utm.my; hazirahzakria@gmail.com

**Abstract**. Polyvinylidene fluoride (PVDF) was chosen in this study as the main material in fabricating membrane due to its excellent chemical resistance and good thermal stability. Combination of triethyl phosphate (TEP) with DMAc produce better structure of membrane which safer and provide high mechanical strength for membrane. Surface modified PVDF hollow fibre membrane (HFM) was prepared using dry-wet spinning technique by varying air gap namely 10 cm, 20 cm and 30 cm. The morphology was evaluated using scanning electron microscopy (SEM), atomic force microscopy (AFM), mercury intrusion porosimetry (MIP), contact angle, tensile test and performing water flux testing. From the characterization data, PVDF HFM with 3 wt.% PEG 400, HFM 3-10 and HFM 3-20 referred as microfiltration membrane with pore size range 0.1-0.8 μm. While, HFM 3-30 act as ultrafiltration membrane with pore size ranging 0.01-0.1 μm. Experimental results revealed that by increasing the air gap from 10 cm to 30 cm, the porosity and finger-like length decreased due to the higher elongational stress that shift the pores from broad to narrow. Thus, PVDF HFM at 10 cm air gap, HFM 3-10 achieve the highest water flux due to the higher porosity, longer finger-like length and hydrophilicity achieved. The modified HFM at shorter air gap was found to be a promising membrane structure for excellence water performance and eco-friendly to environment.

## 1. Introduction

Membrane technology widely used as a separation process to eliminate unwanted contaminants in the water due to its excellent separation system, outstanding capability of high capacity of contaminants removal and low energy consumption [1]. Polymeric membrane especially polyvinylidene fluoride (PVDF) was chosen as one of the attractive organic macromolecule polymers in fabrication of membrane on account of its excellent chemical resistance, high mechanical strength, longer timespan and good thermal stability [2]. Furthermore, PVDF possess high solubility in various organic solvents compared to other materials.

Dimethylacetamide (DMAc) is stated as one of the widely used solvent for dissolving the PVDF due to its high solubility in DMAc. However, the toxic content of DMAc needs to be considered. According to Chang et al [3], triethyl phosphate (TEP) is believed to have much safer solvent due to its low toxic content. In addition, the used of TEP as solvent can help in producing high mechanical strength of membrane [4]. Thus, the membrane produced can withstand high resistance and pressure as well act as green membrane for safety environment.

Although PVDF experiencing excellent characteristics for wastewater treatment, however, it can encounter fouling issue due to its hydrophobic nature which can encounter lower water flux performance. Surface modification was needed to obtain the hydrophilic PVDF membrane. Polyethylene glycol (PEG) found to be one of the most frequently used additives in membrane fabrication. PEG easily dissolves in water and various organic solvents as well as known as non-toxic material and low cost [5]. In addition, it acts as pore former [6] and help in increasing the membrane permeability [5]. Moreover, modification of membrane by varying the air gap during fabrication procedure also affected the properties of produced membrane.

To acquire a good physico-chemical property of membrane in terms of robust and outstanding performance for water treatment, DMAc and TEP were chosen as solvents in dissolving PVDF in order to produce high mechanical strength of membrane. In addition, lower molecular weight of PEG (PEG 400) was used as pore former and different air gap were applied to increase the hydrophilicity of the polymeric membrane. There are various configuration of polymeric membrane including flat sheet, hollow and tubular. Flat sheet is widely used configuration and the easiest to be fabricated. However, it tends to have lower surface area. Hollow fibre configuration was chosen in this study due to its favor condition such as high surface area that can be fabricated using dry-wet spinning technique.

## 2. Methodology

### *2.1 Materials*
Polyvinylidene fluoride (PVDF, Kynar 760 Series-powder, Solvay Specialty Polymers France) was used as polymer. Polyethylene glycol (PEG 400) as pore former purchased from Evergreen Engineering & Resources. Dimethylacetamide (DMAc, Sigma Aldrich) and triethyl phosphate (TEP, Merck) were used as solvent to dissolve a polymer.

### *2.2 Fabrication of PVDF hollow fibre membrane*

*2.2.1 Polymer dope preparation.* PVDF was dried in a 60 $^{0}$C vacuum oven for 24 h to remove moisture. The compositions of single layer dope solutions are shown in Table 1. TEP and DMAc respectively were mixed and stirred at 240 rpm inside a Scott bottle at 80°C until homogenous. 15 wt.% of PVDF was added into the solution and stirred at 80 °C for 24 hours until completely dissolved. 3% of PEG 400 was added into the solution as pore former and stirred for another 24 hours as stated in Table 1.

**Table 1.** Composition of PVDF hollow fibre membrane.

| PVDF (wt%) | PEG 400 (wt%) | TEP (wt%) | DMAc (wt%) |
|---|---|---|---|
| 15 | 3 | 50 | 32 |

*2.2.2 Dry/wet Spinning technique.* Normal tube and orifice spinneret were used to produce hollow fibre membranes. Dope solution was filled into dope reservoir with spinning conditions as shown in Table 2 to fabricate the hollow fibre membranes [7]. Then, as- spun hollow fibre was immersed in a deionized water tank for 1 day to extract the residual diluent. The post treatment was done by soaking the membrane into 50% of ethanol for 1 hour followed by immersing the membrane into 100% ethanol for another 1 hour. Finally, the membrane was dried at room temperature for 1 day.

**Table 2.** Parameters for spinning conditions.

| Sample | Dope flow rate (ml/min) | Bore fluid flow rate (ml/min) | Bore fluid | Spinneret dimension (mm) | Air gap (cm) |
|---|---|---|---|---|---|
| HFM 3-10 | | | | | 10 |
| HFM 3-20 | 7.81 | 8 | Distilled water | 0.8/1.2 | 20 |
| HFM 3-30 | | | | | 30 |

*2.3. Characterization*

*2.3.1 Scanning Electron Microscopy (SEM)* SEM (Model: TM3000, Hitachi) was used to observe the finest structural morphology of the cross-section and the surface of PVDF HFM. The sample was coated with platinum for 10 s under vacuum to ensure the micrograph produced was clear and reduce charging effect. The micrograph was obtained with magnification of 60x, 600x and 10Kx.

*2.3.2 Mercury Intrusion Porosimetry (MIP).* The porosity of membrane was analyzed using Automated Mercury Porosimeter (Model: AutoPore IV Series, Micromeritics). The membrane was cut with size 1 cm and weight around of 0.25 g. This technique was controlled with pressure and able to measure the diameter of pore [6].

*2.3.3 Atomic Force Microscopy (AFM).* Atomic force microscopy (Model: SE-100 Park System) was used to investigate the roughness of PVDF HFM surface [6]. A micro cantilever is located at one end while tiny probe is equipped at the other end. Firstly, sample membrane was put on the object stage by keeping the surface in horizontal plane. Then, the studied area with the lop of the optical microscope was selected. Finally, the instrument scan the tested area and the surface information was obtained [8].

*2.3.4 Contact angle.* The changes in contact angle of PVDF HFM surface with water was measured using contact angle goniometer (Model: OCA15EC, Dataphysics). Deionized water was used as a contact liquid. Multiple points of the sample were measured to obtain the result.

*2.3.5 Tensile strength.* The tensile strength of the PVDF HFM was tested by loading the 500 N (Zwick/Roell) load cell with the test standard of ASTM D 2256/D 2256 M at room temperature. The tensile strength was calculated using equation 1 [6]:

$$\text{Tensile strength } (Pa) = \frac{\text{Load at break (N)}}{\text{Cross sectional area (m}^2)} \tag{1}$$

*2.3.6 Water permeation analysis.* The changes in water flux of PVDF HFM was conducted using dead-end cross-flow permeation cell of the membrane by pumping the feed across the membrane and allow the permeate to flow out through lumen side of membrane. All the analysis was carried out with pressure 1 bar at 25°C with the length of membrane of 10 cm. The flux, F (L/m$^2$.h) was calculated using the following equation 2.

$$F=\frac{V}{At} \quad (2)$$

Where V (L) is the permeate volume collected. While t (h) is time and A (m$^2$) is the exposed membrane area.

## 3. Results and discussion

### *3.1 Physical properties*

The morphology structure of the fabricated PVDF hollow fibre membrane (HFM) were investigated using SEM as shown in Figure 1. Average outer diameter of HFM 3-10, HFM 3-20 and HFM 3-30 were 1.53 mm, 1.50 mm and 1.41 mm respectively. The reduction of the diameter was due to the higher air gap that possess higher elongational stress which make the membrane's wall stretched. The decrease of membrane's wall diameter also effected from the presence of PEG 400 that increase the thermodynamic instability of dope with water because of its non-solvent effect as a phase separation enhancer [9]. However, HFM 3-20 and HFM 3-30 exhibit shorter average finger-like length of 52.37 μm and 31.30 μm respectively at outer and inner layer membrane with sponge-like structure at intermediate space. The spongy structure developed at intermediate space was called as sandwich-like structure on account of pension-coagulant interface instability happened during phase inversion procedure [10;6]. The formation of shorter finger-like may be due to the higher percentage of PEG 400 that lead in low viscosity of dope solution not suitable to be interacted with higher air gap. Macrovoids can be seen near inner finger-like structure and the size increased when air gap increased due to the stretching process where dope solution experienced longer exposure to air before it reaches water to form membrane. The length of finger-like structure was crucial for determination of higher permeability that highly required for water treatment process [11].

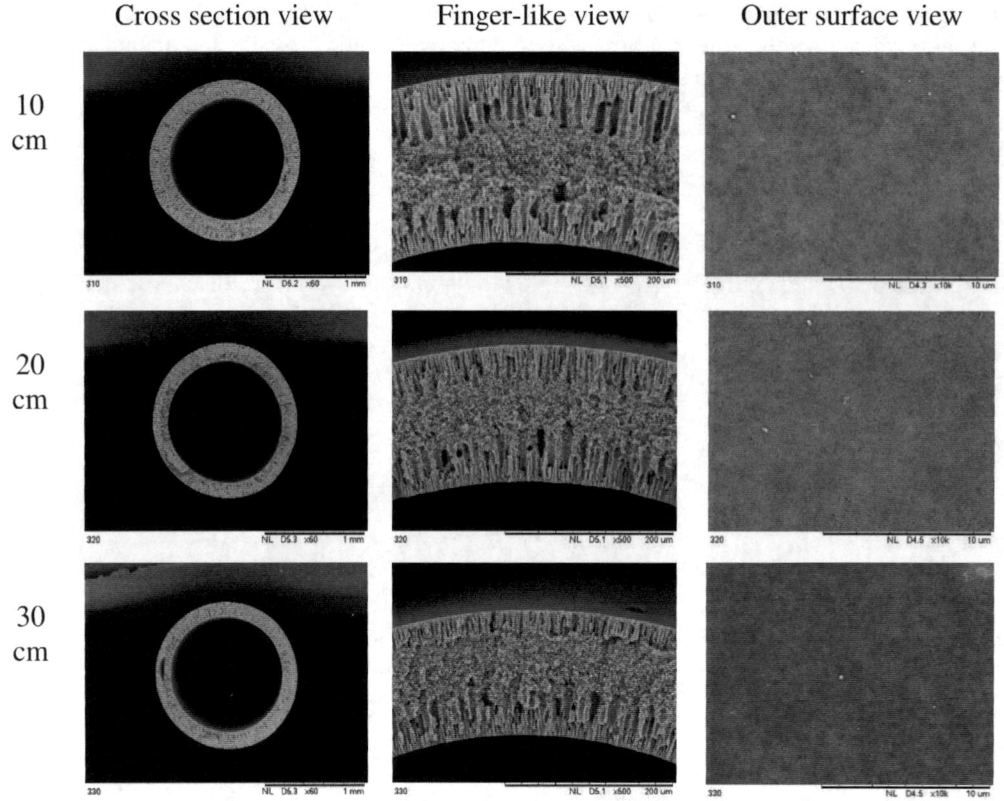


**Figure 1**. SEM images of PVDF HFM with 3% PEG 400 at different air gap.

Figure 2 and Figure 3 depicts the porosity and pore size distribution respectively decreased when air gap increased. As can be seen in Figure 3, the broad peak of each membrane indicates that the membrane composed of sponge-like structure, represented by uniform pore formation throughout the membrane [11]. The intense pore size distribution from average sponge-like structure at the first peak of HFM 3-10 and HFM 3-20 lies in the range between 0.1-0.8μm referring to microfiltration membrane [12]. While, pore size distribution of HFM 3-30 reduced and numerous below 0.1 μm that nominated them as ultrafiltration membrane. Additionally, the mercury intrusion intensity decreased upon the reduction of air gap. This indicate that pore size became smaller when increasing the air gap during spinning procedure. This was due to the greater molecular orientation and a tighter molecular packing. Moreover, the elongation-induced pore deformation occurred when increasing the air gap. A round pore might be elongated as an elliptical pore with narrow fluid channel and the pore size diameter may shift from broad to narrow [13]. Besides, no significantly differences observed at the pore size distribution between 100-1000 μm which corresponded to the formation of finger-like structure.

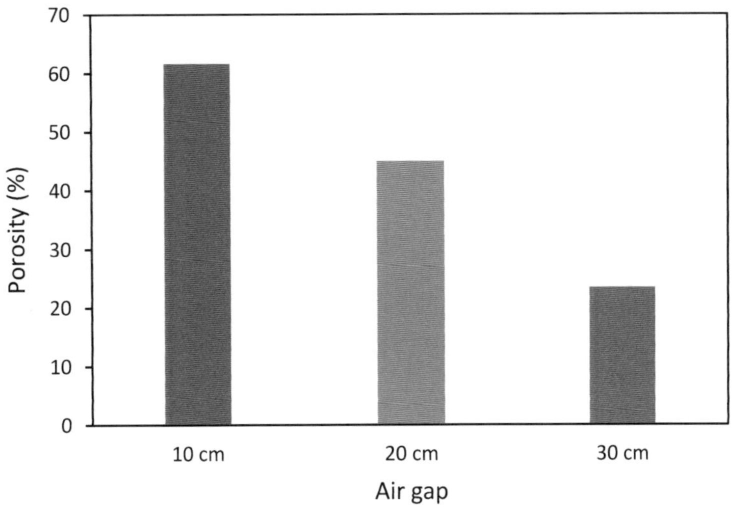


**Figure 2**. Porosity of PVDF HFM at different air gap.

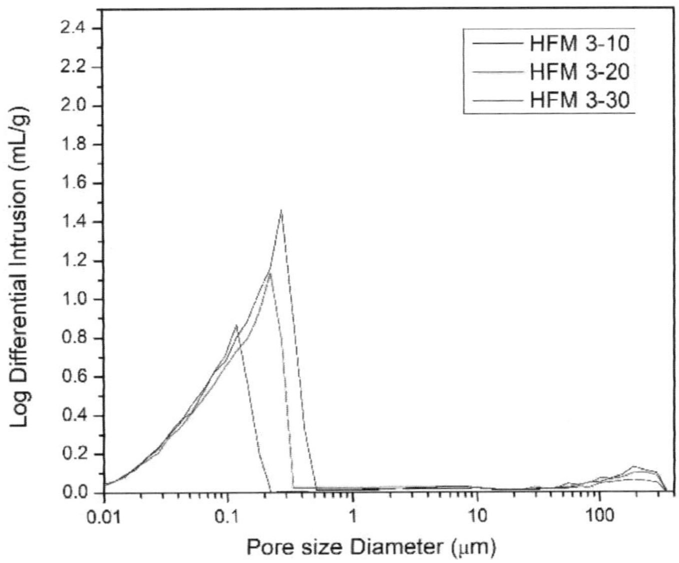


**Figure 3**. Pore size distribution of PVDF HFM at different air gap.

Surface roughness of the membrane was observed by AFM analysis. From observation in Figure 4 and Figure 5, the surface roughness was decreased from 8.78 nm to 4.18 nm when air gap increased from 10 cm to 30 cm. This may be due to the polymer aggregates are merging with each other when increasing the air gap [14]. Moreover, it could be due to the longer contact time of dope suspension exposed to air at higher air gap before it touches the water to form membrane. Thus, stretching the membrane and produce smooth surface. Surface roughness of membrane was crucial for selecting a suitable one for coating process. Higher surface roughness of membrane possessed higher selectivity as support for coating procedure as it provides higher available surface area for particle attachment and expected to minimize the leaching problem during water treatment process [15]. This is because, the higher the roughness, the stronger the attachment of material onto the membrane surface [16].

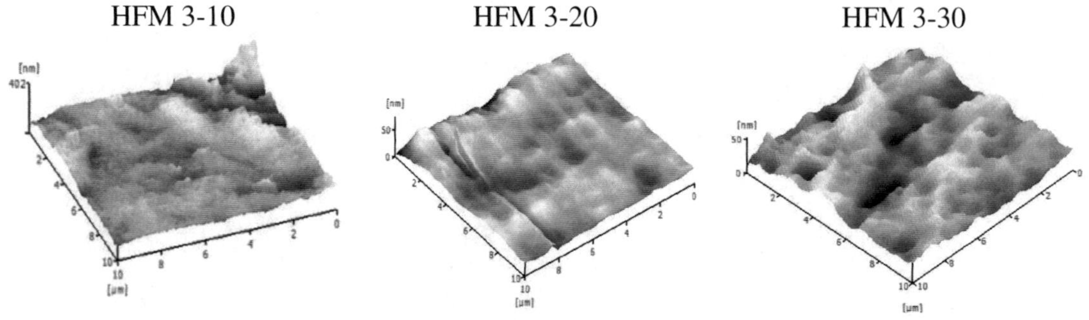


**Figure 4**. Surface micrograph of PVDF HFM at different air gap.

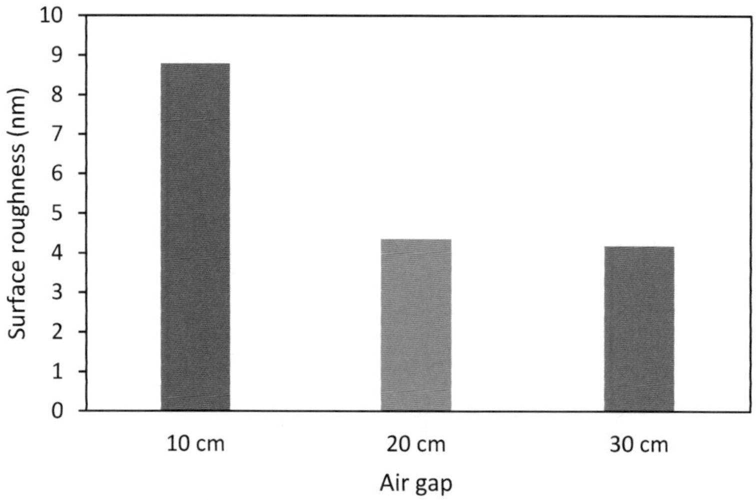


**Figure 5**. Surface roughness of PVDF HFM at different air gap.

Figure 6 depicts the contact angle value of hollow fibre membrane with different pore former loading and air gap. The contact angle tends to increase from $77.01^0$ to $84.20^0$ when increasing the air gap. This was due to the reducing of surface roughness which reduce the surface area causing the membrane tend to exhibit hydrophobic nature. Moreover, this could be on account of the longer period for dope suspension exposed to the air before reaching the water resulting smooth surface of membrane. The hydrophobicity was crucial for determination of fouling issue. The higher the contact angle, the higher the hydrophobicity nature that lead to fouling issue. The particles may clog into the pores of membrane and reduce the water flux performance [17].

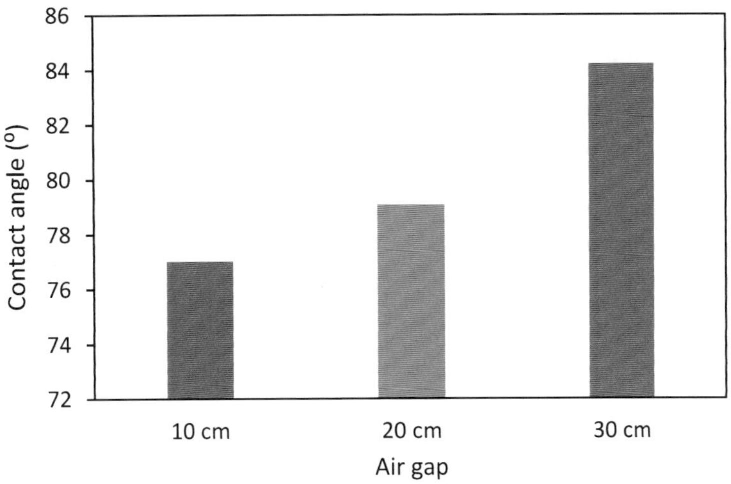


**Figure 6**. Contact angle of PVDF HFM at different air gap.

Figure 7 summarizes the results of tensile strength of PVDF HFM 3 at different air gap. The tensile strength increased when increasing the air gap to 30 cm. This could be due to the role of TEP that help in increasing the tensile strength of membrane. The spongy structure at intermediate space of membrane with interconnected porosity produced by TEP as well as shorter finger-like structure at higher air gap help in developing the stronger structure of membrane [4]. Furthermore, all the membrane in this study shows outstanding tensile strength which indicate higher than 2 MPa compared with previous study by Dzinun et al [7]. The tensile strength of single layer PVDF hollow fibre membrane fabricated by using DMAc only as solvent shown only 1.82 MPa [7], lower compared to this study when using combination of DMAc and TEP. It can be proved that the interconnected bicontinuous pores developed with dense skin layer formed by using TEP help to increase the tensile strength of membrane.

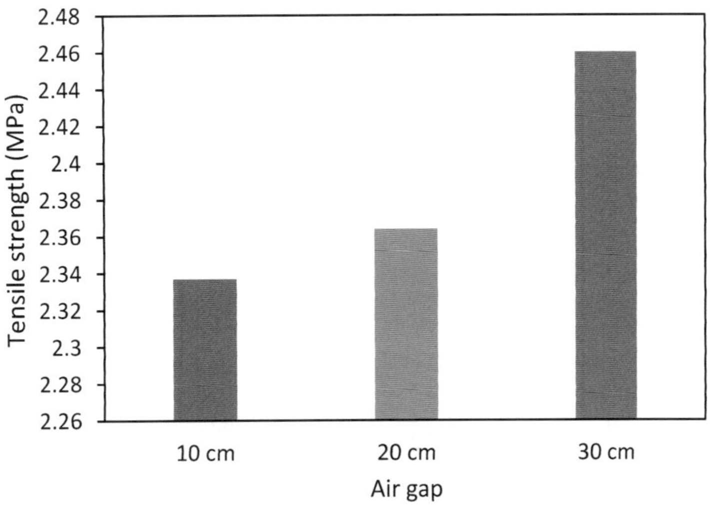


**Figure 7**. Tensile strength of PVDF HFM at different air gap.

*3.2 Water flux performance*

Water permeation analysis was conducted to indicate the water flux performance of PVDF HFM. The water flux trend shown in Figure 8 correlated with the morphology of membrane. As the porosity and finger-like length decreased at higher air gap, the water flux performance also decreased on account of shorter finger-like that lead a difficulty for the water to pass through the pores. HFM 3-10 exhibit the highest water flux performance at 2338.88 L/m$^2$.h compared to the membrane from previous literature by Abed et al [4] which only possessed at 114.59 L/m$^2$.h. Although water flux at 30 cm air gap reduced to 1334.18 L/m$^2$.h, it revealed as an outstanding performance compared to PVDF hollow fibre membrane produced from previous literature written by Shi et al [18] that exhibit only 350 L/m$^2$.h under similar air gap.

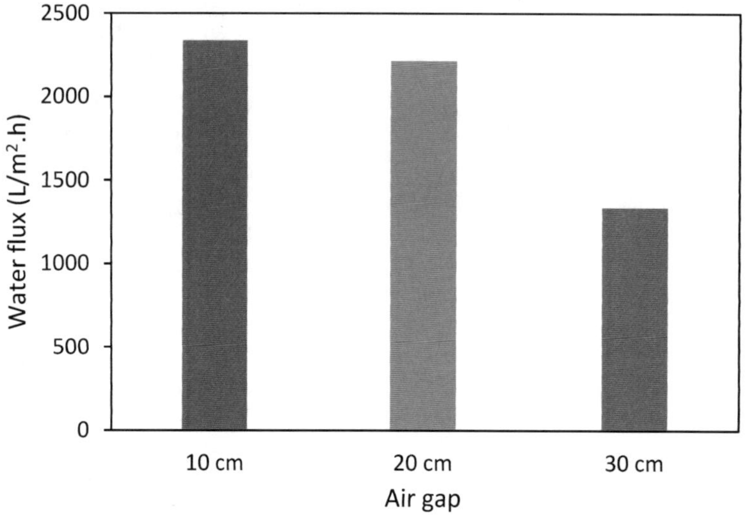


**Figure 8**. Water flux of PVDF HFM at different air gap.

## 4. Conclusions

PVDF hollow fibre membranes with three different air gap, 10 cm, 20 cm and 30 cm have successfully prepared via a dry wet spinning technique. The PVDF HFM 3 membranes have finger-like structure at the outer and inner layer with a sponge-like structure at the intermediate space. Different air gap during spinning procedure significantly affected the finger-like structure, porosity, pore size distribution, membrane surface roughness and increases of tensile strength as well as contact angle which give rise to hydrophobicity. PEG 400 at 3 wt.% loading affected the finger-like structure which form a longer finger-like that help to enhance in water flux performance. The characterization study revealed the HFM 3-10 exhibit the highest porosity of 61.23% and longest finger-like structure which resulting the highest water flux performance at 2338.88 L/m$^2$.h. Furthermore, HFM 3-10 also possessed highest surface roughness which lead the membrane nominated as a good support for coating. Thus, reduce the leaching issue during the water treatment. Besides, HFM 3-10 also experiencing hydrophilicity nature which can minimize the fouling problem. The used of low toxicity of TEP as a solvent also assist in producing a green membrane with outstanding mechanical strength to withstand high pressure. For overall, HFM 3-10 can be nominated as suitable microfiltration membrane for separation process in water treatment.

## 5. Conflict of interest

The authors declare that they have no conflict of interest

## 6. Acknowledgement

The authors gratefully acknowledge the financial support from the Ministry of Higher Education Malaysia under the Malaysia Research University Network (MRUN) Grant (Project number: R.J130000.7809.4L867) and the Higher Institution Centre of Excellence Scheme (Project Number: R.J090301.7809.4J430), and also Universiti Teknologi Malaysia under the Transdisciplinary Research Grant (TDR) (Project number: Q.J130000.3509.05G75), and UTM Award Grant (Project number: R.J130000.7709.5M003). The authors would also like to thank the Research Management Centre (RMC), Universiti Teknologi Malaysia for the technical support and services provided.

## References

[1] Bassyouni M, Abdel-Aziz M, Zoromba MS, Abdel-Hamid S, Drioli E 2019 *Journal of Industrial and Engineering Chemistry* **73** 19-46.
[2] Wang L, Huang D, Wang X, Meng X, Lv Y, Wang X 2015 *Desalination* **361** 25-37.
[3] Chang J, Zuo J, Zhang L, O'Brien GS, Chung T-S 2017 *Journal of Membrane Science* **539** 295-304.
[4] Abed MRM, Kumbharkar SC, Groth AM, Li K 2012 *Journal of Membrane Science* **407-408** 145-154.
[5] Syawaliah, Arahman N, Mukramah, Mulyati S 2017 *IOP Conference Series: Materials Science and Engineering* **180** 1-8
[6] Kamaludin R, Mohamad Puad AS, Othman MHD, Kadir SHSA, Harun Z 2019 *Polymer Testing* **78** 10 1-18
[7] Dzinun H, Othman MHD, Ismail AF, Puteh MH, Rahman MA, Jaafar J 2015 *Chemical Engineering Journal* **269** 255-261
[8] Zhao S, Li Y, Wang Y, Ma Z, Huang X 2019 *Fuel* **244** 78-90.
[9] Li Y, Jin C, Peng Y, An Q, Chen Z, Zhang J 2019 *Journal of the Taiwan Institute of Chemical Engineers* **95** 487-494.
[10] Dzinun H, Othman MHD, Ismail AF, Puteh MH, Rahman MA, Jaafar J 2015 *Journal of Membrane Science* **479** 123 131.
[11] Adam MR, Othman MH, Sheikh Abdul Kadir SH, Mohd Sokri MN, Tai ZS, Iwamoto Y 2020 *Membranes* **10** 1-18
[12] Koyuncu I, Sengur R, Turken T, Guclu S, Pasaoglu ME 2015 Advances in water treatment by microfiltration, ultrafiltration, and nanofiltration *Advances in Membrane Technologies for Water Treatment* ed Basile A, Cassano A, Rastogi NK (Oxford: Woodhead Publishing) chapter 3 pp 83-128.
[13] Khayet M. *Chemical Engineering Science* **58** 3091-3104.
[14] Khulbe KC, Feng CY, Matsuura T, Mosqueada-Jimenaez DC, Rafat M, Kingston D 2007 *Journal of Applied Polymer Science* **104** 710-721.
[15] Paxton NC, Wong CS, Desselle MR, Allenby MC, Woodruff MA 2020 Bone morphogenetic protein-assisted bone regeneration and applications in biofabrication *Biomaterials for Organ and Tissue Regeneration* ed Vrana N, Knopf-Marques H, Barthes J (Oxford: Woodhead Publishing) chapter 14 pp 363-391.
[16] Purkait MK, Sinha MK, Mondal P, Singh R 2018 Introduction to membranes *Stimuli Responsive*

*Polymeric Membranes* (Elsevier) chapter 1 pp 1-37

[17] Khan SJ, Hankins NP, Shen L-C 2016 Submerged and Attached Growth Membrane Bioreactors and Forward Osmosis Membrane Bioreactors for Wastewater Treatment *Emerging Membrane Technology for Sustainable Water Treatment* ed Hankins NP, Singh R (Boston: Elsevier) chapter 11 pp 277-96.

[18] Shi H, Liu F, Xue L 2013 *Journal of Membrane Science* **437** 205-215.

# Titanium Dioxide Incorporated Polyamide Thin Film Composite Photocatalytic Membrane for Bisphenol A Removal

Noor Syahida Mat Anan[1], Juhana Jaafar[1,*], Shuichi Sato[2], Rohimah Mohamud[3]
[1]Advanced Membrane Technology Research Centre (AMTEC), School of Chemical and Energy Engineering, Faculty of Engineering, Universiti Teknologi Malaysia, 81310 Skudai, Johor, Malaysia
[2]Department of Electronic Engineering, Tokyo Denki University, 5 Senju-Asahi-cho,Adachiku, Tokyo 120-8551, Japan
[3]Department of Immunology, School of Medical Sciences, Health Campus, Universiti Sains Malaysia, 16150 Kubang Kerian, Kelantan.
*Corresponding author: juhana@petroleum.utm.my, Tel:+607-5535352, Fax: +607-5535925

**Abstract.** Polyamide thin film composite immobilized with $TiO_2$ is prepared using M-phenylene diamine and Trimesoyl chloride via interfacial polymerization. To determine the characteristics and performance of the membranes, effect of $TiO_2$ loading is studied. 2 (w/v) % of MPD solution and 0.2 (w/v) % of TMC/n-hexane solution were used to prepare PA TFC membrane with different amount of $TiO_2$ loading (0 wt. %, 0.1 wt. %, 0.2 wt. %, 0.3 wt. %, 0.4 wt. % and 0.5 wt. %).PA/$TiO_2$ composite membranes are characterized by FESEM, EDX, FTIR and water contact angle. The properties of the membranes have improved upon $TiO_2$ loading. With over 90% of BPA rejection exhibited by all the fabricated membranes, it is proved that the membranes especially PA/$TiO_2$ TFC membrane with 0.4 wt. % $TiO_2$ (with the highest rejection and degradation value of 99.9% and 14% respectively) are potentially employed as the hybrid photocatalytic membrane for BPA removal.

*Keywords: polyamide, thin film composite, Bisphenol A, Titanium Dioxide, photocatalytic*

## 1. Introduction

Worldwide development in the industrial sector has become one of the main factors of the global water crisis by means of destroyed clean water sources [1]. Thus, reuse of wastewater has become one of the solutions to augment freshwater resources. However, to win public confidence on the matter is a great challenge. Demonstration to the public about the reliability of the safe reusable water production technology is crucial [2]. This water crisis has induced researchers to explore and develop numerous water and wastewater treatment technologies to treat harmful pollutants and transform the water into reusable or even drinkable water. One of the pollutants found in wastewater that has become a great concern to researchers is endocrine-disrupting compounds (EDC). Generally, chemicals that can affect the endocrine system of animals including humans are called EDCs [3]. In a specific way, the word "endocrine disrupters" generally means the synthetic chemicals and natural plant compounds that may influence the endocrine system which is the communication system of glands, hormones, and cellular receptors in charge of the body's internal functions. Many of these compounds have been related to developmental, reproductive, and other health issues in wildlife and laboratory animals [4].

One of the EDCs found in large abundance in wastewater is Bisphenol A (BPA). This leads to the issues of the impact of BPA exposures daily on humans due to its high-volume production. The estrogenic properties of BPA, a ubiquitous synthetic monomer that can leach into the food and water supply, have prompted considerable research into exposure-associated health risks in humans [5]. This chemical is one of the EDCs found in water and likely to perform as an endocrine disruptor even at environmentally appropriate amounts.

Fortunately, research in membrane technology provides many great processes that can be used in water and wastewater treatment. Among them, there is one that has good separation and permeation characteristics [6] and has been reported to achieve high water flux in the membrane separation process [7] which is polyamide (PA) thin film membrane. Thin film that acts as an active layer is commonly produced via interfacial polymerization (IP) method that can provide good selectivity and high-water permeation rate. MPD and TMC are usually used in the fabrication of PA thin film by IP since these chemicals provide aromatic cross-linked PA layer and subsequently produce high water flux [8]. The dense structure of PA layer is the major aspect that contributes to the high separation performance meanwhile the porous support layer contributes to the high water flux. Other than that, PA layer acts as photocatalyst site where $TiO_2$ photocatalyst is embedded. $TiO_2$ is an excellent photocatalyst with applications in various fields. The main advantages of $TiO_2$ are its high chemical stability when exposed to acidic and basic compounds, its nontoxicity, its relatively low cost and its highly oxidizing power, which makes it a competitive candidate for many photocatalytic applications [9].

The main objective of this study is to investigate the effect of $TiO_2$ loading on the characteristics and performance of PA/$TiO_2$ TFC membrane. In this study, polysulfone (PSf) flat sheet membrane is used as the porous substrate of PA thin film meanwhile $TiO_2$ is immobilized into PA layer thus providing the dual effect as the photocatalytic membrane. PA/$TiO_2$ TFC photocatalytic hybrid membrane prepared in this study was believed to achieve better separation and photodegradation performance in eliminating BPA. The $TiO_2$ photocatalyst was exposed directly to light source since it was embedded on the top surface layer of membrane and this approach might enhance the light penetration towards the $TiO_2$ and consequently could improve the photocatalytic activity. The physicochemical properties and photocatalytic and membrane separation performances are determined as a function of $TiO_2$ loading to the PA TFC membranes.

## 2. Materials and methods

### *2.1 Fabrication of PA/$TiO_2$ thin film composites membranes by interfacial polymerization*

The commercial PSf flat sheet membranes were used in this experiment as support or substrates membranes. These substrates membranes were soaked in 2% (w/v, i.e., 2g MPD/100 mL water) of MPD solution for 1 min. $TiO_2$ nanoparticles were immobilized in the MPD/water solution with different concentrations (0-0.5 wt.%). After removing the excess MPD solution on the substrate surface, 100 mL of 0.2% (w/v) TMC/n-hexane solution was gently poured onto the substrate surface. N-hexane was selected as the organic solvent. The TMC was allowed to react with the MPD for a fixed time (1 minute) to form a PA selective layer, followed by draining off the excess TMC solution from the membrane surface. Thermal treatment of MPD/TMC membranes involving placing fresh PA TFC membranes in ventilation oven at 80°C at ambient pressure for 5 min. The flat sheet based TFC membrane was stored in DI water for further use. The prepared membranes were recognized differently based on the amount of $TiO_2$ loading accordingly as PA/0.1 $TiO_2$, PA/0.2 $TiO_2$, PA/0.3 $TiO_2$, PA/0.4 $TiO_2$, PA/0.5 $TiO_2$ TFC membranes while PSf and PA substrate are control membranes.

### *2.2 Characterization of PA/$TiO_2$ thin film membranes*

#### *2.2.1 Morphological analysis by Field-Emission Scanning Electron Microscope (FESEM) and energy dispersive X-ray (EDX)*

The morphological structure of PSf substrate membrane and PA/$TiO_2$ TFC membrane was analyzed using Field Emission Scanning Electron Microscopy (FESEM) image. The membranes sample was broken in liquid nitrogen, and the fractured cross sections and the bottom surface of the membranes were observed. An energy dispersive X-ray analyzer (EDX, OXFORD INSTRUMENTS, USA) was used to quantify elemental analysis and dispersion of $TiO_2$ on top of the PSf PA $TiO_2$ thin film composite membrane.

*2.2.2 Fourier Transforms Infrared Spectroscopy (FTIR)*
FT-IR spectra were carried out on a Perkin Elmer infrared spectrometer using attenuated total reflection (ATR) accessory to provide a non-destructive measurement method. All samples were directly put in contact with ATR diamond by clamping each sample onto the diamond surface, and then pressure was applied to ensure good optical contact between the sample and the diamond. All samples were scanned within the wave range of 650–4000 $cm^{-1}$ at room temperature.

*2.2.3 Contact Angle Analysis*
The relative hydrophilicity of the membrane surface was determined by measuring the contact angle of a water drop (1µL) deposited onto the membrane surface. The contact angle was measured by a static sessile drop method with a goniometer. Deionized water was used as the probe liquid in all measurements. To reduce the evaporation effect, measurements were made as quickly as possible (less than 10 seconds) while 5 measurements were taken for each sample and the average calculated.

*2.3 Preparation of Synthetic Pharmaceutical Wastewater (BPA solution)*
In this study, BPA was chosen as the organic compound to evaluate the effectiveness of the fabricated PA/$TiO_2$ TFC membranes. The standard stock solution of the BPA >99% purchased from Sigma Aldrich was prepared in methanol, freshly before being spiked into the feed solution. The stock solution was diluted to the required concentration (100mg/L) for further use.

*2.4 Performance Study of PSF/PA $TiO_2$ TFC Membrane*

*2.4.1 Pure Water Flux*
The pure water flux ($J$) for the prepared membranes was carried out by using a cross flow filtration system at 5 bar pressure at room temperature. The prepared membranes were pre-pressurized to minimize the compaction effect by using deionized water for 30 min at the pressure of 5 bar before measurement. The pure water flux ($J$) was calculated using the following equation;

$$J = V / A.t$$

where $J$ is the pure water flux (L•$m^{-2}h^{-1}$), $V$ is the permeated water volume (L), $t$ is the permeation time (h), and $A$ is the membrane area.

*2.4.2 BPA rejection*

The rejection of BPA was carried out through the cross flow filtration system. 1 liter of 100ppm BPA solution was used as the feed solution. The concentration of the feed and the permeate solution was identified by using high performance liquid chromatography (HPLC) analyzer (Agilent Technology 1200 Series). The BPA rejection is expressed in terms of percentage of rejection according to the following equation;

$$\text{Rejection of BPA} = \frac{Cf - Cp}{Cf} \times 100$$

where $C_f$ is the feed concentration and $Cp$ is the permeate concentration.

*2.4.3 BPA Degradation (Photocatalytic activity)*

The photocatalytic activity of PA/$TiO_2$ fabricated membranes was run in a photocatalytic membrane reactor placed in the closed container made from stainless steel box with the dimension of (60 x 35 x 42) cm as shown in Figure 1. As the light source, UV lamp (Vilber lourmat, λ=312nm, 30 watt) manufactured by Wuhan Co-shine Technology Co., Ltd, China was installed in the photocatalytic reactor. The prepared membrane was immersed in 1000 mL of 100 ppm BPA solution and was oxygenated in the dark for 120 minutes to achieve the adsorption/desorption equilibrium. The sample was collected every 30 minutes and was analyzed to determine the membranes adsorption. After 120 minutes has ended, the BPA solution along with the membrane was exposed to UV light for 5 hours and the sample was collected at regular time intervals (30 minutes) during irradiation within 5 hours. The degradation sample was quantified by using HPLC analyser (Agilent Technology 1200 Series) coupled with a programmable UV detector. The photocatalytic activity was expressed in the percentages of BPA degradation according to the following equation.

$$\text{Degradation of BPA} = \frac{C0-Ct}{Ct} \times 100$$

where $C_0$ was the initial BPA concentration and $C_t$ was the BPA concentration at time,t.

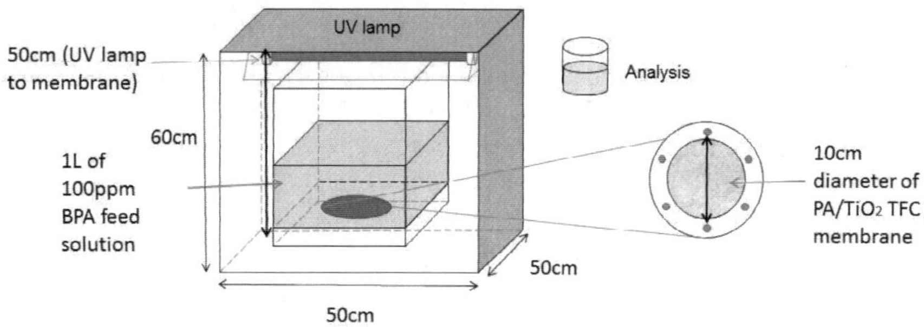


Figure 1: Schematic diagram of photocatalytic reactor set up of PA/$TiO_2$ TFC membrane

## 3. Results and discussion

*3.1 Morphological analysis by Field-Emission Scanning Electron Microscope (FESEM), Fourier Transforms Infrared Spectroscopy Analysis (FTIR) and energy dispersive X-ray (EDX)*

*3.1.1 Field-Emission Scanning Electron Microscope (FESEM) Analysis*

The morphological structure of PSf substrate and PA/$TiO_2$ TFC membranes has been analyzed using FESEM. FESEM image was used to verify the existence of the selective layer or PA layer. Figure 2 shows the comparison of surface morphological structure between PSf substrate and PA/$TiO_2$ TFC membranes with different loadings of $TiO_2$. Image for top of PSf substrate has shown smooth surface while PA TFC possessed a rough surface. The iconic "ridge and valley" structure that represents the PA structure tends to form this rough surface and it can be observed through the PA TFC membrane surface. The formation of the "ridge and valley" structure is an important indicator that proved the formation of PA layer on the PSf substrate [7]. Meanwhile, another study by Arash Mollahosseini and Ahmad Rahimpour [10] also stated that typical "ridge and valley" structure was formed due to the interaction between MPD and TMC and also with the formation of PA on the surface of PSf support membrane. IP process of MPD and TMC is believed to involve an incipient fast stage that forms a dense core barrier layer followed by a slow-growth stage that is MPD diffusion-limited. The first fast reaction generates the dense core layer that is significantly thinner than the extended loose layer formed later as monomer diffuses through the dense core layer. In addition, the PA contains a crosslinked portion (X) and a more hydrophilic linear moiety (Y) containing free carboxylic acid groups [11]. Figure 3 has shown the PA structure formed from an IP reaction between MPD and TMC which has a crosslinked portion (X-fraction) and linear moiety (Y-fraction), which has an unreacted acid chloride group that subsequently, hydrolyzes to a carboxylic acid group.

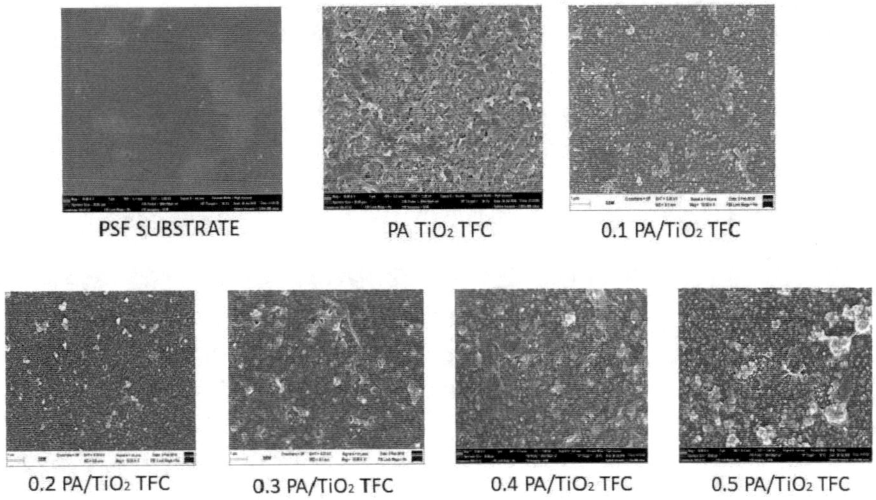


Figure 2: FESEM image of PSF substrate and PA/$TiO_2$ TFC membrane with different loading of $TiO_2$ (0-0.5%)

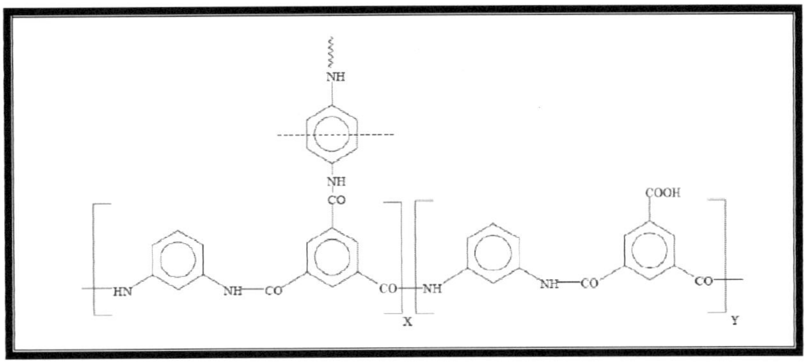


Figure 3: The polyamide structure formed from an IP reaction between MPD and TMC has a crosslinked portion (X-fraction) and linear moiety (Y-fraction), which has an unreacted acid chloride group that subsequently, hydrolyzes to a carboxylic acid group

As shown in Figure 2, compared to PA TFC image with clear "**ridge and valley**" structure, the structure in PA/$TiO_2$ TFC membrane gets less visible since $TiO_2$ particles have covered it. This might happen due to the immobilization of $TiO_2$ in PA layer has affected the PA layer image structure. More than that, the addition of nanoparticles during IP process will affects PA integrity and reduces its cross linking degree that significantly affect the ridge and valley structure formed [12]. Meanwhile, same pattern of FESEM results image has been recorded which has been studied by Thu and co-researchers [13]. In addition, white dots have been detected on all PA/$TiO_2$ TFC membranes which are strongly believed are $TiO_2$ particles. The white spot observed around the features was attributed to $TiO_2$ nanoparticles because of their more "spherical shape" [14]. As shown in this figure, $TiO_2$ particles have been distributed well on all PA/$TiO_2$ TFC membranes. This is supported by EDX mapping image in Figure 6 which has shown the distribution of $TiO_2$ in PA layer.

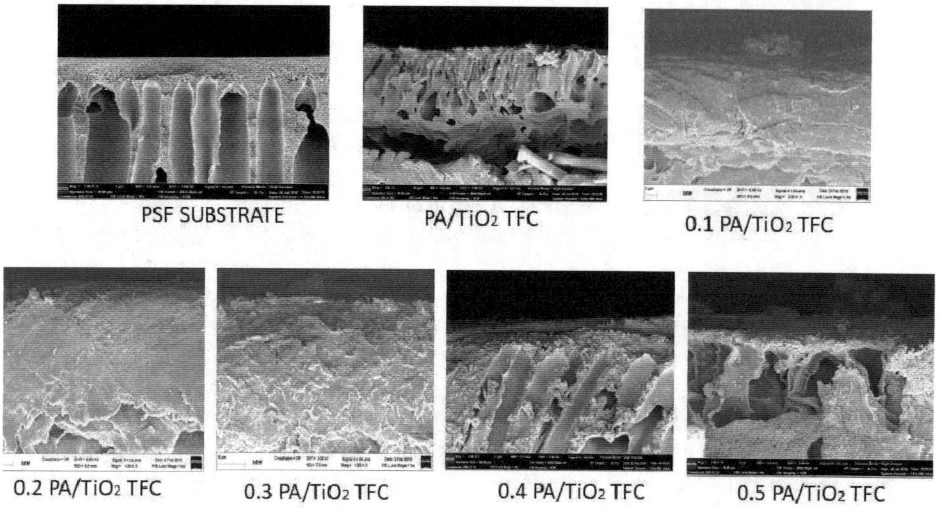


Figure 4: Cross-sectional FESEM image of PSF substrate and PA/$TiO_2$ TFC membrane with different loading of $TiO_2$ ( 0 wt.% - 0.5 wt.%)

Figure 4 shows the comparison of cross-sectional FESEM image between PSf substrate membrane and PA/$TiO_2$ TFC membrane with different loading of $TiO_2$. As can be seen in Figure 4, the PSf substrate membrane had a smooth surface with finger pore like structure at the bottom. Meanwhile, image that represents the PA TFC membrane, the surface of PSf substrate membrane has been covered with another rough layer which is believed as PA layer. According to the image of PA/$TiO_2$ TFC membrane, the formation of a dense selective layer of PA in TFC membrane has been detected which is absent in PSF substrate image. The same features were displayed by TFN membranes prepared by Wang and co-researchers when they were fabricating thin film nanocomposites membranes [15]. Through observation based on these two images, it can be stated that the fabrication of PA layer is a success on top of PSf substrate. As seen from this FESEM image, the cross-sectional image was not clear due to the error during the preparation of sample. However, the image has shown the rough surface and dense structure of PA layer which exists on top of PSf substrate membrane compared to the smooth surface of PSf substrate membrane. This image has proven that PA layer exists as the top layer in all PA/$TiO_2$ TFC membranes even when the “ridge and valley” structure of PA in Figure 2 has been covered by immobilization of $TiO_2$.

### *3.1.2 Fourier Transforms Infrared Spectroscopy Analysis*

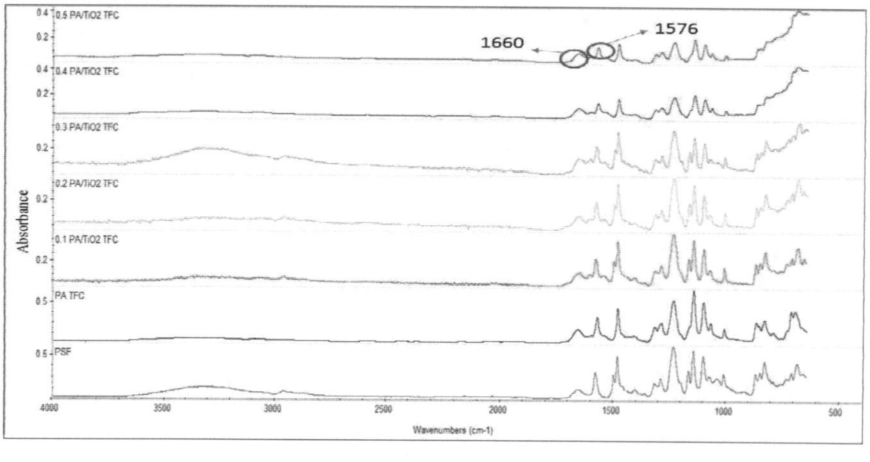

Figure 5: FTIR peak frequencies for PSf substrate membrane and PA TFC membrane and PA/$TiO_2$ TFC membrane

Figure 5 displays the comparison of FTIR analysis data between PSf substrate, PA TFC and PA/$TiO_2$ TFC membranes. The results clearly show the existence of amides functional group at the frequency of 1660 $cm^{-1}$ attributed to the stretching vibration of C=O for PA TFC membranes which this corresponding peak was found absent in PSf substrate membrane. Besides that peak, there is one more peak that represents the amide functional group in PA TFC membranes which is at 1576 $cm^{-1}$ representing the N-H bending in PA bonding. All these bands are categorized in amide group produced by the polymerization process in PA structure formation, as stated by Li and co-researcher [16]. This functional group of amides that existed has verified the formation of PA layer on top of PSf substrate membrane which supported the morphological structure studied. The FTIR spectra of PA/$TiO_2$ TFC membrane are almost identical to PA TFC membrane, indicating the in-situ polymerization has occurred in the existence of $TiO_2$ photocatalyst. As seen from the image, the intensity of peaks for 0.4 PA/$TiO_2$ TFC and 0.5 PA/$TiO_2$ TFC membranes were found lower as compared to others membrane. This might be due to the higher content of $TiO_2$ in the membrane that have created more interaction with the PA.

### *3.1.3 EDX Analysis*

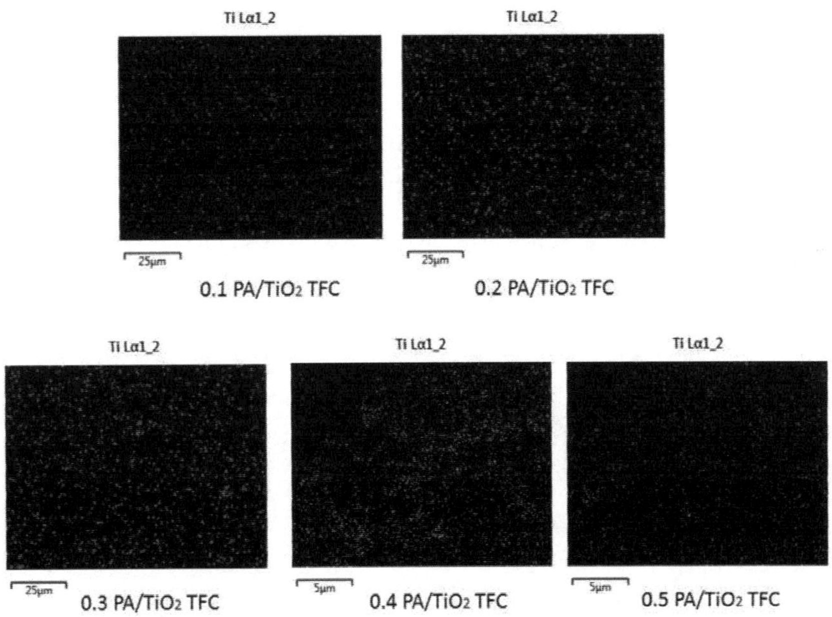


Figure 6: EDX mapping image of PA/$TiO_2$ TFC membrane with various loading (0.1-0.5wt.%)

**Table 1.** The EDX data of PA/$TiO_2$ TFC membrane with various loading (0.1-0.5wt.%)

| Type of membrane | C (wt%) | O (wt%) | Ti (wt%) |
|---|---|---|---|
| PA/0.1 TiO2 | 73.2 | 25.0 | 1.8 |
| PA/0.2 TiO2 | 74.4 | 22.6 | 3.0 |
| PA/0.3 TiO2 | 74.1 | 21.9 | 4.0 |
| PA/0.4 TiO2 | 44.9 | 27.7 | 27.4 |
| PA/0.5 TiO2 | 66.8 | 27.8 | 5.4 |

To investigate the dispersion quality of $TiO_2$ nanoparticles, the EDX mapping analysis was also carried out throughout the hybrid polyamide thin films. Figure 6 shows the percentage of Ti detected which represents the quantity of $TiO_2$ distributed in PA/$TiO_2$ TFC membranes along with the mapping image. As shown in Figure 6, the amount of $TiO_2$ has been increased from 1.8% (PA/0.1 $TiO_2$ TFC membrane) up to 27.4% (PA/0.4 $TiO_2$ TFC membrane) then dropped to 5.4% (PA/0.5 $TiO_2$ TFC membrane). The dramatic drop might be due to the leaching of $TiO_2$ particles from PA layer when over amount of $TiO_2$ particles were added to the PA/TFC membrane. As can be see through Figure 6, the amount of colored dots that represent $TiO_2$ particles has become more obvious and clear from 0.1wt.% to 0.4wt.% $TiO_2$ loading but in 0.5wt.% $TiO_2$ loading, the dots have been reducing. The lowest density dispersion of $TiO_2$ nanoparticles has been obtained in the case of PA/0.1 $TiO_2$ TFC membrane. However, more concentrated dispersion of $TiO_2$ has been observed from PA/0.2 $TiO_2$ TFC membrane until PA/0.4 $TiO_2$ TFC membrane. Figure 6 has clearly shown that the maximum amount of $TiO_2$ identified in PA/TFC membranes is 27.4% which is in 0.4wt.% $TiO_2$ PA/TFC membrane. These results show that 0.4wt.% $TiO_2$ loading has become the optimum loading since the loading higher than that value has initiated the leaching problem. The thin film layer of aromatic polyamide TFC membranes is composed of the crosslinked form of three amide linkages and the linear form with pendant free carboxylic acid and two amide linkages. The behavior of $TiO_2$ on the polymer with COOH group is in two different adsorption ways. One way is that $TiO_2$ was bound with two oxygen atoms of carboxylate group via bidentate coordination to $Ti^{4+}$ cations. The other way was to form a H-bond between carbonyl group and the surface hydroxyl group of $TiO_2$ [17].

### *3.2 Contact Angle Analysis*

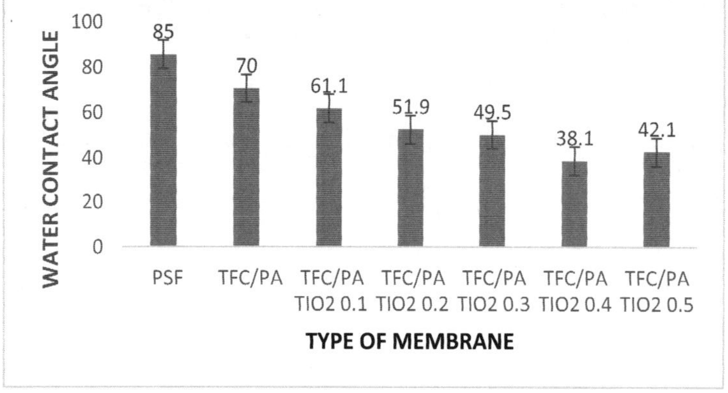


Figure 7: Water contact angle data

The water contact angle (WCA) is commonly applied to study the hydrophilicity of membranes; thus the hydrophilicity features of PSf substrate, PA TFC membrane and PA/$TiO_2$ TFC membranes were investigated and presented in Figure 7. Figure 7 shows that the hydrophilicity of the membrane surface has become much better with the presence of PA layer. As shown in the figure, the WCA value for PA TFC membrane is much lower than PSf substrate. The value of WCA was decreased from 85° to 70° for PSf substrate and PA TFC membranes respectively. The improved hydrophilicity properties make PA TFC membrane desirable for wastewater treatment application. Good hydrophilic polar amides functional group that exists in PA layer structure is the part that makes the value of WCA in PA TFC lower compared to PSf substrate. More than that, membrane with lower WCA value can pull more water molecules onto its surface and improve the transport of water across the membrane [18]. Based on other study, the value of WCA of their PSF membrane is around 75° [19]. The hydrophilicity of PSF membrane might varies depending on the method and parameters used during the membrane fabrication. Meanwhile, the hydrophilicity of PA TFC membranes studied by Rajaeian and co-researcher was 79° compare to its substrate which is 87.3°. Their finding in wetting properties of these two types of membrane is almost identical to this study.

Figure 7 also shows the WCA values for PA TFC membrane and PA/$TiO_2$ TFC membranes at different loading of $TiO_2$ immobilized. As shown in the graph, the value of WCA of PA TFC membrane without $TiO_2$ particles is much higher compared to PA/$TiO_2$ TFC membranes. The value of 70° has been gradually reduced with the existence of PA/$TiO_2$ TFC layer. This proves that $TiO_2$ has further improved the hydrophilicity of PA TFC membranes. The addition of $TiO_2$ has contributed to the hydrophilicity of PA/$TiO_2$ TFC membrane due to the hydrophilic nature of $TiO_2$ particles. This was believed to be due to the presence of the rich hydroxyl group in the $TiO_2$ photocatalyst that has high ability to attract water molecules onto the PA/$TiO_2$ TFC membrane surface, which causes an increase in the degree of hydrophilicity [20]. As expected, the WCA value for PA/0.5 $TiO_2$ TFC is higher than PA/0.4 $TiO_2$ TFC which means that the hydrophilicity of the membrane has been decreased. This belief was correlated to the leaching of $TiO_2$ particles from PA layer due to the overwhelming of $TiO_2$ immobilization since the amount and dispersion of $TiO_2$ particles is less in PA/0.5 $TiO_2$ TFC membrane as shown in EDX mapping image in Figure 6.

### *3.3 Performance Study of PA $TiO_2$ TFC Membrane*

#### *3.3.1 Pure Water Flux*

Figure 8 shows the difference of water flux value in L/m$^2$.h for PSf substrate ,PA TFC and PA/TiO2 TFC membranes. As shown in Figure 8, the water flux value of PA TFC membrane is much lower than PSf substrate. This result is relevant since NF membranes have much lower water flux value compared to UF membranes. Generally, the morphological structure of each membrane is affecting its performance such as water flux value. Based on the morphological study as stated before, PSf substrate membrane has porous structure while PA TFC has denser structure than PSf substrate. This denser structure has slowed down the water flow through the membrane and significantly reduced the water flux value. The typical water flux value of TFC membranes is about 10 L/ m$^2$.h [21]. As stated in Figure 8, the water flux value for PA TFC membrane is lower than TFC typical water flux value. This may be due to the fact that PSf substrate used already has low water flux value compared to the typical UF membrane's flux value which is in the range of 100-2000 L/m$^2$.h [22].

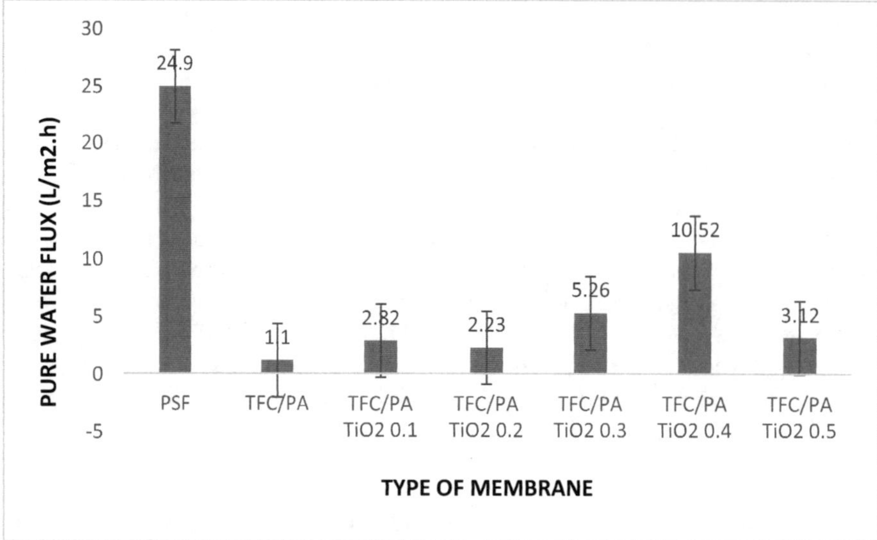


Figure 8: Pure water flux data in L/m$^2$.h at 5 bar

As shown in the figure, the lowest value is for PA TFC membrane without $TiO_2$ particles. Meanwhile, PA/0.4 $TiO_2$ TFC membrane exhibited the highest pure water flux value. Generally, all PA/$TiO_2$ TFC membranes in this study generated low water flux value due to the dense PA layer structure. However, the addition of $TiO_2$ particles also affects the pure water flux value since it is related to the hydrophilicity of the membrane. Since improving membrane hydrophilicity can increase the membrane flux, the highest water flux at 0.4% $TiO_2$ loading might be due to the high hydrophilicity properties of that membrane. The improvement in hydrophilicity properties might also be due to the higher proportion of oxygen-containing functional groups on the PA surface from the nucleophile neutralization of acyl chloride molecule during interfacial polymerization [23]. As expected, the WCA value for 0.5 PA/$TiO_2$ TFC is higher than 0.4 PA/$TiO_2$ TFC which means that the hydrophilicity of the membrane has been decrease. This believed was correlated to the leaching of $TiO_2$ particles from PA layer due to overwhelming of $TiO_2$ immobilization since the amount and dispersion of $TiO_2$ particles is less in 0.5 PA/$TiO_2$ TFC membrane as shown in EDX mapping image.

*3.3.2 BPA rejection*

Figure 9 shows the comparison of BPA removal efficiency between PSf substrate, PA TFC and PA/$TiO_2$ TFC membranes. Through the BPA rejection test, PSf substrate membrane can only remove about 19% of BPA from its initial feed value, 100ppm. This might happen due to the large pore size of PSf substrate membrane and the majority of BPA particles can pass through the membrane. This proves that UF substrate membrane is not practical for BPA removal. Meanwhile, PA TFC membrane has rejected about 98.5% BPA which is much more efficient compared to PSf substrate membrane. Based on the morphological study, PSf substrate has been detected to have porous structure, which explained the poor rejection of BPA since BPA is small in size and can easily pass through the membrane. Meanwhile, the dense layer of PA layer on top of PSf substrate is believed to be the part that contributes to the high BPA rejection.

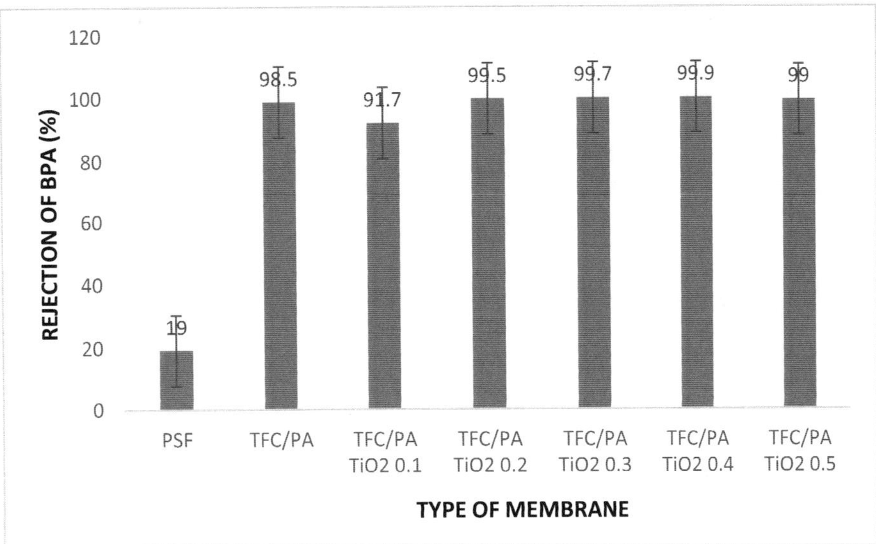


Figure 9: Rejection percentage of BPA

As shown in the figure, the lowest rejection performed is by PA/$TiO_2$ TFC membrane with 0.1% $TiO_2$ loading which is 91.7%. Meanwhile, the highest rejection achieved is by PA/0.4 $TiO_2$ TFC membrane with 99.9% BPA rejection. All fabricated membranes has high BPA rejection which is 90% and above due to the structure of PA layer formed. The dense structure of PA layer contributes to high rejection of BPA. The solution-diffusion process is the most common mechanism for mass transport through nonporous membrane like PA TFC membrane which consist of five fundamental processes. The BPA molecules must first be transported or diffused through the liquid film. Then the BPA solution is in the upstream surface of the membrane and followed by diffusion through the membrane matrix. Then, desorption of solute molecules in the downstream side (permeated side) of the membrane and lastly the diffusion through the boundary layer of the permeate phase [24]. The variation in BPA removal is concluded associated to the physical characteristics of the membrane and also the physicochemical characteristics of the BPA [25,26]. The illustrations of possible BPA rejection mechanism through PSF and its nanocomposite membranes are shown as in Figure 10.

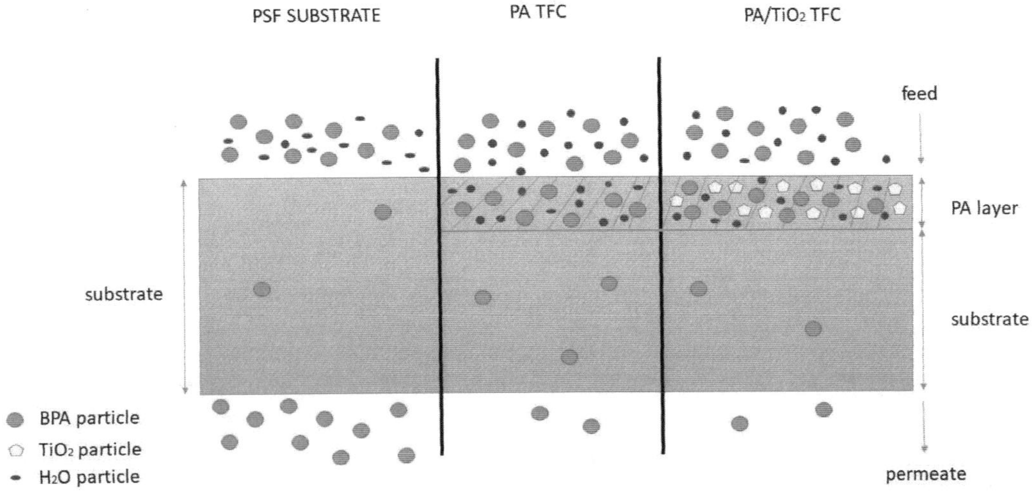

Figure 10: Illustration of possible BPA rejection mechanism through PSF and its nanocomposite membranes

### *3.3.3 Adsorption and Photocatalytic performance of PA/$TiO_2$ TFC membranes*

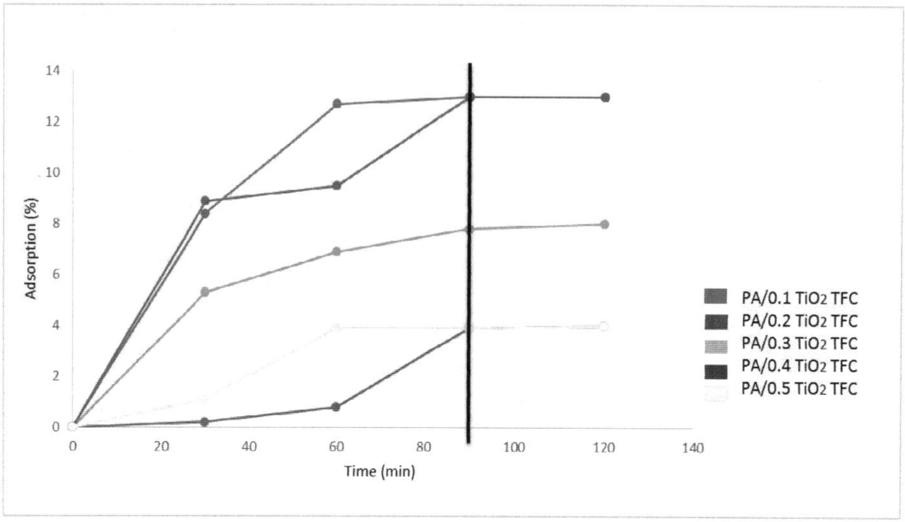


Figure 11: Kinetic adsorption of BPA in the dark

The removal efficiency of the BPA by the prepared PA/$TiO_2$ TFC membranes were measured based on adsorption and photocatalytic degradation of BPA solution with an initial concentration of 100 ppm. The kinetic adsorption of the membrane is displayed in Figure 11. Figure 11 shows that most BPA adsorptions occur on the membrane surface in 90 minutes. Hence, the membrane is kept in the dark for 120 minutes ahead of UV irradiation. It is to confirm that the photodegradation under UV irradiation started after adsorption reach the equilibrium state [27]. After 120 minutes of BPA adsorption, 0.1 PA and 0.4 PA $TiO_2$ TFC membranes showed the maximum BPA adsorption, which is around 13%. The adsorption of 0.2 PA, 0.3 PA and 0.5 PA $TiO_2$ TFC membranes has adsorption value of 4%, 8% and 4% respectively. Similar study on adsorption of $TiO_2$ photocatalytic membrane towards BPA has reported that the adsorption value of BPA by their membrane is 15% which is quite identical to this study [28].

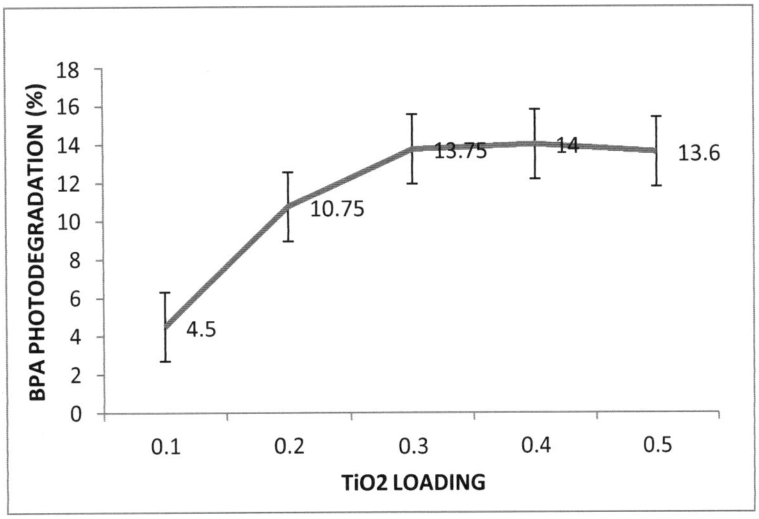


Figure 12: BPA photodegradation performed by PA/$TiO_2$ TFC photocatalytic membrane with different loading

Figure 12 shows the photodegradation percentage performed by PA/$TiO_2$ TFC photocatalytic membranes with different loading of $TiO_2$. Based on the data obtained, the highest percentage of photodegradation of BPA achieved by this type of membrane is 14% which is performed by 0.4 PA/$TiO_2$ TFC photocatalytic membrane. Meanwhile, the lowest photodegradation is performed by 0.1 PA/$TiO_2$ TFC photocatalytic membrane which is 4.5%. 0.4 PA/$TiO_2$ TFC membrane achieved the highest photodegradation percentage value and this might be due to high $TiO_2$ particles embedded in this membrane compared to other PA/$TiO_2$ TFC membranes as shown in FESEM and EDX image before.

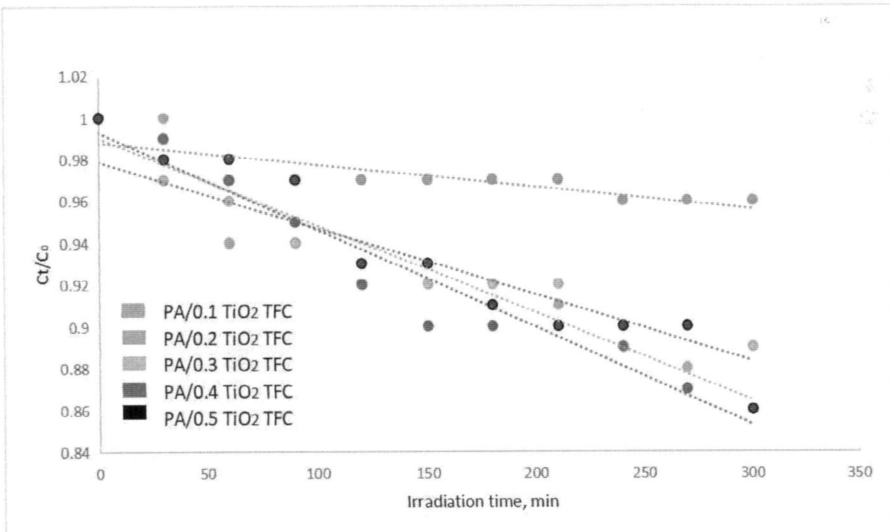


Figure 13: Photodegradation of BPA under UV irradiation

Meanwhile, based on Figure 13, it can be seen that degradation is still proceeding even after 300 minutes of irradiation time except for 0.2 PA/$TiO_2$ TFC membrane. Even after 300 minutes exposure to the UV light source, the degradation of BPA is low. Meanwhile, for 0.2 PA/$TiO_2$ TFC membrane, the degradation has stopped after 240 minutes of irradiation time. This can happened when there is no longer active site available for reaction to take place. However, for others membrane, based on the data, the degradation of BPA might continue if the irradiation time is longer and the percentage of BPA degradation might be higher. It is proposed that the incorporation of the $TiO_2$ photocatalyst to the whole PA layer has produced a wide surface area of $TiO_2$ exposed to UV light and high porosity against the direct interaction with the BPA pollutants, hence promoting excellent BPA degradation process to take place [29]. Moreover, the existence of a uniform dispersion $TiO_2$ catalyst in the PA layer is believed to give the high degradation of the BPA particles. The low degradation of BPA in this study is highly caused by the insufficient amount of $TiO_2$ and the setup of the experiment. The large volume and high concentration of BPA without any movement to get a fast transportation to and from the catalyst other than diffusion during the photocatalytic process hinders a fast degradation of BPA, then results in low degradation of BPA.

Several studies on the degradation of BPA have been done using different types of photocatalytic membranes which resulted in higher degradation of BPA. It was well known that there were a number of variables and parameters that affect the photocatalytic performance which include intensity of the light source, type of catalyst and loading of catalyst. However, since the main variable considered in this study is the $TiO_2$ loading, it is worth comparing the photocatalytic performance of the hybrid membrane by only taking into account the $TiO_2$ content in the membrane. A study by Kamaluddin and team [30] has found that about 86.7% of BPA can be degraded by their dual layer hollow fiber (DHLF) membrane comprised of 7.5 wt% of $TiO_2$ photocatalyst. The $TiO_2$ has been incorporated into DHLF using co-extrusion phase inversion method in the outer layer of the membrane. The high loading of $TiO_2$ used (7.5 wt%) might also be the main contributor to the high BPA degradation. In another study, up to 88% of BPA can be degraded by the ceramic hollow fiber photocatalytic membrane comprised of 1.0 wt% of Ag/$TiO_2$ photocatalyst [31].

Based on these studies, the low degradation of BPA by PA/$TiO_2$ TFC membrane might be caused by the low content of $TiO_2$ available for photocatalysis to take place. However, though at lower loading of $TiO_2$ (0.4 wt.%), PA/TiO2 TFC membrane manages to degrade up to 14% of BPA and hence it is anticipated that higher loading of $TiO_2$, could degrade more BPA. On top of that, the method of immobilizing $TiO_2$ in the membrane is an interesting subject that should be further studied in providing higher effective surface area of $TiO_2$ for photocatalysis activities.

## 4. Conclusion

In this study, PA/$TiO_2$ TFC membrane was successfully fabricated through IP process with immobilization of $TiO_2$ particles on top of PSf substrate. Through FESEM and EDX analysis, the $TiO_2$ particles spotted on PA TFC membrane increase as the $TiO_2$ loading increase but as it reaches 0.5wt.% $TiO_2$ loading, the $TiO_2$ particles have decreased due to the leaching issue. From the FTIR analysis, C=H stretching and N-H bending are present in PA TFC peaks which represents the amides group present in polyamide structure. From contact angle analysis, the hydrophilicity of the membranes improves as the $TiO_2$ loading increase but WCA becomes increased at 0.5wt.% $TiO_2$ loading since loss of $TiO_2$ particles happened. Based on results obtained through the porosimeter test, average pore diameter of PA/$TiO_2$ TFC is not much different for all loading but the average pore diameter of PA TFC membrane is much smaller than PSf substrate. In addition, from the performance study of these membranes, the water flux of PA/0.4 $TiO_2$ TFC membrane is the highest among other loadings while the lowest is at PA TFC membrane without $TiO_2$. From the rejection test, the percentage of removal by these membranes is not much different but the highest percentage of BPA removal is performed by PA/0.4 $TiO_2$ TFC membrane. Meanwhile, PA TFC membranes are much more applicable for BPA removal since PSf substrate only removes about 19% BPA. It is worth mentioning that more $TiO_2$ loading in the PA/$TiO_2$ TFC membranes could contribute to higher photocatalytic activities without compromising the uniformity of $TiO_2$ distribution in the membrane dense skin layer.

## 5. References

[1] Ze-Xian Low, Qi Liu, Ezzatollah Shamsaei, Xiwang Zhang and Huanting Wang, *Preparation and Characterization of Thin-Film Composite Membrane with Nanowire-Modified Support for Forward Osmosis Process.* Membranes, 2015. **5**: p. 136-149.

[2] H.K. Shon, S. Phuntsho and S. Vigneswaran, *Effect of photocatalysis on the membrane hybrid system for wastewater treatment.* Desalination, 2008. **225**: p. 235-248.

*[3]* Singhal, N., Song, Y.,Johnson, A. and Swift, S., *Estrogenic Endocrine Disrupting Compounds.* Auckland Regional Council, 2010. **005**: p. 1-66.

[4] John Helland, E*ndocrine Disrupters as Emerging Contaminants in Wastewater.*House Research Department, 2006. p. 1-4.

[5] Darcie D. Seachrist, Kristen W. Bonk, Shuk-Mei Ho, Gail S. Prins, Ana M. Soto and Ruth A. Keri, *A review of the carcinogenic potential of bisphenol A.* Reproductive Toxicology, 2016. **59**: p. 167-182.

*[6]* Eun-Sik Kim, Young Jo Kim, Qingsong Yu and Baolin Deng, *Preparation and characterization of polyamide thin-film composite (TFC) membranes on plasma-modified polyvinylidene fluoride (PVDF).* Journal of Membrane Science, 2009. **344**: p. 71-81.

[7] Miao Tian, Changquan Qiu, Yuan Liao, Shuren Chou and Rong Wang, *Preparation of polyamide thin film composite forward osmosis membranes using electrospun polyvinylidene fluoride (PVDF) nanofibers as substrates.* Separation and Purification Technology, 2013. **118**: p. 727-736.

[8] Adel Soroush, Jalal Barzin, Mahdi Barikani and Mahdi Fathizadeh, *Interfacially polymerized polyamide thin film composite membranes: Preparation,characterization and performance evaluation.* Desalination, 2012. **287**: p. 310-316.

[9] Marta Castellote and Nicklas Bengtsson. Principles of $TiO_2$ Photocatalysis. *Application of Titanium Dioxide Photocatalysis to Construction Materials.* 2011. p. 5-10.

[10] Arash Mollahosseini and Ahmad Rahimpour, *Interfacially polymerized thin film nanofiltration membranes on $TiO_2$ coated polysulfone substrate.* Journal of Industrial and Engineering Chemistry, 2014. **20**: p. 1261-1268.

[11] Abdel Hameed Mostafa El- Aassar. Polyamide Thin Film Composite Membranes Using Interfacial Polymerization: Synthesis, Characterization and Reverse Osmosis Performance for Water Desalination. *Australian Journal of Basic and Applied Sciences*, 2012. p. 382-391.

[12] G.S. Lai, W.J. Lau, P.S. Goh, A.F. Ismail, Y.H. Tan, C.Y. Chong, R. Krause- Rehberg and S. Awad. Tailor-made Thin Film Nanocomposite Membrane Incorporated with Graphene Oxide using Novel Interfacial Polymerization Technique for Enhanced Water Separation. *Chemical Engineering Journal*, 2018. p. 1-27.

[13] Thu Hong Anh Ngo, Dung The Nguyen, Khai Dinh Do, Thu Thi Minh Nguyen, Shinsuke Mori and Dung Thi Tran. Surface modification of polyamide thin film composite membrane by coating of titanium dioxide nanoparticles. *Journal of Science: Advanced Materials and Devices*, 2016. **1**: p. 468-475.

[14] Majid Peyravi, Mohsen Jahanshahi , Ahmad Rahimpour, Ali Javadi and Saeideh Hajavi. Novel thin film nanocomposite membranes incorporated with functionalized $TiO_2$ nanoparticles for organic solvent nanofiltration. *Chemical Engineering Journal*, 2014. **241**: p. 155–166.

[15] Wang, J., Wang, Y., Zhu, J., Zhang, Y., Liu, J., and Van der Bruggen, B. Construction of $TiO_2$@graphene oxide incorporated antifouling nanofiltration membrane with elevated filtration performance. *Journal of Membrane Science*, 2017.**533**: p. 279–288.

[16] Li, H., Shi, W., Zhang, Y., Du, Q., Qin, X., and Su, Y. Improved performance of poly(piperazine amide) composite nanofiltration membranes by adding aluminum hydroxide nanospheres. *Separation and Purification Technology*, 2016. **166**: p. 240–251.

[17] Sung Ho Kim, Seung Yeop Kwak, Byeong Hyeok Sohn and Tai Hyun Park. Design of $TiO_2$ Nanoparticle Self Assembled Aromatic Polyamide Thin Film Composite Membrane as an

Approach to Solve Biofouling Problem. *Journal of Membrane Science*, 2003. **211**: p. 157-165.

[18] Bano, S., Mahmood, A., Kim, S.-J. and Lee, K.-H. Graphene oxide modified polyamide nanofiltration membrane with improved flux and antifouling properties. *J. Mater. Chem. A*, 2015. **3**(5): p. 2065–2071

[19] S.M. Mousavi, E. Saljoughi, Z. Ghasemipour and S.A. Hosseini. Preparation and Characterization of Modified Polysulfone Membranes with High Hydrophilic Property Using Variation in Coagulation Bath Temperature and Addition of Surfactant. *Polymer engineering and science,* 2012. p. 2196-2205.

[20] G. Akpan and B.H. Hameed, *Parameters affecting the photocatalytic degradation of dyes using TiO2-based photocatalysts: A review.* Journal of Hazardous Material, 2009. **170**: p. 520-529.

[21] W.J. Lau, Stephen Gray, T. Matsuura, D. Emadzadeh, J. Paul Chen and A.F. Ismail, *A review on polyamide thin film nanocomposite (TFN) membranes: History, applications, challenges and approaches.* Water Research, 2015. **80**: p. 306-324.

[22] H. K. Shon, S. Phuntsho, D. S. Chaudhary, S. Vigneswaran and J. Cho, *NanofAiltration for water and wastewater treatment – a mini review.* Drinking Water Engineering and Science, 2013. **6**: p. 47-53.

[23] Kedchaikulrat P, Vankelecom IFJ, Faungnawakij K and Klaysom C. Effects of colloidal $TiO_2$ additives on the interfacial polymerization of thin film nanocomposite membranes. *Colloids and Surfaces A: Physicochemical and Engineering Aspects,* 2020.

[24] Amira Abdelrasoul, Huu Doan, Ali Lohi and Chil-Hung Cheng. Mass Transfer Mechanisms and Transport Resistances in Membrane Separation Process. *Mass Transfer- Advancement in Process Modelling*, 2015.

[25] Heo, Jiyong. "Removal of Micropollutants and NOM in Carbon Nanotube-UF Membrane System from Seawater." *Water Science & Technology*. 2011. 63(11):2737.

[26] Kim, Jae Hyuk, Pyung Kyu Park, Chung Hak Lee, and Heock Hoi Kwon. Surface Modification of Nanofiltration Membranes to Improve the Removal of Organic Micro-Pollutants (EDCs and PhACs) in Drinking Water Treatment: Graft Polymerization and Cross-Linking Followed by Functional Group Substitution. *Journal of Membrane Science,* 2008. **321**(2): p. 190–98.

[27] Dzinun, Hazlini et al. Photocatalytic Degradation of Nonylphenol by Immobilized $TiO_2$ in Dual Layer Hollow Fibre Membranes. *Chemical Engineering Journal*, 2015. **269**: p. 255–61.

[28] N. A. M. Nor, Juhana Jaafar, A.F. Ismail, Mukhlis A. Rahman, M.H.D. Othman, T. Matsuura, F. Aziz, N. Yusof, W.N.W. Salleh and M.N. Subramaniam. Effects of heat treatment of TiO2 nanofibers on the morphological structure of PVDF nanocomposite membrane under UV irradiation. *Journal of Water Process Engineering*, 2017. **20**: p. 193-200.

[29] Mohamed, Mohamad Azuwa, W. N. W. Salleh, Juhana Jaafar, a. F. Ismail, and Nor Azureen Mohamad Nor. Photodegradation of Phenol by N-Doped $TiO_2$ Anatase/rutile Nanorods Assembled Microsphere under UV and Visible Light Irradiation. *Materials Chemistry and Physics,* 2015. **162**: p. 113–23.

[30] Roziana Kamaludin, Zatilfarihiah Rasdi, Mohd Hafiz Dzarfan Othman, Siti Hamimah Sheikh Abdul Kadir, Noor Shafina Mohd Nor, Jesmine Khan, Wan Nor I'zzah Wan Mohamad Zain, Ahmad Fauzi Ismail, Mukhlis A Rahman and Juhana Jaafar. Visible Light Active Photocatalytic Dual Layer Hollow Fiber (DLHF) Membrane and Its Potential in Mitigating the Detrimental Effects of Bisphenol A in Water. *Membranes*, 2020. **32**: p. 1-18.

[31] Usman Shareef, Mohd Hafiz Dzarfan Othman, Ahmad Fauzi Ismail and Asim Jilani. Facile Removal of Bisphenol A from Water through Novel Ag-Doped TiO2 Photocatalytic Hollow Fiber Ceramic Membrane. *Journal of the Australian Ceramic Society*, 2019. p. 1-11.

**Acknowledgments**
We would like to thank AMTEC staff for the help regarding the characterization process of membrane samples. We would also like to show gratitude to Nurafidah Arsat for assistance in conducting lab work until it was fully completed. This research was supported by HICOE grant from the Ministry of Higher Education (MOHE) with grant number: R.J090301.7851.4J433.

Funding: This work was supported by HICOE grant from the Ministry of Higher Education (MOHE) with grant number: R.J090301.7851.4J433

# Ionic Conductivity and Structural Properties for CMC - AN in Propylene Carbonate as Plasticizer in Solid Biopolymer Electrolytes

N.H. **Ahmad**[1]* and H. **Nurhafizah**[2]
[1, 2] Department of Physics, Faculty of Science,
Universiti Teknologi Malaysia, 81310
Skudai, Johor Bahru, Johor, Malaysia.

*Corresponding author's phone: +60162057179
E-mail: nurhidayahahmad@utm.my*

**Abstract.** The present study aims to investigate the structural and ionic conductivity of carboxymethyl cellulose (CMC) - ammonium nitrate (AN) – propylene carbonate (PC) as proton conducting polymer electrolytes. The CMC-AN-PC were prepared via solution casting technique. The complexion of polymer electrolyte films has been confirmed via FTIR studies. Fourier transform infrared (FTIR) spectra showed the occurrence of complexation between the polymer electrolyte components. Impedance studies reveals that the highest ionic conductivity of CMC-AN-5wt% PC 6.5 x $10^{-3}$ $Scm^{-1}$ was observed in polymer electrolyte system obtained at ambient temperature.
**Keyword.** polymer electrolyte, solution casting technique, ionic conductivity, Fourier transform infrared, Impedance

## 1. INTRODUCTION

Ionically conductive polymer electrolytes have potential applications in electrochemical devices such as batteries [1, 3], sensors [2], electrochromic windows [3], supercapacitors [4] and etc. Due to their capability and versatility, solid polymer electrolytes (SBEs) have been introduced in this system to improve the conductive of polymer electrolytes. SBE plays an important role in solid state ionics due to their distinctive properties, such as ease of fabrication, good electrode–electrolyte contact and good dimensional, thermal stability, able to accommodate a wide range of ionic salt doping compositions and high ionic conductivity [5].

CMC is a cellulose derivative and appears as a white or slightly yellow flocculent fiber powder or white powder. It is odourless, tasteless, non-toxic, and is soluble in cold water or hot water to form a certain stick degree of the transparent solution. CMC is a natural organic polymer, the most abundant bio-macromolecule, biocompatible and biodegradable [3, 6]. It contains a hydrophobic polysaccharide backbone and many hydrophilic carboxyl groups and shows an amphiphilic characteristic [3]. CMC is produced by etherification of the hydroxyl groups with sodium monochloroacetate (SMCA) in the presence of aqueous alkali as shown in Figure 1.

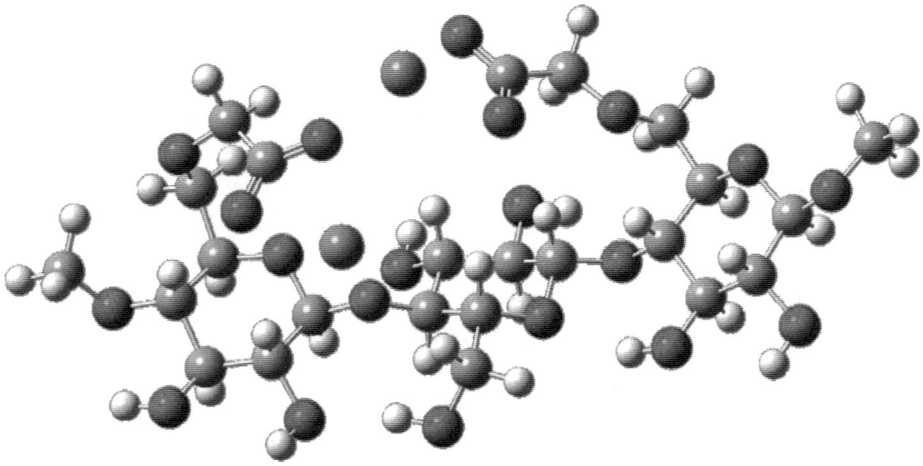

**Figure1** Structure of pure CMC from GaussView 5.0 (red – oxygen, dark grey – carbon, white – hydrogen, purple - sodium).

The addition of ionic dopant to the polymer was reported to increase the ionic conductivity of SBEs by increasing the number of free mobile ions [7]. Thus, the choice of a suitable doping material is essential to ensure the SBE is able to dissociate well in a non-aqueous medium and also provide the pathway for proton to travel with high mobility between electrodes.

In addition, the main purpose of using plasticization methods in polymer-ionic dopant is to enhance the ionic conductivity [7]. A good plasticizer should possess several properties, such as low viscosity, low volatility, high molecular weight, good miscibility between polymer-salt complexes and a high dielectric constant [8].

A variety of plasticizers has been used in polymer electrolytes, such as polyethylene glycol, PEG, imethylformamide, DMF, ethylene carbonate, EC, dibutyl phthalate, DBP, glycerol, dimethyl sulfoxide, DMSO. These plasticizers have been chosen due to their good properties such as, high dielectric constant, help to enhance the ionic dissociation, low viscosity, decrease the glass transition and increase the amorphous content [9]. In this research, propylene carbonate (PC) has been chosen as the plasticizer since it has high dielectric constant, $\varepsilon = 64.4$ [10]. In addition, PC is highly soluble with CMC and AN.

The present work aimed to investigate the electrical and structural of the SBEs system by using electrical impedance spectroscopy (EIS) and Fourier Transform Infrared (FTIR) Spectroscopy, respectively. These experiment combination techniques give in deep understanding on the SBE behavior.

## 2. Experimental

2.0 g of carboxymethyl cellulose (CMC) from Acros Organic Co. was added into distilled water and stirred until it was completely dissolved. The varied amount of ammonium nitrate (AN) was added to the CMC solution and stirred again until the solutions became homogeneous. The mixture was cast into petri dish and heated onto the oven at constant temperature of 50°C for almost 14 hours. Then, the solutions were left to dry at room temperature (28°C) for film formation. For the plasticized sample, the highest conductivity of CMC-AN SBE was chosen to be added with plasticizer. The same method was used to produces CMC-AN-PC SBEs by adding different amount of propylene carbonate (PC) as plasticizer into CMC-AN mixture as captured in Table 1.

**Table 1** Designation of CMC-AN-PC SBEs.

| Designation | Formulation | | | | |
|---|---|---|---|---|---|
| | CMC (g) | AN (wt.%) | AN (g) | PC (wt.%) | PC (g) |
| 0P | 2 | 45 | 1.64 | 0 | 0.00 |
| 5P | | | | 5 | 0.16 |
| 10P | | | | 10 | 0.34 |
| 15P | | | | 15 | 0.53 |
| 20P | | | | 20 | 0.76 |
| 25P | | | | 25 | 1.10 |

### *2.1 Electrical Impedance Spectroscopy*

Electrical impedance spectroscopy (EIS) was utilized to determine the ionic conductivity of SBEs in the frequency range of 50 Hz to 1MHz. The range of the frequency has been used because of the bulk resistance, $R_b$ exists at that range. Electrical measurements were performed on a HIOKI 3532-50 LCR Hi-Tester by ambient temperature at 303K. SBE was placed between the blocking stainless steel electrodes of a sample cell holder connected EIS. The bulk resistance ($R_b$) of SBE was obtained from the plot of imaginary part ($-Zi$) versus real part ($Zr$) of impedance.

### *2.2 Fourier Transform Infrared Spectroscopy*

The complexation of SBEs can be determined by using Fourier Transform Infrared spectroscopy. A Thermo Nicolet 380 FTIR spectrometer equipped with an Attenuated Total Reflection (ATR) accessory utilizing a germanium crystal. The spectrum was recorded in the frequency ranging between 4000 and 700 $cm^{-1}$ with a resolution of 4 $cm^{-1}$.

IOP Conf. Series: Materials Science and Engineering 1142 (2021) 012016 doi:10.1088/1757-899X/1142/1/012016

## 3. Discussion

### *3.1 Cole-Cole plot*

Figure 2 showed the Cole-Cole plot of CMC-AN-PC SBEs at ambient temperature. From the Cole-Cole plot, the value of bulk resistance, $R_b$ was determined. The $R_b$ can be established by measuring the intercept of the tilted line with the x-axis. The Cole-Cole plot of the SBEs containing 0, 5, 10, 15, 20 and 25 wt% of PC showed straight line trend. This straight-line graph or spike represent formation of double layer capacitance at the blocking electrode-electrolyte interface which causes the inclination and is better known as the polarization effect due to migration of ions [11].

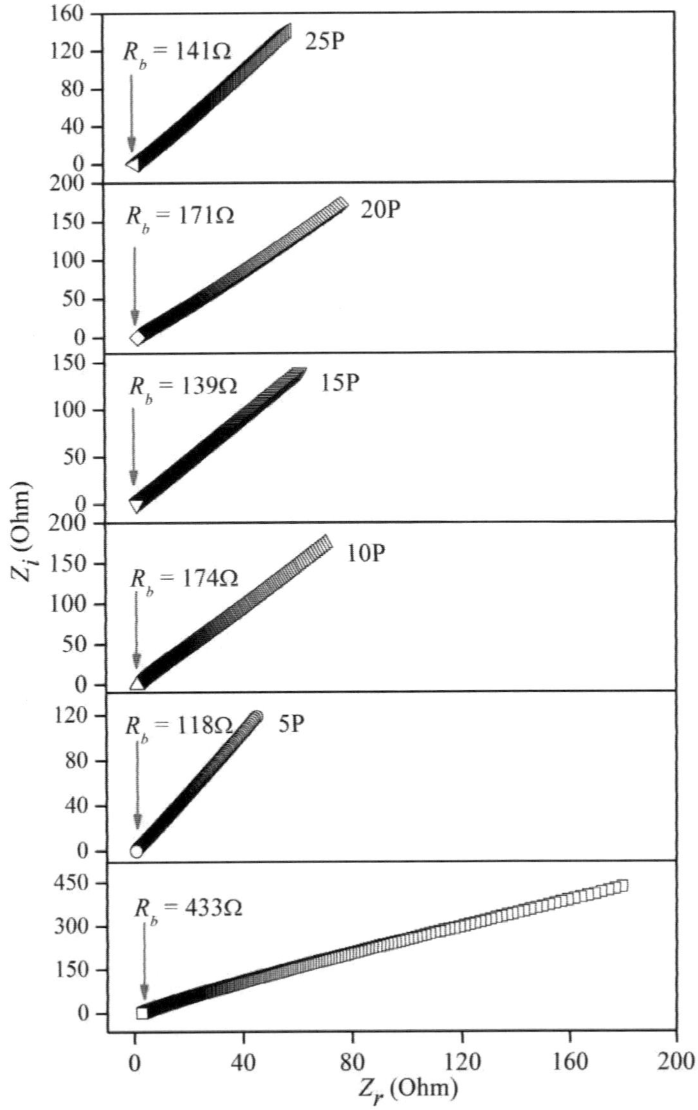


**Figure 2** Cole-Cole plot of CMC-AN-PC SBEs at room temperature.

### *3.2 Ionic conductivity at room temperature*

The value of the bulk resistance, Rb from the Cole-Cole plot in Figure 2 can be substituted into the ionic conductivity ($\sigma$) equation below,

$$\sigma = \frac{t}{R_b A}$$

where, $t$ is the thickness, $R_b$ is the bulk resistance, $A$ ($cm^2$) is the electrode-electrolyte contact area of SBEs.

The ionic conductivity of this system was divided into three regions (Figure 3), where Region I is between 0 – 5 wt.%, Region II between 5 – 20 wt.% and Region III 20 – 25 wt.%. In Region I, the ionic conductivity of unplasticized SBE is $(2.11 \pm 0.03) \times 10^{-3}$ $Scm^{-1}$ and increases to $(6.53 \pm 0.01) \times 10^{-3}$ $Scm^{-1}$ for CMC-AN-5 wt.% PC. After 5 wt.% PC, the ionic conductivity starts to decrease until it reaches the minimum ionic conductivity value of Region II, 20 wt.% PC $(2.63 \pm 0.01) \times 10^{-3}$ $Scm^{-1}$. Beyond 20wt.% PC, the ionic conductivity increases to $(3.63 \pm 0.01) \times 10^{-3}$ $Scm^{-1}$.

As reported by previous work [10], the addition of plasticizer with high dielectric constant such as PC can assist the dissociation of ion / aggregate of salt and producing more free $H^+$ ion to facilitate in the conduction process. However, the insufficient of PC added into CMC-AN SBE may not assist in salt dissociation and causes a reduction in ionic conductivity (Region II in Figure 2). In Region III, the addition of sufficient PC concentration is believed to introduce a shorter pathway between H+ ion and PC, causing more ion dissociation. The number of free ion and ionic mobility are expected to increase and thus improves in ionic conductivity. This result is similar to the work reported by [7] in their plasticized polymer-salt system. Based on these results, it can be said that the addition of AN and EC has influenced the ionic conductivity variation in this system.

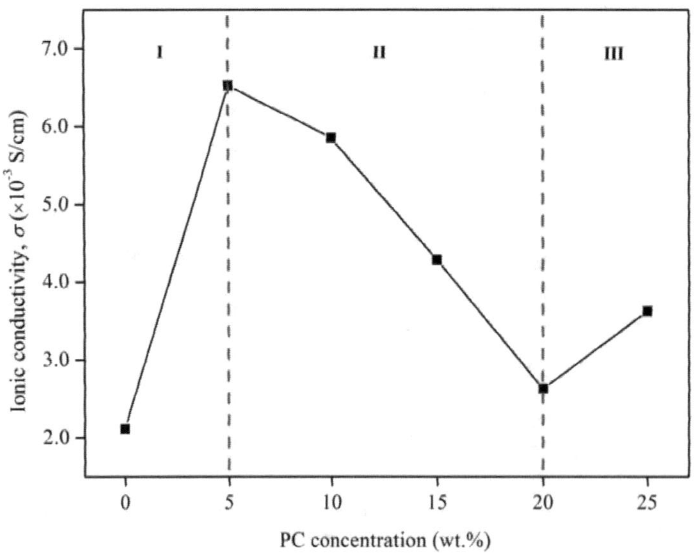


**Figure 3** Ionic conductivity plot of PC concentration in SBEs.

### *3.3 FTIR analysis*

In order to study the interaction between CMC-AN-PC, Figure 4 is plotted as (a) pure PC; (b) pure AN and (c) pure CMC. Figure 5 represents the FTIR spectra of CMC-AN-PC between wavenumber of 4000 and 700 $cm^{-1}$. A broad absorption peak of O-H stretching at 3680 $cm^{-1}$ to 2680 $cm^{-1}$. Pure CMC shows a strong peak of symmetric stretching mode of carboxylic group, $\upsilon_a$ (C=O) at 1591 $cm^{-1}$. The bands located at 1417 $cm^{-1}$ and 1321 $cm^{-1}$ corresponds to $CH_2$ scissoring, δ(C-H) and bending mode of C-OH bonding, δ(C-OH), respectively. The doubled stretching mode of C-O-C were observed at 1105 $cm^{-1}$ and 1057 $cm^{-1}$. This result is agreement as reported by several researchers [12-13].

Pure AN exhibits intense vibration bands in the range of 4000$cm^{-1}$ to 900$cm^{-1}$ as showed in Figure 4. As compare to Figure 5, at wavenumber of 3250 $cm^{-1}$, 3074 $cm^{-1}$ and 2351 $cm^{-1}$ corresponds to stretching mode of ammonium ($NH_4^{+)}$ ion. The mixture of CMC-AN-PC SBEs were observed the asymmetric stretching and deformation modes of both $\upsilon(NH_4^+)$ and $\upsilon(NO_3^-)$ to appear between 1800$cm^{-1}$ to 1500$cm^{-1}$[7]. Zhao et al (2008) reported the asymmetric deformation mode of ammonium ions, $\upsilon(NH_4^+)$ occurs at 1470$cm^{-1}$. The others of the vibration bands from pure AN at the lower wavenumber belongs to the stretching $\upsilon(NO_3^-)$ and bending δ(NO3-) modes of nitrate ions [15-17].

In Figure 5, a peak corresponds to C=O stretching of PC was appeared at 1770 $cm^{-1}$ at 5P to the lower wavenumber of 1750 $cm^{-1}$ upon the addition of 5wt.% of PC in CMC-AN SBE. The peaks were prominent and the intensity increased as PC composition increased. The increase in peak intensity at C=O is believed due to increase of PC supplied in the polymer matrix and proving the interaction between CMC and PC. At the peaks for C-H bending deformation at 1422 $cm^{-1}$ at 5P seems decreased to the lower wavenumber to 1405 $cm^{-1}$ at 25P. It can be summarized that, there is no obvious increment of intensity changes between CMC-AN and PC concentration from the IR spectrum. That is the reason, the ionic conductivity in not increase upon the addition of 10wt.% until 20wt.% of PC concentration from the IR spectrum.

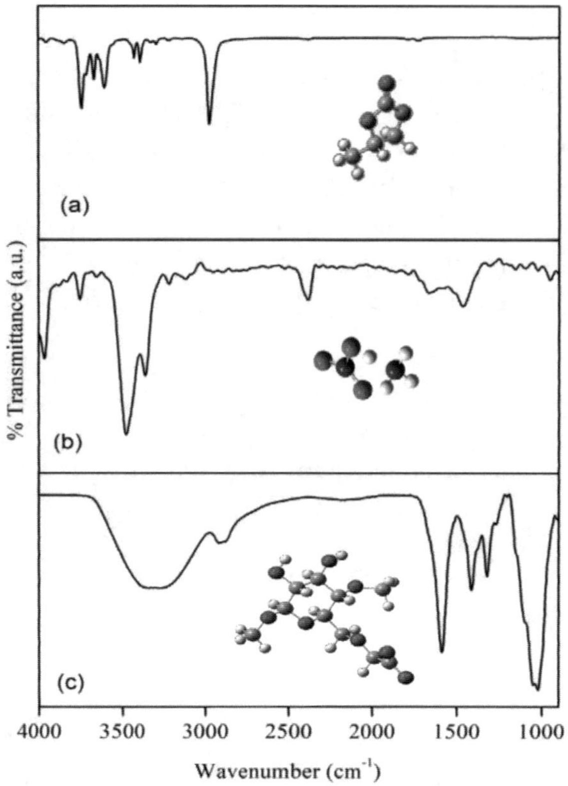


**Figure 4** (a) Pure PC (b) pure AN and (c) pure CMC.

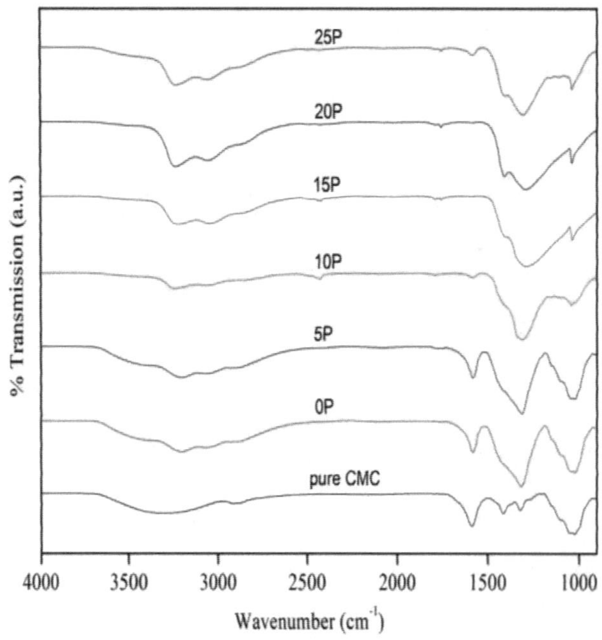


**Figure 5** IR spectra of 0P, 5P, 10P, 15P, 20P and 25P SBEs.

## 4. Conclusion

The CMC-AN-PC SBEs have been successfully prepared by solution casting technique. The highest ionic conductivity obtained at 6.5 x $10^{-3}$ $Scm^{-1}$ at 5 wt. % of PC at ambient temperature. FTIR studies confirmed that the complexation of CMC-AN-PC SBEs. The changes of wavenumber showed that the interaction was completely occurred when 5wt.% of PC was added to CMC-AN system. The IR spectrum shows the amount of PC in the CMC-AN is not sufficient enough to increase the ionic conductivity and to dissociate in this system.

## Acknowledgement

We gratefully acknowledge the financial support from Ministry of Higher Education (MoHE) through grant UTMER Vot. 19J62 of Universiti Teknologi Malaysia and the experimental assistance from Faculty of Science.

## References

[1] Hongfei L et al 2018 *Energy Environ. Sci.* **11**, 941-951.
[2] Higier A et al 2017 *Int. J. Hydrogen Energy* **42**, 4706-4709.
[3] Rani M et al 2014 *Polymers* **6**(9), 2371–2385.
[4] Vidyanand V et al 2017 *Journal of Materials Chemistry A* **5**, 8461-8476.
[5] Shujahadeen B A et al 2020 *Materials* **13**, 3675.
[6] Sohaimy M I H and Isa M I N 2015 *Applied Mechanics and Materials* **719–720**, 67–72.
[7] Hafiza M N and Isa M I N 2020 *Journal of Membrane Science* **597**, 117176.
[8] Hailan K et al 2018 *RSC Adv.* **8**, 11643-11651.
[9] Dhatarwal P and Sengwa R J 2017 *Indian Journal of Pure and Applied Physics* **55**, 7-18.
[10] Ahmad N H and Isa M I N 2016 *Carbohydrate Polymers* **137**, 426–432.
[11] Azli A A et al 2020 *Ionics* **26,** 5773–5804.
[12] Sugama T and Pyatina T 2015 *Cement and Concrete Composites* **55**, 281-289.
[13] Yadollahi M and Namazi H 2013 *Journal of Nanoparticles Research* **15**, 1-9.
[14] Zhao et al 2008 *Carbon* **47**, 331-333.
[15] Kamarudin K H and Isa M I N 2013 *International Journal of Physical Sciences* **8**, 1581-1587.
[16] Kadir et al 2011 *Electrochemica Acta* **55**, 1475-1482.
[17] Wu et al 2007 *Aerosol Science and Technology* **41**, 581-588.

# Textile dye Reactive Black 5 (RB5) removal by visible light photocatalyst and its characterization

**S H Mohamed Noor[1], M H D Othman[1*], J Jaafar[1], M A Rahman[1], A F Ismail[1]**

[1]School of Chemical and Energy Engineering, Faculty of Engineering, Universiti Teknologi Malaysia (UTM), 81310 Skudai, Johor, Malaysia

hafiz@petroleum.utm.my; sitihawamnoor@gmail.com

**Abstract**. Visible light photocatalysis is now a subject of interest for researchers to explore further in the treatment of wastewater as it can save costs and be environmentally friendly. Cadmium sulphide (CdS) is one of the photocatalysts of visible light that has excellent properties and a low band gap. The toxicity and carcinogenicity of textile industry effluent containing predominantly textile dyes disturbs the environment. In this study, the removal of reactive black 5 (RB5), the most used dyes with CdS as visible light photocatalyst, was tested. Using X-ray diffraction (XRD), scanning electron microscope (SEM), ultraviolet-visible-near infrared (UV-Vis-NIR) spectrophotometers and photocatalytic tests under visible light, the physical and chemical properties of CdS were characterised. CdS has an irregular shape and this can be demonstrated through SEM. The band gap obtained was 2.12 eV and this was associated with the degradation efficiency of CdS under visible light as it can degrade RB5 after 360 minutes of exposure by up to 80 percent. This study proves that CdS is a strong photocatalyst of visible light that has a small band gap and crystalline particles.

## 1. Introduction

In March 2019, a case reported in Sungai Kim Kim, Pasir Gudang, Johor, Malaysia about industrial factories that eliminate their chemical wastes into the river [1]. The river affected the community in large radius from the river. After the air contaminated with the poisonous gases, more than 5000 people get affected with difficulties to breath, nausea and vomit symptoms [2]. This case shows that our environment is sick and the awareness among us about the important to keep our earth healthy is low.

For textile industry specifically, it consume huge quantity of water as it hold as second in the world`s largest water consumer and more than 90% disposed [3]. It also discharged their effluents into environment. Their effluents are made up of metals, dyes and other pollutants [4]. 15-20% of wastewater and 17-20% of industrial effluent composed of textile dyes. While during the colouring process, 90% of the dyes used are released as effluents [5]. Dyes could affects the ecosystem as it quite high in pH, colour, chemical oxygen demand, suspended solids, metals, temperature, salts and biochemical oxygen demand [4]. It also a toxic substance, has carcinogenic and mutagenic properties and not only affecting water bodies, but also cause diseases towards humans and animals [6].

There is a law under Malaysia Environmental Quality Act 1974, Quality of Environment (Scheduled Wastes) Regulations 2005 that specify chemicals waste from industrial are scheduled waste. There are five types of waste under this law, SW1-SW5 which consist of metal wastes, inorganic substances, organic substances, waste with organic or inorganic substances and other wastes. Thus, any type of industries needs to obey and dispose their waste through this scheduled waste channel. However, our environment still affected by the irresponsible party.

There are several conventional method to remove textile dyes such biological methods, physical methods and chemical methods [7]. Ozonation, photochemical and oxidative process grouped under chemical methods, decolorization by white-rot fungi and adsorption by microbes are under biological methods while ion exchange, irradiation, membrane filtration and adsorption by activated carbon are some of physical methods. However, these methods gives out common drawbacks to the process such formation of secondary compound, not effective for all dyes, intense sludge formation and not cost effective [8].

Photocatalysis is an alternative to save the environment as currently our environment needs our help due to excessive exposure especially from industry effluents. Photocatalysis is a process that require catalyst and light to accelerate a chemical reaction. This process has been used in organic pollutant degradation especially in wastewater, hydrogen production, air purification and as disinfectant. Photocatalysis made up of two distinct process which are homogenous process and heterogeneous process [9].

Homogenous process needs metal compound and under certain condition, hydroxyl radicals generated by the high metal ions oxidation state react with the organic matter and degrade the organic matters. Heterogenous process require semiconductor materials as the catalyst. Titanium dioxide ($TiO_2$) and zinc oxide (ZnO) are the most common used semiconductor materials. The process needs just a light pressure and temperature and low in maintenance cost. To be a great catalyst, the material needs to fulfil few criteria which can utilize UV or visible light, inert in terms of biologically and chemically, photocorrosion stable, low cost and safe [9]. However, $TiO_2$ and ZnO can only activated under UV light and a low band gap photocatalyst is preferable as it can activated under visible light.

Cadmium sulfide (CdS) is an amazing substance. It has multifunction and has been applied in various area. The combination of cadmium and sulfur forming cadmium sulfide exist in vivid yellow to brown colour. It has molecular weight of 144.46 g/mol and specific gravity of 4.82 g/cm$^3$. It exists in solid but in two polymorphs, cubic hawleyite and hexagonal greenockite and has a sublimation point of 980 °C. The speciality of this substance is it does not soluble in water either its hot or cold [10]. CdS is grouped into II–VI semiconductor and has a 2.42 eV bandgap at 515 nm wavelength [11]. It has been applied in photocatalysis, gas sensor, infrared and laser detector, solar cells which act as buffer layer in Copper-Indium-Gallium-Selenide solar cells, catalyst in photochemical, nonlinear optical materials and devices of luminescence and optoelectronic [10].

Textile industry effluent which mainly contains the textile dyes are disturbing the ecosystem with its toxicity and carcinogenic properties. It also may threaten aquatic life habitat and endangered species may facing extinction. Thus, this study was conducted to remove of Reactive Black 5 (RB5) which is the most frequent used dyes in textile industry using visible light photocatalyst CdS. The physical and chemical properties of CdS was characterized using Scanning Electron Microscope (SEM), X-ray Diffraction (XRD), Ultraviolet-Visible-Near Infra-Red (UV-Vis-NIR) spectrophotometer and the photocatalytic test under visible light.

## 2. Methodology

### *2.1 Materials*
Cadmium sulfide (CdS, Sigma Aldrich, USA) was used as main component. Textile dye Reactive Black 5 (RB5, Sigma Aldrich, USA) used as contaminant to be degraded by photocatalyst.

### *2.2. Characterization*

*2.2.1 Scanning Electron Microscopy (SEM).* The morphological and structural characteristic of the CdS surface were examined using scanning electron microscopy (SEM; TM3000, Hitachi) analysis. Each sample was placed on a stub and coated with gold for 3 minutes under vacuum condition. The SEM image for the sample was taken for different magnifications 2000x and 5000x.

*2.2.2 X-ray Diffraction (XRD).* CdS crystallinity and phase identification was analysed by X-ray Diffraction (XRD). The analysis was carried out at 40 kV and 30 mA. The sample also employed CUK-β radiation at wavelength of 0.15418 nm at an angular incidence of $2\theta = 20\text{-}80°$ with a scan step speed of 1°/min.

*2.2.3 Ultraviolet-Visible-Near Infra-Red (UV-Vis-NIR) spectrophotometer.* UV-Vis-NIR spectrophotometer was used to determine CdS band gap value and the optical property on light absorption capability under UV and visible light irradiation was examined through this method. The sample was clamped to the sample holder and scanned between 200-1000 nm using the spectrophotometer. A graph projected from the analysis and the band gap can be determined from the graph.

### *2.3 Photocatalytic test under visible light*
Photocatalytic activity of CdS to degrade RB5 as textile dye measured under visible light irradiation. 0.1 g CdS powder added in several concentration of RB5 which are 10 ppm, 15 ppm and 20 ppm. The suspension then stirred in dark place for 30 min to reach adsorption-desorption equilibrium. 10 ml of suspension taken and filtered, and the absorbance measured using Perkin Elmer UV-Visible Spectrophotometer at 597 nm for every 30 minutes interval in 360 minutes. The results then calculated using equation below to obtain the degradation of RB5 percentage [12].

$$\text{Degradation of organic contaminants} = \frac{\text{Co-Ct}}{\text{Ct}} \times 100 \qquad (1)$$

Equation of photocatalytic activity percentage where $C_0$ is the initial concentration at t=0 and $C_t$ is concentration at time 30, 60, 90, 120, 150, 180, 210, 240, 270, 300, 330 and 360 minutes.

## 3. Results and discussion

### *3.1 Physical properties*

Figure 1 (a) and (b) are the image of CdS powder under magnification of 2000x and 5000x. The shape showing that CdS has crystalline shape although the image quite not clear. It has a range of 17-22 nm size of particles. As compared with [10] Sahare (2018) research, CdS have a crystalline structure and irregular particle which in more detail, it has hexagonal wurtzite and sphalerite structure and with more larger particle size range 30-50 nm. The crystalline, irregular structure and smaller particle size helps the CdS having large total surface area and higher photocatalytic performance as larger surface area exposed towards the source of light.

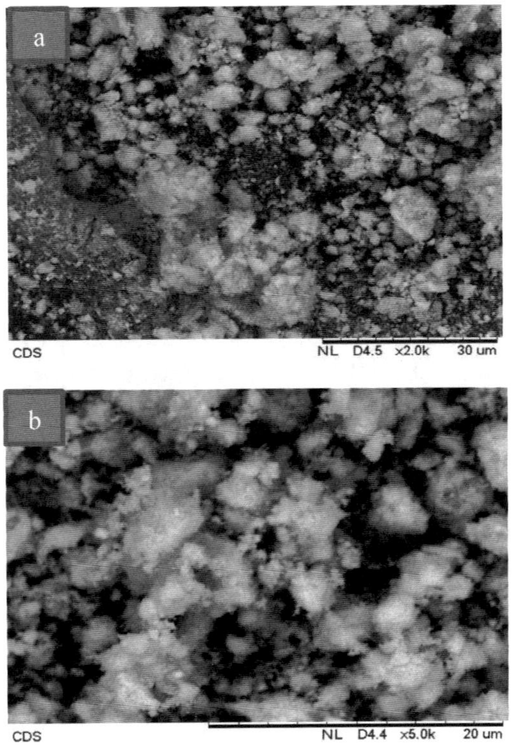


**Figure 1**. Image from SEM of CdS under (a) 2000x and (b) 5000x magnification

The result showed that the substance is 100% contained CdS particle and proved the CdS exist in crystalline shape as XRD only can detect crystalline solid. Besides, it was found the diffraction peak at 25.03°, 43.58° and 51.78° as in Figure 2. The DB Card Number is 01-089-0440. Compared to [13] Aggarwal (2016) research, the peaks quite similar to this study which are $2\Theta$ = 26°, 44° and 52° confirming the powder contained fully crystalline CdS particle.

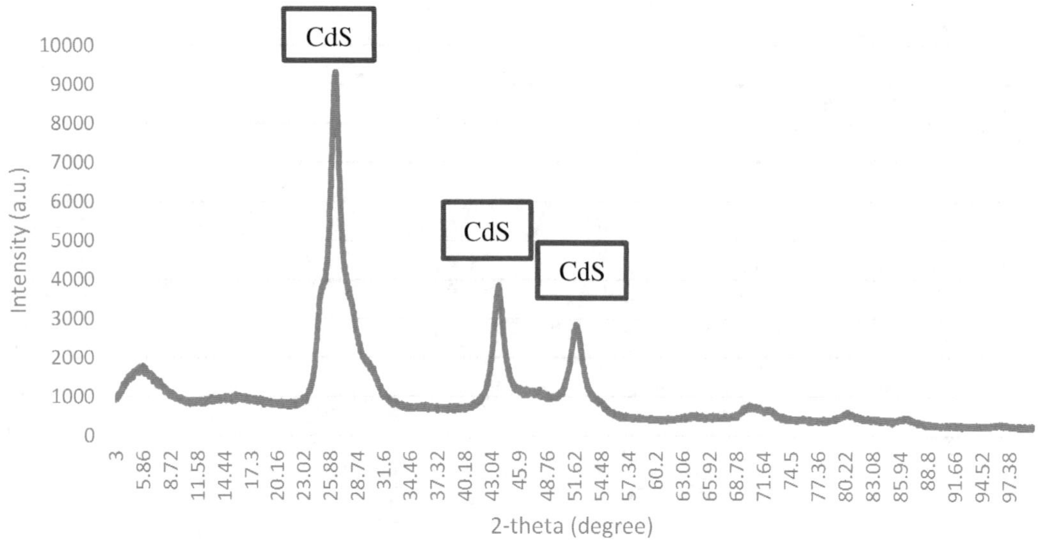


**Figure 2**. Graph from XRD of CdS

Band gap determination from UV-Vis-NIR spectrophotometer proving CdS is photocatalyst with a band gap value 2.12 eV (Figure 3). Band gap value 2.12 eV portrayed that the CdS can works as photocatalyst the best under visible light and can work under ultraviolet but not at its best performance. CdS can excites its electron from the valence band to conduction band without the aid of ultraviolet lamp thus it saves the cost. The visible light region which in the range of 400 nm to 765 nm wavelength are the spectrum that can be seen by human eyes. Thus, as the CdS can be activated with the spectrum up to 585 nm, it only needs normal visible light and by using visible lamp during night. Compared with a study by [14] Pan and Zhu (2015), CdS has a band gap of 2.42 eV which a little bit higher than found in this study. This happened may due to the CdS used in this study was the commercialized CdS and the CdS used by [14] Pan and Zhu (2015) was synthesized CdS. Thus, the band gap may change a little bit due to distance between conduction band and valence band may differ.

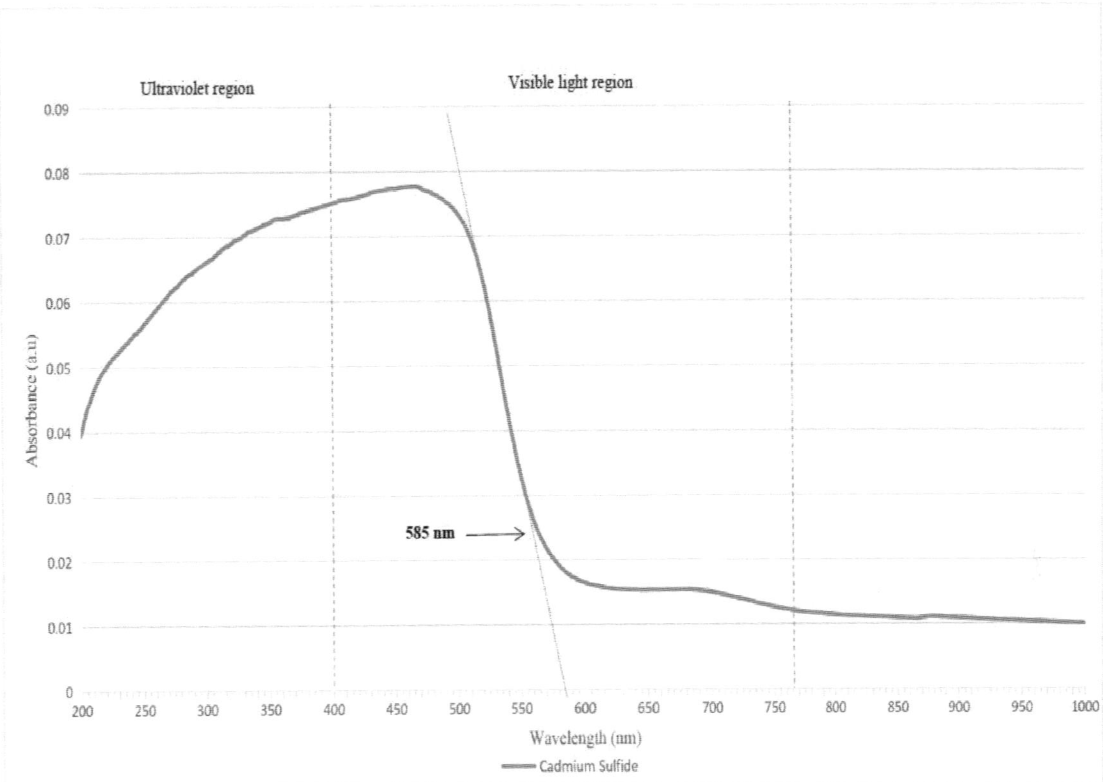


**Figure 3**. UV-Vis-NIR graph of CdS

Photocatalytic activity of CdS to degrade Reactive Black 5 (RB5) as textile dye under visible light irradiation was analysed. The absorbance of the RB5 for every 30 minutes interval in 360 minutes was measured using Perkin Elmer UV-Visible Spectrophotometer at spectrum of 597 nm. Figure 4 showed the effectiveness of CdS powder which able to degrade textile dye RB5 up to 93.72 % at 10 ppm, 92.57% at 15 ppm and 90.66% at 20 ppm in 360 minutes under visible light irradiation.

In comparison with a study by [12] Ismail *et al.* (2019), raw bauxite without any modification able to degrade 67% of RB5 under visible light in 360 minutes. CdS showed better performance as have higher degradation percentage in same degradation time. Other than that, a study by [15] Lucas *et al.* (2013) also focused on degradation of textile dye RB5 by using other substance, the hybrid nanoparticle of $Fe_3O_4$@$SiO_2$@$TiO_2$. The nanoparticle successfully degraded 91% of RB5 after 60 minutes of exposure but the drawback of this study is the duration of time needed to prepare the nanoparticle as it is not commercially available compared to CdS that is buyable. Besides, RB5 also been degraded by n-doped $TiO_2$ with degradation percentage of 88.4% after six hours of exposure [16]. Thus, making CdS is better than n-doped $TiO_2$, bauxite and $Fe_3O_4$@$SiO_2$@$TiO_2$ with better degradation of textile dye RB5.

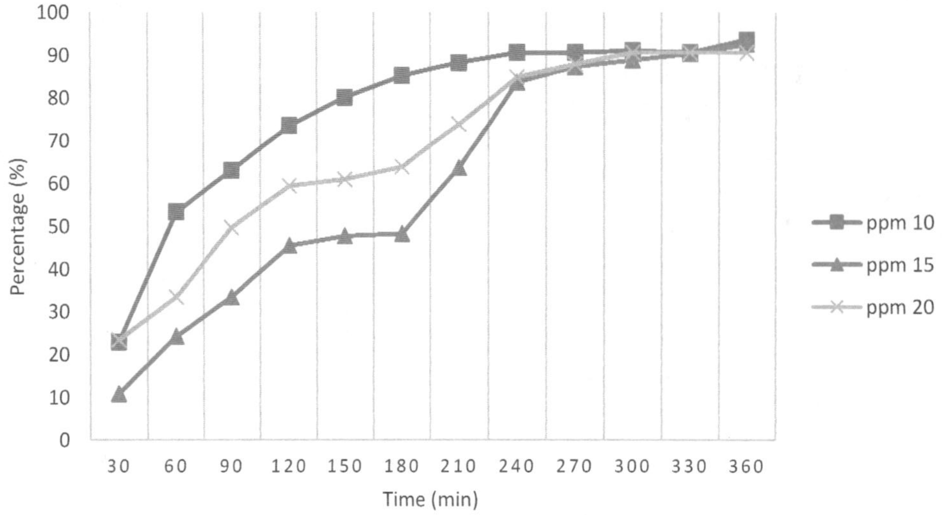


**Figure 4**. Percentage of RB5 degradation graph by CdS

## 4. Conclusions

Cadmium sulfide has very low band gap and crystalline structure. It also a good visible light photocatalyst in removing 93.72% 10 ppm textile dye Reactive Black 5 (RB5) in 360 minutes under visible light exposure. Thus, CdS can be use as a photocatalyst to remove dyes in textile industry wastewater as the photocatalysis process itself is safe, degrading the pollutants into carbon dioxide and water.

## 5. Conflict of interest

The authors declare that they have no conflict of interest

## 6. Acknowledgement

The authors gratefully acknowledge the financial support from the Ministry of Higher Education Malaysia under the Malaysia Research University Network (MRUN) Grant (Project number: R.J130000.7809.4L867) and the Higher Institution Centre of Excellence Scheme (Project Number: R.J090301.7809.4J430), and also Universiti Teknologi Malaysia under the Prototype Research Grant ICC (PRGS-ICC) (Project number: R.J130000.7709.4J483), and UTM High Impact Research Grant (Project number: Q.J130000.2409.08G34). The authors would also like to thank the Research Management Centre (RMC), Universiti Teknologi Malaysia for the technical support and services provided.

## References

[1] Yap, C.K., Peng, S.H.T. and Leow, C.S., 2019. Contamination in Pasir Gudang Area, Peninsular Malaysia: What can we learn from Kim Kim River chemical waste contamination? *Journal of Humanities and Education Development*, 1(2), pp.82–87.

[2] Ismail, S.N.S., Abidin, E.Z. and Rasdi, I., 2020. A Case Study of Pasir Gudang Chemical Toxic Pollution: A Review on Health Symptoms, Psychological Manifestation and Biomarker Assessment. *Malaysian Journal of Medicine and Health Sciences*, 16(SUPP11), pp.2636–9346.

[3] Aditya, W., Notodarmodjo, S. and Helmy, Q., 2019. Decolorization of Reactive black-5 Dye by UV based photocatalytic with immobilized ZnO nanoparticles onto ceramic plate surface. *IOP Conference Series: Materials Science and Engineering*, 536(1).

[4] Yaseen, D.A. and Scholz, M., 2019. *Textile dye wastewater characteristics and constituents of synthetic effluents: a critical review*, Springer Berlin Heidelberg.

[5] Al-Sakkaf, B.M., Nasreen, S., Ejaz, N. and Feng, F., 2020. Degradation Pattern of Textile Effluent by Using Bio and Sono Chemical Reactor. *Journal of Chemistry*, 2020.

[6] Lellis, B., Fávaro-Polonio, C.Z., Pamphile, J.A. and Polonio, J.C., 2019. Effects of textile dyes on health and the environment and bioremediation potential of living organisms. *Biotechnology Research and Innovation*, 3(2), pp.275–290.

[7] Karthik, V. et al., 2014. An overview of treatments for the removal of textile dyes. *Journal of Chemical and Pharmaceutical Sciences*, 7(4), pp.301–307.

[8] Sivarajasekar, N. and Rajoo, B., 2015. Agriculture waste biomass valorisation for cationic dyes sequestration : A Agriculture waste biomass valorisation for cationic dyes sequestration : A concise review. *Agriculture waste biomass valorisation for cationic dyes sequestration : A Agriculture waste biomass valorisation for cationic dyes sequestration : A*, 7(October), pp.737–748.

[9] Saravanan, R., Gracia, F. and Stephen, A., 2017. Basic Principles, Mechanism, and Challenges of Photocatalysis. In: *Nanocomposites for Visible Light-induced Photocatalysis*. pp. 19–40.

[10] Sahare, P., 2018. Synthesis and Characterization of CdS Nanoparticles. *International Journal of Engineering Research & Technology (IJERT)*, Volume 6(Issue 1). Available at: https://www.ijert.org/synthesis-and-characterization-of-cds-nanoparticle.

[11] Dumbrava, A., Badea, C., Prodan, G. and Ciupina, V., 2010. Synthesis and characterization of cadmium sulfide obtained at room temperature. *Chalcogenide Letters*, 7(2), pp.111–118.

[12] Ismail, N.J., Othman, M.H.D., Kamaludin, R. and Mohamad, I., 2019. Characterization of Bauxite as a Potential Natural Photocatalyst for Photodegradation of Textile Dye. *Arabian Journal for Science and Engineering*, 44(12), pp.10031–10040. Available at: https://doi.org/10.1007/s13369-019-04029-9.

[13] Aggarwal, S., 2016. Photo Catalytic Degradation of Methyl Orange by Using CdS Semiconductor Nanoparticles Photo catalyst Photo Catalytic Degradation of Methyl Orange by Using CdS

Semiconductor Nanoparticles Photo catalyst. *International Research Journal of Engineering and Technology (IRJET) e-ISSN:*, 3(4 April), pp.2–6.

[14] Pan, A. and Zhu, X., 2015. *Optoelectronic properties of semiconductor nanowires*, Elsevier Ltd.

[15] Lucas, M.S. et al., 2013. Photocatalytic degradation of Reactive Black 5 with TiO 2 -coated magnetic nanoparticles. *Catalysis Today*, 209, pp.116–121. Available at: http://dx.doi.org/10.1016/j.cattod.2012.10.024.

[16] Kamaludin, R. et al., 2018. Visible-Light-Driven Photocatalytic N-Doped TiO2 for Degradation of Bisphenol A (BPA) and Reactive Black 5 (RB5) Dye. *Water, Air, and Soil Pollution*, 229(11).

# Removal of direct red dye using polysulfone-chitosan ultrafiltration membrane

**J Hamdan[1], N S M Sabri[1], H Hasbullah[1*], M N M Sokri[1], S A Rahman[2] and F Kadirkhan[3]**

[1] Advanced Membrane Technology Research Centre (AMTEC), School of Chemical and Energy Engineering (SCEE), Universiti Teknologi Malaysia, 81310, Skudai, Johor, Malaysia

[2] Faculty of Chemical & Natural Resources Engineering, Universiti Malaysia Pahang, Lebuhraya Tun Razak, Kuantan, Malaysia

[3] Gas Sustainability Technology (GaSTech), R&D Department, Group Research & Technology (GR&T), Project Delivery & Technology (PD&T), Petronas Research Sdn Bhd, Malaysia

*Email: hasrinah@utm.my

**Abstract**. Textile industries are one of the largest pollution contributor to the environment through high amount of wastewater discharge that contain high concentration dyes. The present state of application of using membrane technology in the removal of dyes from textile wastewater has shown positive pathway for this research. Chitosan are biopolymer that are naturally occurring. It is used as a hydrophilic additive in the fabrication of polysulfone (PSF) and chitosan (CS) membrane. Different loadings of CS at 0.1, 0.3, 0.5, 0.7 and 1.0wt% were used. The fabricated PSF-CS membrane by phase inversion method were characterized using SEM for its morphology, FTIR for its functional group and contact angle for membrane hydrophilicity. As the addition of CS increase the hydrophilicity of the PSF-CS membrane, excessive CS precipitation was observed at chitosan loading at 1.0wt% creating a large macro void causing lower permeation performance. It is found that PSF-CS membrane with 0.7wt% has the best contact angle and dye rejection performance of 65.49° and 99.78%.

## 1. Introduction

Textile industry is one of the vital and vast emerging industry worldwide including Malaysia. Despite being a contributor to the economic growth, textile industry has been categorized as one of the industry that produces massive pollution problem especially with respect to liquid effluent. Wastewater effluent produced from textile factory have high possibility to cause environmental pollution and health-risk issues to rise due to the large scale production and extensive dye usage in dyeing and finishing involved[1]. Textile wastewater effluent that contain dyes has high intensity in color even in low concentration and removing dye itself is a challenging problem [2]. When wastewater contain untreated dye is released into the sewer, the high intensity color will affect the aquatic life because dye particles will block sunlight from penetrating through causing the photosynthetic activity to decrease [3]. Existing techniques that currently applied by the industry is either ineffective in removing dye as they pass through the municipal wastewater treatment facilities or they are very expensive such as activated sludge adsorbent.

Therefore, other alternative such as membrane technology could be a promising technique in treating textile effluent, effectively. Several studies have been conducted in matter of removing, treating or purifying textile wastewater effluent through membrane technology. In 2012, Abid [4] uses reverse osmosis and nanolfiltration to treat several synthetic dye which are acid red, reactive black and reactive blue. They achieved up to 99% for reverse osmosis technique while the nanolfiltration achieved up to 97% rejection with all the synthetic dye. Karisma [5] studied nanolfiltration polyetherimide membrane for the removal of reactive red (RR120), obtained 81% rejection performance operating at 50 psi and Gholami [6] also studied nanolfiltration polyamid and cellulose acetate blend running pressure at 0.5-3 bar, were able to achieved 92.5% rejection for disperse dye. Recently, Kadhim [7] studied ultrafiltration of the removal of two toxic dye which are acid black and rose bengal using graphene oxide mixed matrix membrane. As a result, the membrane displayed high rejection rate of 99% for both dyes. Therefore, membrane technology is an efficient way in the treatment of textile wastewater in order to tackle the existing challenges.

Polymeric membrane used in this study is polysulfone (PSF) where it exhibits high mechanical strength, good chemical resistance, thermal stability and excellent aging resistance making it a suitable material for the application in textile wastewater effluent treatment [8]. However, for the application of PSF in the industry comes down to a few major downsides which are PSF is naturally hydrophobic [9] and over a period of time protein or particles presence in the wastewater effluent got absorbed onto the membrane surface can block the surface pores and causes serious fluctuation in flux and low rejection performance [10]. Therefore, hydrophilic additive or nanoparticles can be included with the PSF membrane to produce a hydrophilic membrane. In this study, chitosan is used as an additive to be incorporated inside the PSF polymer to formed a mixed matrix membrane for the removal of direct red (DR80) dye.

Chitosan originated from chitin through deacetylation process. It is a hydrophilic material due to presence of two crosslinkable groups of hydroxyl (-OH) and amine ($-NH_3$) [11]. It became soluble in aqueous acidic solution when the degree of deacetylation of chitin reaches 50% and it is widely used in different field used for removing pollutants through coagulation, flocculation, filtration and membrane filtration [12]. Chitosan is a pseudonatural cationic polymer and is unique enough to be used in different applications mainly as solutions, gels, films or fibres [13]. Furthermore, chitosan has been used in other previous studies such as PVDF [14], PES [15], films [16] and coatings [17]. The preparation of the PSF-CS mixed matrix membrane using phase inversion which is relatively simple process. The produce membrane was thoroughly characterized and tested for DR80 removal performance.

## 2. Materials and Methods

### *2.1. Chemicals*

Polysulfone pellet (PSF, Udel1700) with average molecular weight of 69 500 $gmol^{-1}$ was purchased from Solvay Advanced Polymers. Other following chemicals were directly supplied by Sigma Aldrich, USA, such as chitosan low molecular weight (degree of deacetylation > 75%), polyvinylpyrrolidone (PVP, K40) with average molecular weight of ~40,000 $gmol^{-1}$ was used as an additive for the dope solution, while 1-methyl-2-pyrrolidone (NMP, Purity = 99.5%) was used as the organic solvent to dissolve dope solution. The soluble dye used for the filtration performance was Direct Red 80 (DR80) with average molecular of 1373.07 $gmol^{-1}$. Tap water was used as the coagulation bath.

### *2.2. Membrane fabrication*

Prior to the fabrication of PSF-CS dope solution, PSF pellet need to be oven-dried for 24 hours at 80°C to remove any moisture present. The method of blending chitosan inside polymeric dope solution consist of suspending chitosan powder in N-methyl-1-pyrrolidone solution together with polyvinylpyrrolidone as the pore former at 50°C until homogenous solution was formed. Fix amount of PSF pellet was slowly added into the solution forming a homogenous dope solution. Dope solutions were mixed using a magnetic stirrer and sonicated for 1 hour to remove any air bubbles formed during mixing. The membranes were prepared using phase inversion method. Flat sheet membranes with thickness approximately ~200µm were casted onto a glass plate and was immersed into the coagulation bath at 24 °C. The prepared PSF-CS membrane was kept in the coagulation bath for two (2) days to remove any excess solvent and let dried in the oven at 35 °C. The prepared membrane was kept sealed in an air tight container until further used.

### 2.3. *Membrane characterization*

The functional group of the membrane was examined using a ATR-Fourier Transform Infrared Spectroscopy (ATR-FTIR, Nicole 5700 Thermo Electron) to analyse the chemical structure of pristine PSF membrane and modified PSF-CS membrane. The morphology for the surface and cross section of the membrane was characterize using Scanning Electron Microscopy (SEM, Hitachi TM3000). The membrane sample was fractured in liquid nitrogen and spurred coated with gold before carrying out the analysis. The hydrophilicity of the membrane was determined using water contact angle which was measured with Contact Angle System OCA. Approximately 1µl of deionized water was carefully dropped on the membrane surface at 5 different points and the average of measured values are recorded as the final water contact angle.

### 2.4. *Permeation performance*

All permeation performance for pristine and modified membrane including pure water flux and rejection performance were carried out using crossflow filtration system. The effective membrane area was $2.46 \times 10^{-3}$ m$^2$ at 25 °C and pressure of 1 bar. The membrane was soaked in deionized water for 24-hour prior for testing to ensure even water distribution before being placed in the cell holder. Feed tank was filled with deionized water and permeate sample collection started after 30 minutes of membrane compaction time at pressure of 2 bar. Sample collection continued at every 10-minute interval. The pure water flux (PWF) was calculated using the following equation:

$$J_W(L/m^2h) = \frac{Q}{\Delta t\ A}$$

Where $J_w$ is the pure water flux expressed as L/m$^2$h, while Q is the volume of permeate sample in Litre (L), t is the time interval in hour and A is the total membrane area in m$^2$.

The permeate sample concentration was measured using UV-Spectrophotometer (HACH DR5000). For direct red dye rejection, 50 ppm of direct red dye solution was prepared and maximum wavelength was 532nm. The permeate sample was collected every 30 minutes under the pressure of 1 bar. The system was allowed to run for total of 180 minutes. The rejection (R%) was calculated using the following equation:

$$R(\%) = \left(1 - \frac{C_P}{C_F}\right) \times 100$$

Where R is the rejection in percent (%), while $C_P$ and $C_F$ is the concentration of permeate and feed sample in ppm, respectively.

## 3. Results and Discussions

### 3.1. *ATR-FTIR analysis*

The ATR-FTIR analysis was carried out on the prepared membrane to confirm the presence of chitosan on the surface of the PSF membrane. Different loadings of chitosan (0.1, 0.3, 0.5, 0.7 and 1.0wt%) were prepared. Excessive precipitation chitosan loadings above 1.0wt% in the dope solution can be observe. Figure 1 shows the FTIR spectra for pure chitosan, pristine PSF and PSF-CS membrane. All different loadings of chitosan showed similar FTIR peaks. In Region I, it can indicate the presence of O-H stretching at 3345 cm$^{-1}$, that belongs to hydroxyl group of chitosan that contribute to the hydrophilicity of the PSF-CS membrane. Region II, peaks at 2970 $cm^{-1}$ and 2872 $cm^{-1}$ that belongs to aromatic C-H stretch and aliphatic C-H stretch, respectively proving that chitosan has been successfully incorporated inside the PSF membrane. In Region III, at peak 1659 $cm^{-1}$ depicts the presence of amine group $-NH_3$ originated from chitosan [18].

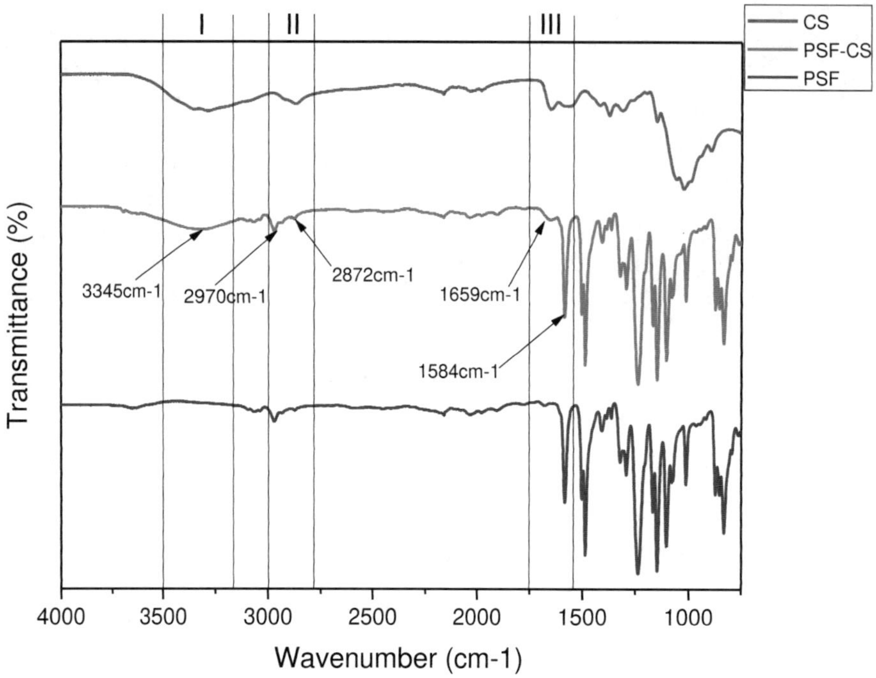


**Figure 1.** ATR-FTIR spectra for CS, PSF-CS and PSF membrane.

### *3.2. Morphology analysis*

The difference on the morphology of PSF-CS membrane were studied by SEM. Figure 2 shows the cross-sectional and surface images of pristine PSF and PSF-CS membrane. From the SEM images, the pristine PSF membrane and PSF-CS membrane have a typical asymmetric structure with top and support layer with finger-like channel [19]. Upon the addition of chitosan, the finger-like channel became wider and short in length. This phenomenon may be cause by the intrusion of non-solvent (water) through the membrane as the chitosan loadings increases therefore, bigger pore can be observed [20]. At the same time, bigger macro void can be observed in all PSF-CS membrane as the chitosan loadings increase from 0.1wt% to 1.0wt%. The pore size of the membrane upon the addition of chitosan was approximately 100nm. As higher loadings chitosan is blend directly with the dope solution, excessive chitosan precipitation can be observed inside the dope solution. This can also be proved from its cross-section where a large "lump" of chitosan can be observed where it settles under the surface of the membrane. Initially, the addition of chitosan was to increase the membrane hydrophilicity but due to this behavior, excessive present chitosan has eventually decreases its hydrophilicity and its rejection performance. Figure 2 also shows that as the chitosan loadings increases, more porous surface can be observed but even more chitosan particles can be seen which will lead to pore blocking.

PSF11

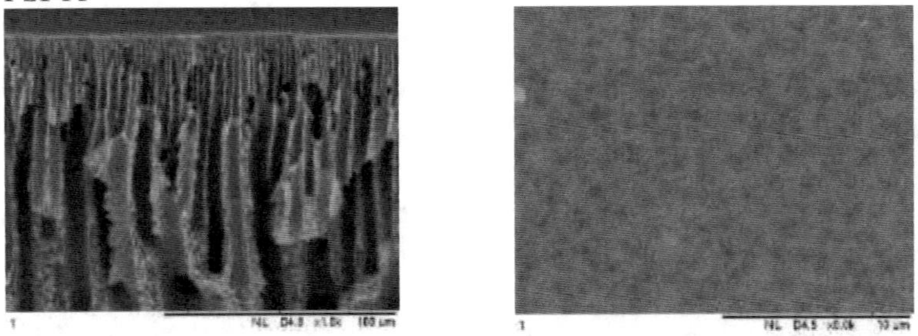

PSF11 CS0.1

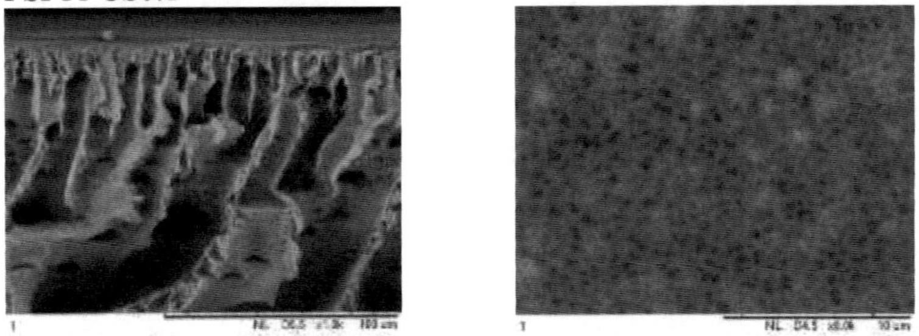

PSF11 CS0.3

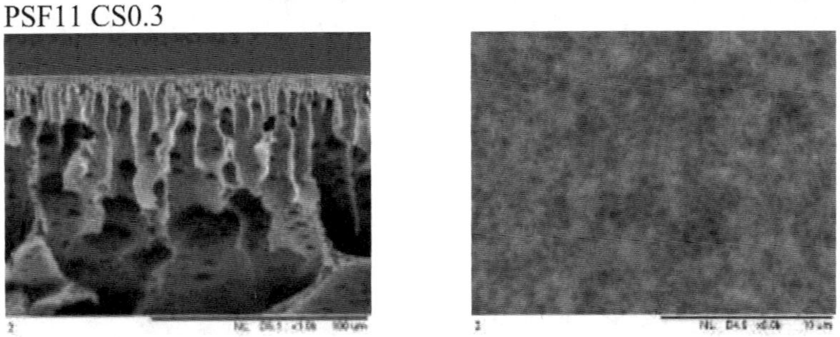

PSF11 CS0.5

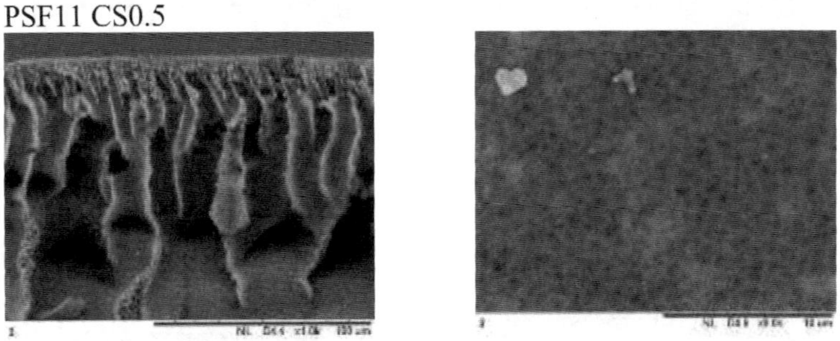

PSF11 CS0.7

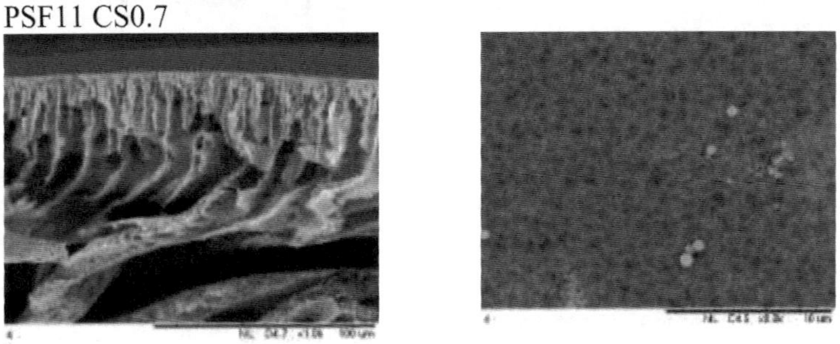

PSF11 CS1.0

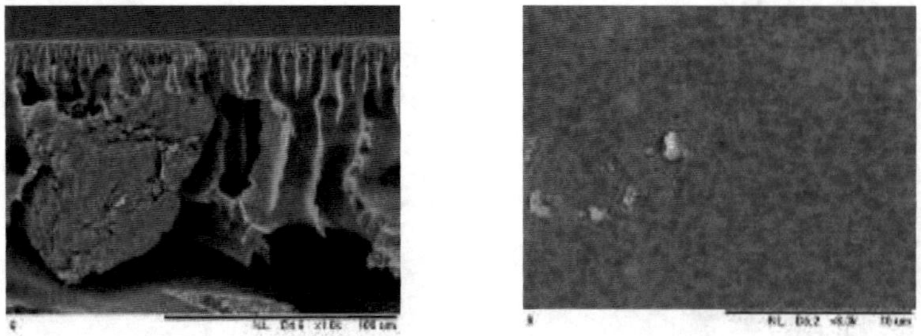

**Figure 2.** SEM images of cross-sectional and surface morphology of pristine PSF and PSF-CS membrane.

### *3.3. Water contact angle analysis*

Water contact angle analysis is carried out to determine the membrane hydrophilicity. Table 1 represents the water contact angle analysis of the PSF and PSF-CS membrane. it is generally known that any membrane that have contact angle <90° is considered as hydrophilic [21]. Pristine membrane is hydrophobic but upon the addition of chitosan, the PSF-CS membrane gradually became hydrophilic. PSF-CS membrane with chitosan loading of 0.7wt% was the most hydrophilic in comparison with the other PSF-CS membrane. As for PSF-CS membrane with chitosan loading of 1.0wt%, the membrane started to become hydrophobic due to the fact that the chitosan particles have started to clog the pores of the membrane.

**Table 1.** Water contact angle analysis for pristine PSF and PSF-CS membrane.

| Membrane | PSF | CS0.1 | CS0.3 | CS0.5 | CS0.7 | CS1.0 |
|---|---|---|---|---|---|---|
| WCA (°) | 95.40 | 92.90 | 81.79 | 78.09 | 65.49 | 82.57 |

### *3.4. Permeation performance*

The pure water flux performance of the membrane was carried out using cross-flow filtration system. As expected, all PSF-CS membrane have higher pure water flux in comparison with pristine PSF membrane. Previously, it is proven that the addition of chitosan increases the membrane hydrophilicity [22]. As the hydrophilicity increases, the pure water flux will also increase. Figure 3 shows the pure water flux analysis, where PSF-CS membrane with chitosan loading of 0.7wt% has the highest amount at 160.249 L/m$^2$h. the pure water flux started to decrease when chitosan loading is at 1.0wt%. this may due to the fact that excessive loading of chitosan has blocked the pore surface of the PSF-CS membrane which reduce the surface are of the membrane and the formation of large macro void that was formed during the fabrication of the membrane.

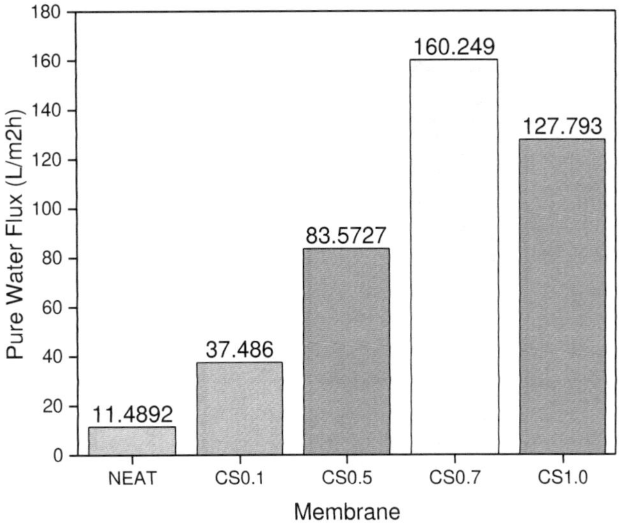


**Figure 3.** Pure water flux performance for PSF-CS membrane.

### *3.5. Dye rejection analysis*

Pristine PSF and PSF-CS membrane that was developed could be used to remove textile effluents that is usually charged. Direct red (DR80) dye (MW = 1373.07 g/mol) is anionic charged dye which is suitable for this developed PSF-CS membrane. The developed PSF-CS membrane has been analyzed to be negatively charged upon the addition of chitosan. Since, both the membrane and DR80 dye particles are both negatively charged, the membrane is able to repel the DR80 dye particles as DR80 is relatively a large dye particle. Since the pore size of the PSF-CS membrane is larger (~100nm) than the size of the dye particles, the dye particles can be removed through charge repulsion mechanism [23]. From Figure 4, PSF-CS with chitosan loading of 0.7wt% has the best DR80 rejection of 99.78% followed by membrane with loading of 0.3wt% with DR80 rejection of 99.56%. From these results, it is worth to note that the removal of dye does not only depend on sieving mechanism [24]. From the high rejection observed, it is believed that the charged repulsion mechanism between the PSF-CS membrane and DR80 dye, together with the morphological structure of the PSF-CS membrane, DR80 can be successfully removed without any further treatment such as backwash or even using polyelectrolyte or micellar assisted ultrafiltration.

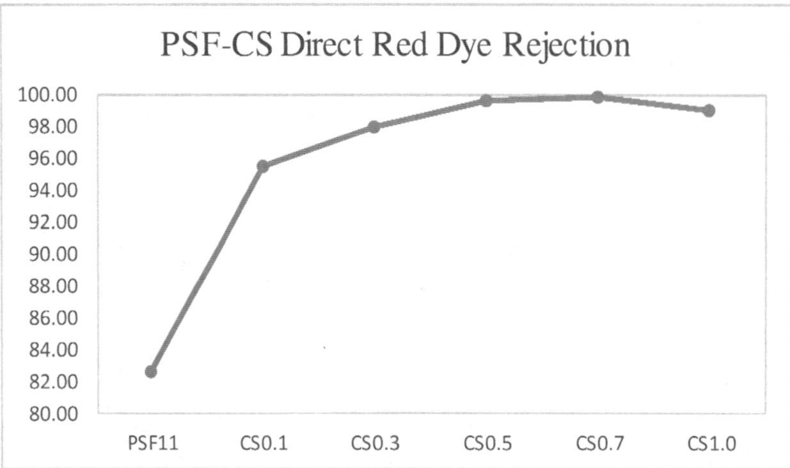


**Figure 4.** Direct red dye rejection performance for PSF-CS membrane.

## 4. Conclusion

In this study, chitosan has been successfully incorporated with PSF membrane producing a PSF-CS membrane. Blending of chitosan inside the dope solution has turned PSF membrane that is naturally hydrophobic into a hydrophilic membrane due to the presence of hydroxyl groups of chitosan. The presence of chitosan also increases the membrane permeation. However, the amount of chitosan used affect the morphological structure of the membrane due to the formation of large macro void and excessive amount of chitosan will eventually reduce the pure water flux performance and its hydrophilicity due to pore blockage. Filtration test performance of direct red (DR80) was performed on the PSF-CS membrane with chitosan loading of 0.7wt% has the highest rejection of 99.78%.

## 5. Acknowledgement

The authors gratefully acknowledge the financial support from the Ministry of Higher Education Malaysia under the Fundamental Research Grant Scheme (R.J130000.7851.5F380) and the Higher Institution Centre of Excellence Scheme (R.J090301.7851.4J434), and also Universiti Teknologi Malaysia under UTMShine Grant (Q.J130000.2451.09G17), and UTM Tier 1 (Q.J130000.2546.18H34 and Q.J130000.2546.16H10). The authors would also like to thank the Research Management Centre (RMC), Universiti Teknologi Malaysia for the technical support and services provided.

## References

[1] Ahmad, A.L., S. Puasa, and S. Abiding, *Crossflow Ultrafiltration for Removing Direct-15 Dye from Wastewater of Textile Industry.* ASEAN Journal on Science and Technology for Development, 2017. **23**: p. 207.

[2] Rauf, M.A., M.A. Meetani, A. Khaleel, and A. Ahmed, *Photocatalytic degradation of Methylene Blue using a mixed catalyst and product analysis by LC/MS.* Chemical Engineering Journal, 2010. **157**(2): p. 373-378.

[3] Yuan, L., D. Huang, W. Guo, Q. Yang, and J. Yu, *TiO2/montmorillonite nanocomposite for removal of organic pollutant.* Applied Clay Science, 2011. **53**(2): p. 272-278.

[4] Abid, M., M. Zablouk, and A. Abid-Alameer, *Experimental Study of Dye Removal from Industrial Wastewater by Membrane Technologies of Reverse Osmosis and Nanofiltration.* Iranian journal of environmental health science & engineering, 2012. **9**: p. 17.

[5] Karisma, D., G. Febrianto, and D. Mangindaan, *Removal of dyes from textile wastewater by using nanofiltration polyetherimide membrane.* IOP Conference Series: Earth and Environmental Science, 2017. **109**: p. 012012.
[6] Gholami, M., S. Nasseri, M. Alizadehfard, and A. Mesdaghinia, *Textile Dye Removal by Membrane Technology and Biological Oxidation.* Water Qual Res J Canada, 2003. **38**.
[7] Kadhim, R.J., F.H. Al-Ani, M. Al-shaeli, Q.F. Alsalhy, and A. Figoli, *Removal of Dyes Using Graphene Oxide (GO) Mixed Matrix Membranes.* Membranes, 2020. **10**(12).
[8] Afsari, M., S. Kheyrieh, and M. Asghari, *Application and modification of polysulfone membranes in wastewater treatment - A review.* Reviews in Chemical Engineering, 2017.
[9] Tiron, L.G., Ş.C. Pintilie, M. Vlad, I.G. Birsan, and Ş. Baltă, *Characterization of Polysulfone Membranes Prepared with Thermally Induced Phase Separation Technique.* IOP Conference Series: Materials Science and Engineering, 2017. **209**: p. 012013.
[10] Nguyen, H.T.V., T.H.A. Ngo, K.D. Do, M.N. Nguyen, N.T.T. Dang, T.T.H. Nguyen, V. Vien, and T.A. Vu, *Preparation and Characterization of a Hydrophilic Polysulfone Membrane Using Graphene Oxide.* Journal of Chemistry, 2019. **2019**: p. 3164373.
[11] Tsai, H.-S. and Y.-Z. Wang, *Properties of hydrophilic chitosan network membranes by introducing binary crosslink agents.* Polymer Bulletin, 2008. **60**(1): p. 103-113.
[12] Islam, S., M.R. Bhuiyan, and M. Islam, *Chitin and chitosan: structure, properties and applications in biomedical engineering.* Journal of Polymers and the Environment, 2017. **25**(3): p. 854-866.
[13] Rinaudo, M., *Chitin and chitosan: properties and applications.* Progress in polymer science, 2006. **31**(7): p. 603-632.
[14] Boributh, S., A. Chanachai, and R. Jiraratananon, *Modification of PVDF membrane by chitosan solution for reducing protein fouling.* Journal of Membrane Science, 2009. **342**(1-2): p. 97-104.
[15] Ghaee, A., M. Shariaty-Niassar, J. Barzin, and A. Ismail, *Chitosan/polyethersulfone composite nanofiltration membrane for industrial wastewater treatment.* Int. J. Nanosci. Nanotechnol., 2013. **9**: p. 213-220.
[16] Uragami, T., T. Matsuda, H. Okuno, and T. Miyata, *Structure of chemically modified chitosan membranes and their characteristics of permeation and separation of aqueous ethanol solutions.* Journal of Membrane Science, 1994. **88**(2): p. 243-251.
[17] Silitonga, R.S., N. Widiastuti, J. Jaafar, A.F. Ismail, M.N.Z. Abidin, I.W. Azelee, and M. Naidu, *The Modification of PVDF Membrane via Crosslinking with Chitosan and Glutaraldehyde as the Crosslinking Agent.* Indonesian Journal of Chemistry, 2018. **18**(1): p. 1-6.
[18] Kil'deeva, N.R., P.A. Perminov, L.V. Vladimirov, V.V. Novikov, and S.N. Mikhaĭlov, *[Mechanism of the reaction of glutaraldehyde with chitosan].* Bioorg Khim, 2009. **35**(3): p. 397-407.
[19] Elizalde, C.N.B., S. Al-Gharabli, J. Kujawa, M. Mavukkandy, S.W. Hasan, and H.A. Arafat, *Fabrication of blend polyvinylidene fluoride/chitosan membranes for enhanced flux and fouling resistance.* Separation and Purification Technology, 2018. **190**: p. 68-76.
[20] Kumar, R., A.M. Isloor, A.F. Ismail, S.A. Rashid, and T. Matsuura, *Polysulfone–Chitosan blend ultrafiltration membranes: preparation, characterization, permeation and antifouling properties.* RSC Advances, 2013. **3**(21): p. 7855-7861.
[21] Mansur, S., m. hafiz, M.H. Othman, A. Ismail, M. Nidzhom, Z. Abidin, N. Said, P. Goh, H. Hasbullah, S. Kadir, and F. Kamal, *Study on the effect of PVP additive on the performance of PSf/PVP ultrafiltration hollow fiber membrane*. 2018.
[22] Jóźwiak, T., U. Filipkowska, P. Szymczyk, J. Rodziewicz, and A. Mielcarek, *Effect of ionic and covalent crosslinking agents on properties of chitosan beads and sorption effectiveness of Reactive Black 5 dye.* Reactive and Functional Polymers, 2017. **114**: p. 58-74.

[23] Arkhangelsky, E., I. Levitsky, and V. Gitis, *Electrostatic repulsion as a mechanism in fouling of ultrafiltration membranes.* Water Sci Technol, 2008. **58**(10): p. 1955-61.

[24] Bouazizi, A., M. Breida, B. Achiou, M. Ouammou, J.I. Calvo, A. Aaddane, and S.A. Younssi, *Removal of dyes by a new nano–TiO2 ultrafiltration membrane deposited on low-cost support prepared from natural Moroccan bentonite.* Applied Clay Science, 2017. **149**: p. 127-135.

# Synthesis of graphene oxide via electrochemical process: A short review towards flexible synthesis method

**I I Edward[a], N Abdul Manaf[b*], S A Tahir Abdul Muthalib[a], M R Musram Rakunman[b], L.S. Tan[b], T Tsuji[b]**

[a]Faculty of Science, Universiti Tunku Abdul Rahman, Kampar, Perak

[b]Department of Chemical and Environmental Engineering, Malaysia-Japan International Institute of Technology (MJIIT), Universiti Teknologi Malaysia, Kuala Lumpur, Malaysia

* Corresponding author: norhuda.kl@utm.my

**Abstract**. Graphene-based material for instance graphene oxide (GO) and reduce graphene oxide (rGO) emerge as unique frontier material that have been widely applied in various application such as energy conversion, nanotechnology as well as oil and gas industry. It properties such as high mechanical properties and enhance electrical conductor have made this material outperforms existing frontier material thus features large-scale commercialization opportunity. At present, electrochemical method has attracted much attention in synthesizing graphene-based material (GO/rGO) due to its green technology (environmentally friendly process), low cost and enhanced efficiency. Nevertheless, based on contemporary studies, different demand in GO/rGO quality requires different parametric setup and characterization analyses subsequently provides significant challenge to identify a standard design of electrochemical that can produce various quality of GO/rGO. This inflexibility affects the progress of commercial scale-up thus entails a systematic review related to various aspects of electrochemical studies for instance parametric analysis, product/material characterization and applications. A parametric analysis includes electrolyte concentration, type of electrolyte, temperature, synthesis time, cathode and anode materials and voltage demand. While, for product characterization, several methods are evaluated for instance Transmission electron microscopy (TEM) and Raman spectroscopy (RAMAN). This review paper is beneficial to the researcher and industries as a rapid guideline towards the flexible mass-production of GO/rGO via electrochemical method. Finally, future research should be conducted on the flexible design of unit operation (i.e. reactor) to ensure process feasibility.

## 1. Introduction

Graphene and graphene oxide (GO) is not notoriously known to the public but it is foreseen to be the wonder material in 21st century just as plastic was considered crucial in the 20th century development. This prevision is manifested by the potential of graphene-based materials for application in different fields due to their unique physical and chemical properties such as its outstanding electronic, mechanical, optical, and thermal properties. A GO and reduce-GO (rGO) are the main research branches and exhibited as a great precursor to synthesis graphene with higher yield and lower synthesising costs. GO is a material that consisted of a single monomolecular layer of graphite with various oxygen contained functionalities, such as OH, C=O and C-O-C groups. While, rGO is a graphene-like sheets by removing the oxygen-containing groups with the recovery of a conjugated structure [1].

The presence of oxygen functionalities in GO makes GO easily disperse in water and organic solvents. This advance property make GO have benefit as an additive on ceramic and polymer manufacturing process in enhancing their electrical and mechanical properties such as tensile strength, elasticity and conductivity [2]. Furthermore, GO/rGO plays a significant role in energy storage and conversion due to its efficient thermal and electrical conductor [3]. Several methods have been explored to determine the best route to synthesis GO/rGO such as the Modified/Improved Hummers method [4-6], chemical vapor deposition (CVD) [7, 8] and microwave irradiation system [9-11].

At present, electrochemical method has attracted much attention in synthesizing graphene-based material (in this review, graphene based materials are described as graphene, GO and rGO and its composites) due to its green technology (environmental friendly process), low cost and enhanced efficiency. Nevertheless, based on contemporary studies, different demand in GO/rGO quality requires different parametric setup and characterization analyses subsequently provides significant challenges to identify a standard design of electrochemical process that can produce various quality of graphene-based material without require huge process modification and extensive characterization studies. Similar comment has been reported by Achee et al. [12], whereas production of graphene-based material features lack of standardization.

This work focuses solely on the recent progress of graphene-based material from electrochemical exfoliation method with the aim to find a measure for the aforementioned limitation/challenge. The literatures analyses involve evaluation of curated articles from year 2010 until 2020, which encompasses of various scientific and technical databases. Quality of graphene based material (product) is discussed based on the optimal operating condition and characterization analyses. Finally, present work proposes facile and flexible design of experiment to synthesis the graphene based material at rapid way.

## 2. Flexible operation of electrochemical exfoliation process

In electrochemical process, graphene-based material is synthesized via graphite conductivity by stimulating the anions/molecules in the electrolyte into graphite subjected to the voltage/current biases, in order to form a graphite intercalation compound. Electrolyzed water which produce oxygenated material will oxidize the graphite intercalation compound to form GO/rGO. Electrochemical exfoliation of GO/rGO from graphite involves two possible alternative routes such as single-step and two-step electrochemical processes, which consists of anodic and cathodic graphite exfoliation mechanisms. Many reviews have been done related to the electrochemical exfoliation process of graphene-based material. Liu et al. [13] have reviewed existing articles related to GO synthesis in term of operating condition of electrochemical exfoliation process, product quality and characterization studies. Their review suggested to emphasize on the fundamental of graphite mechanism in exfoliation process to obtain good quality of graphene-based material, to promote large-scale setup and to determine possible application of graphene-based material in the emerging application. Lee et al.[14] summarized the characterization analyses related to the prevailing synthesis methods of graphene/GO and its derivative with the additional analyses including production cost and yield. Similarly, Le et al. [15] have conducted a review study which combine similar scopes of work as in [13, 14]. Three different methods have been discussed in their review paper, such as two-electrode, three-electrode and electrolyte exfoliation. These trends evident a novelty of this short review article whereas to the best of author knowledge, there is no study proposes a standard electrochemical exfoliation method to accommodate different industrial and product demands. To narrow down the scope of this review, we only select studies involved with platinum as the counter electrode, as tabulated in Table 1. Since platinum is known as a feasible and flexible electrode that can accommodate different type of working electrode.

### *2.1 Single step electrochemical exfoliation process*

Single step electrochemical method involves intercalation of electrolyte ion to produce graphene. Sharif et al. [16] has successfully synthesized a monolayer and bilayer graphene via single-step exfoliation (direct) mechanism. In their work, a minimum defect of as synthesized graphene with high temperature stability in air was produced by using ammonium sulphate as an electrolyte at a voltage of 10 V for 3 h synthesis time. Chen et al. [17] used similar electrolyte solution at a voltage of 10 V with low concentration (0.1 M) to produce graphene from pencil core. Other studies have shown that utilization of ammonium sulphate as the electrolyte is able to produce a thin layer of graphene [17, 18]. Jibrael et al. [19] used three different types of electrolyte solutions (sulfuric acid, nitric acid and distilled water) at voltage of 10 V for 50 min to produce graphene powder. On the other hand, Aghamohammadi et al. [20] used combination of hydrogen fluoride and titanium oxide to produce multi-layered graphene at a voltage of 10 V. They observed that addition of titanium oxide may enhance the exfoliation rate as well as promoting fluorination of graphene sheet as a composite material.

Prakoso et al. [21] produced graphene sheet via graphite extracted from spent Zn–C batteries. The electrochemical exfoliation process was conducted at a voltage of 5 V in a poly (sodium 4-styrenesulfonate) (PSS) solution. It was observed that they managed to produce high quality of graphene flakes using battery waste. Wang et al. [22] synthesized low defect of graphene layer by using graphite electrode coated with parafin. In their work, concentrated sodium hydroxide was used as the electrolyte at voltage of 3 V under the confine space to promote exfolation and to prevent excessing expansion of graphite. Similar concept was adapted by Achee et al. [12] where they demonstrated that utilization of compressed graphite powder can promote continuous intercalation, expansion and exfoliation simultaneously to ease the sytnhesis of graphene. In their work, high yield of graphene was successfully been produced with large lateral size by using sulfuric acid as the electrolyte solution

A well-oxidized monolayer GO have been synthesized by [12, 23, 24] using high concentration of sulphuric acid via exfoliation mechanism. In a different study, addition of surfactant into the electrolye solution is able to produce GO via single step method. Nurhafizah et al. [25] managed to synthesis GO by using a sodium dodecyl sulphate (SDS) surfactant at a voltage of 10 V for 24 h synthesis time. Their as synthesized GO exhibited thinner sheet with wrinkles structures. Alternatively, different concentration of SDS surfactant from 0.001 to 1.0 M was analysed by Md Disa et al. [26]. They observed that the lowest GO production is given by the lowest concentration of SDS used which is 0.001 M. It can be seen, that only few studies have employed single-step electrochemical exfoliation process to obtained GO/rGO. Where, synthesis of graphene is more favourable to use this method.

### *2.2 Two-step electrochemical exfoliation process*

Two-step electrochemical exfoliation process represents the intercalation and oxidation/exfoliation mechanisms in a sequential way. Chernysheva et al. [27] employed two-step mechanism to produce GO. First stage involved an anodic intercalation of graphite anode in sulfuric acid followed by the oxidation/exfoliation of the graphite in ammonium sulphate at a voltage of 10 V for 10 min. Shen et al. [28] found an optimal condition of GO synthesis at voltage of 2 V during intercalation process and 20 V during oxidation stage. Both stages were using high concentration of sulphuric acid (95% and 65%) at 10 min and 1 min synthesis time, respectively. Cao et al. [29] employed similar elctrolyte solutions as in [28] at slightly lower voltage (10 V) to synthesis GO. Their two-step approach led to GO with a high yield (> 70 wt %), good quality (> 90%, monolayer), and reasonable oxygen content (17.7 %). High yield of GO using water electrolytic oxidation of graphite was conducted by Pei et al. [30]. Their study used high concentrated of sulfuric acid without oxidant agent for intercalation process followed by diluted sulfuric acid for oxidation process. It was observed that highly oxidized GO (C/O < 2) was achieved only when the sulfuric acid concentration is in the range of 40 wt.% to 60 wt.%.

Contrary, Kumar et al. [31] used mixture of sulfuric acid and phosphoric acid for intercalation process followed by the oxidation process using potassium permanganate. High quality of GO was obtained based on their experimental condition at 6 h synthesis time with fix intercalation time at 4 min. Sahoo et al. [24] applied different concentration of sulfuric acid at 1 cm gap between two electrodes positions. Initial synthesis involved cathodic pretreatment of working electrode at a negative direct current (DC) of 10 V for 30 s followed by 3 V for 15 min to ensure the expansion of pyrolytic graphite electrode for efficient intercalation process. Their result showed that low concentration of electrolyte (0.5 and 1.0 M sulfuric acid) was capable to stimulate the rate of electrochemical oxidation reactions at the graphite-electrolyte interface which exhibited more oxidation endowment in the product. This condition may contribute to the highest GO yield as correlated by [26].

Other studies have evident that utilization of sulfuric acid, at low synthesizing temperature are able to produce good quality of single/multilayer graphene/graphene based material [30, 32, 33]. For instance, moderate number of GO powder was synthesized using voltage of 2 V for 10 min intercalation process, followed by 60 s in the oxidation process (20 V). In their work, 65% sulfuric acid was used as an electrolyte at room temperature [32]. At the aforementioned condition, the as synthesized GO powder featured a single layer at 0.345 nm. A longer synthesizing time was conducted by Konwar et a. [33], where they managed to produce a highly pure GO in 0.1 M sulfuric acid solution at room temperature at constant potential of 3 V for 4 h. In a separate study, a monolayer of GO sheet with no defects or oxidation-containing functional groups was obtained by Pei et al. [30]. In their study, a concentrated sulfuric acid at 1.6 V (for 20 min) was used followed by the electrochemical reaction in 50% sulfuric acid for 3 min (5 V). Table 1 summarizes the operational condition of electrochemical synthesis of graphene based using platinum and counter electrode. Based on these two methods, it can be predicted that propose standard design to synthesis graphene-based material is via two-step method at 10 V (< 15 min) and 2 V (< 5 min) for intercalation and oxidation processes, using moderate concentration of sulfuric acid.

## 3. Characterization analysis of graphene based material

A GO's complex structure causes it difficult to suggest a standardize characterization method. For certain applications, different characteristics of GO are more significant than in others. Such characteristics are lateral scale, degree of oxidation, level of exfoliation, amount of moisture, categories of functional groups and the existence of impurities from the oxidation phase. These involve stacking of the GO, the lateral proportions of the GO, the functional groups found within/on the GO, and the degree of GO sheet disorder. In characterization analyses, stacking spacing is identified by XRD, while lateral proportions may be measured through methods such as FESEM, FETEM, AFM or optical microcopy, while RAMAN, FTIR or XPS indicate the functional groups present.

In this review, four characterization analyses are reviewed which include FESEM, XRD, FTIR, FETEM and RAMAN. These five characterization analyses are selected based on the frequency distribution study shown in Figure 1. Based on the figure, it can be seen that the common analysis that used to characterize graphene-based material (GO/rGO) are those four.

A crumpled sheets of GO was observed in FESEM analysis when using moderate to high concentration of surfactant electrolyte (0.1 - 1.0 M) while, thick agglomeration of GO was exhibited when low concentration (0.001 M) of surfactant was used [25]. In a different study, a highly-wrinkled of GO sheet was formed under the moderate concentration of sulfuric acid (65%) [35], which corroborate with the previous work done by [25]. Those structure indicates a deformation of the graphene layers due to the linkage of the oxygenated functional groups. Contrary, a large silk veil morphology with slightly folded edges and loose open structure was observed in GO composited with nanofibers when using 0.1 maniline and 1 M sulfuric acid mixed solution [34].

The XRD analysis is used to determine possible changes in the interlayer spacing. For instance, as synthesized GO produced by [35] showed a peak in the range of 9 – 12° corresponding to the (001) reflection plane. An intense and sharp peak at $2\theta = 11.66°$ was revealed which corresponds to the (001) reflection plane [34]. The presence of this characteristic peak exhibited a completely oxidized GO during the electrochemical process.

Study conducted by Deiz-Pascual et al. [32] showed that C/O analysis was corroborate with the EDX analysis, elemental and FTIR analysis which exhibited a homogeneous and good quality of GO sheets, with a clear flake structure and lack of traces from the raw material. The darker areas in the FETEM analysis illustrated a good level of exfoliation, where thickness will be increased when with the increment of C/O ratio at decreasing of oxygen level. A TEM image for a mixed electrolyte solution with the typical 6-fold symmetric diffraction patterns with strong spots indicated the crystalline nature [16]. Similar TEM image pattern can be indicated in the study done by Chen et al. [18]. Where, both GO layers were synthesized using mixed electrolyte. Pei et al. observed that TEM image has consistent result with the AFM analysis [30]. It showed that about 46% of them are monolayer and about 86% are no more than three layers, which are consistent with the AFM measurement results.

Based on Raman spectroscopy, superior quality of graphene-based material (GO/rGO composite/derivative) can be exhibited at D, 2D and G peaks respectively. A typical D peaks can be observed at about 1323 $cm^{-1}$[31], 1354 $cm^{-1}$[35], 1360$cm^{-1}$[34] and 1363 $cm^{-1}$[18] . Essentially, the D band arises from breathing mode of the sp2 [16, 30] and attributed by defects or lattice disorders crystal structure due to the binding of oxygen-functional groups [35]. On the other hand, 2D peak is indicated at peak of 1600 $cm^{-1}$ [34] and 2650 $cm^{-1}$ [31] and usually attributed to second order phonon processes [35]. While, a typical G peak usually at about 1580 $cm^{-1}$[16, 36], 1586 $cm^{-1}$[35], 1590 $cm^{-1}$ [18] and 2700 $cm^{-1}$[34]. It resembles to the in-plane vibration of sp2 carbon atoms [16] with vibrational mode found in a graphite single crystal [37].

## 4. Conclusion and Outlook

This review paper could be used to conclude that electrochemical synthesis of GO is becoming a more common synthesis route to obtain graphene based material. It was also evident that in modern material science, graphene based material is a material that may be of significant use in the future since it can apply in various applications. That being the case, the outlook of this review would be to somehow create an idea for the common electrochemical setup for the synthesis of graphene-based material and also to increase the possibility of making the electrochemical route an ideal one.

**Acknowledgments**
Authors wishing to acknowledge financial support from MTJA-Cess Fund: IF062-2019 and Universiti Teknologi Malaysia, project no. R.K130000.7343.4B463

IOP Conf. Series: Materials Science and Engineering 1142 (2021) 012019 doi:10.1088/1757-899X/1142/1/012019

Table 1: Summary of operational condition of electrochemical synthesis of graphene based material using platinum as the counter electrode

| Ref | Working electrode | Counter Electrode | Power supplies (V) | Electrolyte | Exfoliation time | T (°C) | Characterization analyses | Type of graphene based material |
|---|---|---|---|---|---|---|---|---|
| [16] | Graphite sheet | Platinum | 10 (DC) | $Na_2SO_4$,$NH_4NO_3$, $(NH_4)_2SO_4$,$(NH_4)_2HPO_4$ | 3-4 h | 700 | TEM,RAMAN, XPS, AFM | Graphene |
| [38] | Graphite foil | Platinum | 1.1 or 2<br>2.10 to 30 | H2SO4 | 10 or 30 min<br>30 and 120 s | 25 ± 2 | TEM,SEM, AFM, RAMAN, FTIR | Single-layered GO sheets |
| [39] | Graphite | Platinum | | 1.$H_2SO_4$<br>2.$HBF_4$ + Water<br>3.$HBF_4$ + Methanol | | | | GO |
| [40] | Pencil graphite electrode (PGE) | Platinum | 2 | $H_2SO_4$ + $KNO_3$ + Phosphate Buffer | - | - | SEM | Graphene oxide/cobalt oxide nanocomposite |
| [41] | Graphite | Platinum | -1.2 to +0.3 | 0.1M $H_2SO_4$ | - | - | SEM, EDX, CV | Graphene/Pt and graphene/Au tubular microengines |
| [18] | Graphite rod | Platinum | 15 | 0.1M Oxalic acid + 0.05 $Na_2SO_4$ | - | - | SEM, TEM, RAMAN, FT-IR, XRD | GO |
| [42] | Glassy Carbon Electrode | Platinum | 0.01 | 0.1M Phosphate Buffer Solution | - | - | SEM, XRD, RAMAN, CV, DPV | (RGO)/cobalt oxide ($Co_3O_4$) nanocomposite |
| [43] | Carbon Paste Electrode | Platinum | -0.2 to 0.9 | 0.01M ortho aminophenol + 0.5M $HClO_4$ + 0.1M $LiCO_4$ + 0.005M SDS | - | 25 ± 2 | SEM | Magnetic functional graphene oxide (MFGO) |
| [32] | Flexible graphite foil stuck on tungsten wire with silver glue | Platinum | 1 to 2 | 98 wt% $H_2SO_4$ | 10 to 30 minutes | 25 ± 2 | GC-MS, SEM, TEM, FT-IR, XRD, RAMAN, AFM | GO Powder |

Table 1: Summary of operational condition of electrochemical synthesis of graphene based material using platinum as the counter electrode (cont.)

| Ref | Working electrode | Counter Electrode | Power supplies (V) | Electrolyte | Exfoliation time | T (°C) | Characterization analyses | Type of graphene based material |
|---|---|---|---|---|---|---|---|---|
| [44] | Graphite | Platinum | 10 | $H_2SO_4$ | 4 hours | - | XRD, FT-IR, SEM, 13C-NMR | GO sheets |
| [45] | Glassy Carbon Electrode | Platinum | -0.3 to 0.6 | - | - | - | SEM, TEM | Polypyrrole (PPy), reduced graphene oxide (RGO), and gold nanoparticles (nanoAu) biocomposite |
| [46] | GO sample | Platinum | 0.6 | - | - | - | RAMAN, XPS, SEM, AFM/SKPFM | Reduced graphene oxide (rGO)/copper (Cu) composite films |
| [47] | Modified homemade carbon-paste electrode | Platinum | - | $H_2SO_4$ | - | - | SEM, TEM, FT-IR, XRD | Reduced graphene oxide and poly(m-aminobenzenesulfonic acid, ABSA) nanocomposite (PABSA−rGNO) |
| [48] | Glassy Carbon Electrode | Platinum | - | Propylene Carbonate with $LiClO_4$ | - | - | FESEM, HRTEM, AFM, CV | Graphene quantum dots from graphene oxide |
| [33] | Graphite | Platinum | 3 | 0.1M $H_2SO_4$ | 4 hours | - | RAMAN, FTIR, EIS, CV | GO |
| [49] | Glassy Carbon Electrode | Platinum | -0.8 - +1.4 | CTAB + 0.1M $H_2SO_4$ | | | FESEM, EDX, FTIR, XRD | Poly(CTAB)/GO |
| [50] | Graphite | Platinum | | $H_2SO_4$ | - | - | SEM, FTIR, XRD, TGA, DSC | Multilayer GO |

"-": Data was unavailable

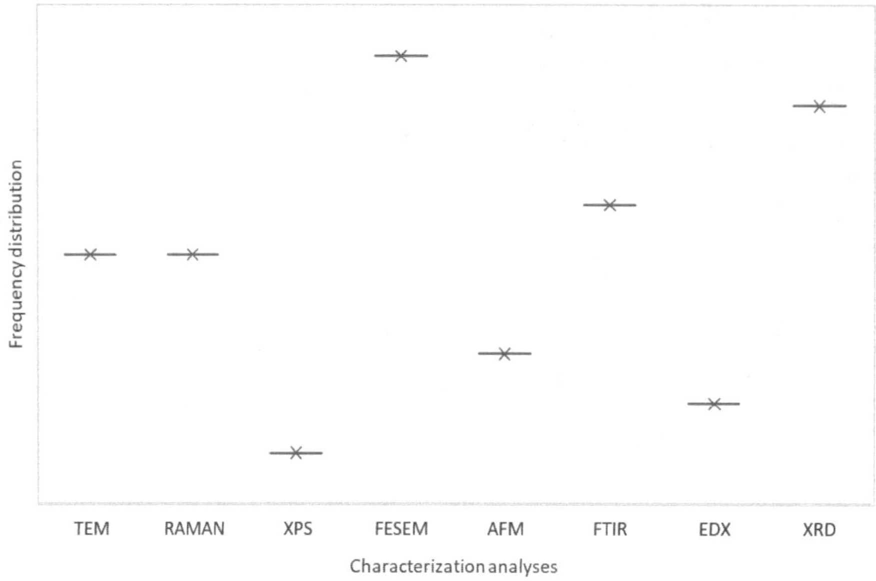


Figure 1: Frequency distribution of characterization analyses of the graphene-based material (GO/rGO) based on the curated articles in Table 1.

**Acknowledgment**

Authors wishing to acknowledge financial support from MTJA-Cess Fund - IF062-2019 and Universiti Teknologi Malaysia, project no. R.K130000.7343.4B463

**References**

1. Pei, S. and H.-M. Cheng, *The reduction of graphene oxide.* Carbon, 2012. **50**(9): p. 3210-3228.
2. Sadegh, H., *Development of graphene oxide from graphite: a review on synthesis, characterization and its application in wastewater treatment.* Reviews on Advanced Materials Science, 2017. **49**(1).
3. Kumar, H., et al., *Recent advancement made in the field of reduced graphene oxide-based nanocomposites used in the energy storage devices: A review.* Journal of Energy Storage, 2021. **33**: p. 102032.
4. Marcano, D.C., et al., *Improved synthesis of graphene oxide.* ACS Nano, 2010. **4**(8): p. 4806-14.
5. Zhou, Q., Y. Lu, and H. Xu, *High-yield production of high-quality graphene by novel electrochemical exfoliation at air-electrolyte interface.* Materials Letters, 2019. **235**: p. 153-156.
6. Siaw, W.C., et al., *Synthesis of graphene oxide from industrial waste.* IOP Conference Series: Materials Science and Engineering, 2020. **778**: p. 012050.
7. Ryu, B.D., et al., *Gallium dopant-induced tunable electrical properties of reduced graphene oxide using metal organic chemical vapor deposition.* Applied Surface Science, 2020. **504**: p. 144500.
8. Mehravar, S., S. Fatemi, and M. Komiyama, *Magnetic property and structural study of nickel supported on reduced graphene oxide prepared by chemical vapor deposition.* Surface and Interface Analysis, 2020.
9. Xie, X., Y. Zhou, and K. Huang, *Advances in Microwave-Assisted Production of Reduced Graphene Oxide.* Frontiers in Chemistry, 2019. **7**(355).
10. Yang, B., et al., *Synthesis of graphene by microwave irradiation for dye adsorption.* RSC Advances, 2014. **4**(110): p. 64771-64780.
11. Wang, X., et al., *Fast and facile microwave-assisted synthesis of graphene oxide nanosheets.* RSC Advances, 2014. **4**(104): p. 60102-60105.
12. Achee, T.C., et al., *High-yield scalable graphene nanosheet production from compressed graphite using electrochemical exfoliation.* Scientific reports, 2018. **8**(1): p. 1-8.
13. Liu, F., et al., *Synthesis of graphene materials by electrochemical exfoliation: Recent progress and future potential.* Carbon Energy, 2019. **1**(2): p. 173-199.
14. Lee, X., et al., *Review on graphene and its derivatives: Synthesis methods and potential industrial implementation.* Journal of The Taiwan Institute of Chemical Engineers, 2019. **98**: p. 163-180.
15. Li, L., et al., *Review—Progress of Research on the Preparation of Graphene Oxide via Electrochemical Approaches.* Journal of The Electrochemical Society, 2020. **167**(15): p. 155519.
16. Sharif, F., et al., *Synthesis of a high-temperature stable electrochemically exfoliated graphene.* Carbon, 2020. **157**: p. 681-692.
17. Chen, K., D. Xue, and S. Komarneni, *Nanoclay assisted electrochemical exfoliation of pencil core to high conductive graphene thin-film electrode.* Journal of colloid and interface science, 2017. **487**: p. 156-161.
18. Chen, D., et al., *Photosynergetic Electrochemical Synthesis of Graphene Oxide.* Journal of the American Chemical Society, 2020. **142**(14): p. 6516-6520.

19. Jibrael, R.I. and M.K.A. Mohammed, *Production of graphene powder by electrochemical exfoliation of graphite electrodes immersed in aqueous solution.* Optik, 2016. **127**(16): p. 6384-6389.
20. Aghamohammadi, H., A. Heidarpour, and S. Ghasemi, *Electrochemical synthesis of fluorinated graphene nanoplatelets in electrolytes containing hydrofluoric acid and TiO2 nanoparticles.* FlatChem, 2020. **22**: p. 100172.
21. Prakoso, B., et al., *Facile synthesis of battery waste-derived graphene for transparent and conductive film application by an electrochemical exfoliation method.* RSC Advances, 2020. **10**(17): p. 10322-10328.
22. Wang, H., et al., *Preparation of graphene sheets by electrochemical exfoliation of graphite in confined space and their application in transparent conductive films.* ACS applied materials & interfaces, 2017. **9**(39): p. 34456-34466.
23. Lowe, S.E., et al., *The role of electrolyte acid concentration in the electrochemical exfoliation of graphite: Mechanism and synthesis of electrochemical graphene oxide.* Nano Materials Science, 2019. **1**(3): p. 215-223.
24. Sahoo, S.K., et al., *Industrial scale synthesis of few-layer graphene nanosheets (FLGNSs): an exploration of electrochemical exfoliation approach.* Journal of Applied Electrochemistry, 2020. **50**(6): p. 673-688.
25. Nurhafizah, M.D., et al., *Effect of voltage applied for graphene oxide/latex nanocomposites produced via electrochemical exfoliation and its application as conductive electrodes.* Diamond and Related Materials, 2020. **101**: p. 107624.
26. Md Disa, N., et al., *The Synthesis of Graphene Oxide via Electrochemical Exfoliation Method.* Advanced Materials Research, 2015. **1109**: p. 55-59.
27. Chernysheva, M.N., et al., *Investigation of sulfuric acid intercalation into thermally expanded graphite in order to optimize the synthesis of electrochemical graphene oxide.* Journal of Electroanalytical Chemistry, 2020. **858**: p. 113774.
28. Shen, Y., et al., *Structure and Properties Evolution of Graphene Oxide Sheets During Low-Temperature Reduction on a Solid Substrate.* The Journal of Physical Chemistry C, 2020.
29. Cao, J., et al., *Two-Step Electrochemical Intercalation and Oxidation of Graphite for the Mass Production of Graphene Oxide.* Journal of the American Chemical Society, 2017. **139**(48): p. 17446-17456.
30. Pei, S., et al., *Green synthesis of graphene oxide by seconds timescale water electrolytic oxidation.* Nature Communications, 2018. **9**(1): p. 145.
31. Kumar, N. and V.C. Srivastava, *Simple Synthesis of Large Graphene Oxide Sheets via Electrochemical Method Coupled with Oxidation Process.* ACS Omega, 2018. **3**(8): p. 10233-10242.
32. Díez-Pascual, A.M., et al., *Tailorable synthesis of highly oxidized graphene oxides via an environmentally-friendly electrochemical process.* Nanomaterials, 2020. **10**(2): p. 239.
33. Konwar, S., et al. *High Purity Graphene Oxide Using Electrochemical Synthesis and Its Application.* in *Macromolecular Symposia.* 2019. Wiley Online Library.
34. Gao, Z., et al., *Electrochemical synthesis of layer-by-layer reduced graphene oxide sheets/polyaniline nanofibers composite and its electrochemical performance.* Electrochimica Acta, 2013. **91**: p. 185-194.
35. Díez-Pascual, A.M., et al., *Determination of riboflavin based on fluorescence quenching by graphene dispersions in polyethylene glycol.* RSC Advances, 2016. **6**(24): p. 19686-19699.
36. Li, S.-J., et al., *Electrochemical synthesis of a graphene sheet and gold nanoparticle-based nanocomposite, and its application to amperometric sensing of dopamine.* Microchimica Acta, 2012. **177**(3): p. 325-331.
37. Tuinstra, F. and J.L. Koenig, *Raman spectrum of graphite.* The Journal of chemical physics, 1970. **53**(3): p. 1126-1130.

38. Shen, Y., et al., *Structure and Property Evolution of Graphene Oxide Sheets during Low-Temperature Reduction on a Solid Substrate.* The Journal of Physical Chemistry C, 2020. **124**(26): p. 14371-14379.
39. Nishina, Y. and S. Eigler, *Chemical and electrochemical synthesis of graphene oxide – a generalized view.* Nanoscale, 2020. **12**(24): p. 12731-12740.
40. Razmi, H., L. Ezzati, and Z. Khorablou, *Direct Electrochemical Synthesis of Graphene Oxide/Cobalt Oxide Nanocomposite on Pencil Graphite Electrode for Highly Sensitive and Selective Detection of Insulin in Pharmaceutical Samples.* Journal of The Electrochemical Society, 2019. **166**(12): p. B961-B968.
41. Martín, A., et al., *Template Electrosynthesis of High-Performance Graphene Microengines.* Small, 2015. **11**(29): p. 3568-3574.
42. Dinesh, B., et al., *In situ electrochemical synthesis of reduced graphene oxide-cobalt oxide nanocomposite modified electrode for selective sensing of depression biomarker in the presence of ascorbic acid and dopamine.* Journal of Electroanalytical Chemistry, 2017. **786**: p. 169-176.
43. Ehsani, A., et al., *Electrosynthesis, physioelectrochemical and theoretical investigation of poly ortho aminophenol/magnetic functional graphene oxide nanocomposites as novel and hybrid electrodes for highly capacitive pseudocapacitors.* Journal of Colloid and Interface Science, 2017. **490**: p. 695-702.
44. Singh, P.K., et al., *Electrochemical synthesis of graphene oxide and its application as counter electrode in dye sensitized solar cell.* Journal of Renewable and Sustainable Energy, 2014. **6**(1): p. 013125.
45. Wu, B., et al., *Electrochemical synthesis of polypyrrole, reduced graphene oxide, and gold nanoparticles composite and its application to hydrogen peroxide biosensor.* Nanomaterials, 2016. **6**(11): p. 220.
46. Xie, G., M. Forslund, and J. Pan, *Direct electrochemical synthesis of reduced graphene oxide (rGO)/copper composite films and their electrical/electroactive properties.* ACS Applied Materials & Interfaces, 2014. **6**(10): p. 7444-7455.
47. Yang, T., et al., *Direct and freely switchable detection of target genes engineered by reduced graphene oxide-poly (m-aminobenzenesulfonic acid) nanocomposite via synchronous pulse electrosynthesis.* Analytical chemistry, 2013. **85**(3): p. 1358-1366.
48. Kalita, H., et al., *Electrochemical synthesis of graphene quantum dots from graphene oxide at room temperature and its soil moisture sensing properties.* Carbon, 2020.
49. Abraham, P., et al., *Electrochemical synthesis of thin-layered graphene oxide-poly (CTAB) composite for detection of morphine.* Journal of Applied Electrochemistry, 2020. **50**(1): p. 41-50.
50. Yakovlev, A., et al., *Electrochemical Synthesis of Multilayer Graphene Oxide by Anodic Oxidation of Disperse Graphite.* Russian Journal of Electrochemistry, 2019. **55**(12): p. 1196-1202.

# AUTHOR INDEX